PHYSICAL/ CHEMICAL PROCESSES

HOW TO ORDER THIS BOOK

BY PHONE: 800-233-9936 or 717-291-5609, 8AM–5PM Eastern Time

BY FAX: 717-295-4538

BY MAIL: Order Department
Technomic Publishing Company, Inc.
851 New Holland Avenue, Box 3535
Lancaster, PA 17604, U.S.A.

BY CREDIT CARD: American Express, VISA, MasterCard

VOLUME 2

PHYSICAL/ CHEMICAL PROCESSES

INNOVATIVE HAZARDOUS WASTE TREATMENT TECHNOLOGY SERIES

EDITED BY
HARRY M. FREEMAN
RISK REDUCTION ENGINEERING LABORATORY
U.S. ENVIRONMENTAL PROTECTION AGENCY
CINCINNATI, OHIO

TECHNOMIC PUBLISHING CO., INC.
LANCASTER · BASEL

Innovative Hazardous Waste Treatment Technology Series—Volume 2
a TECHNOMIC publication

Published in the Western Hemisphere by
Technomic Publishing Company, Inc.
851 New Holland Avenue
Box 3535
Lancaster, Pennsylvania 17604 U.S.A.

Distributed in the Rest of the World by
Technomic Publishing AG

Copyright © 1990 by Technomic Publishing Company, Inc.
All rights reserved

No part of this publication may be reproduced, stored in a
retrieval system, or transmitted, in any form or by any means,
electronic, mechanical, photocopying, recording, or otherwise,
without the prior written permission of the publisher.

This book was edited by Harry Freeman in his private capacity. No official support or
endorsement by the Environmental Protection Agency or any other agency of the federal
government is intended or should be inferred.

Printed in the United States of America
10 9 8 7 6 5 4 3 2 1

Main entry under title:
    Innovative Hazardous Waste Treatment Technology Series—Volume 2/
        Physical/Chemical Processes

A Technomic Publishing Company book
Bibliography: p.

Library of Congress Card No. 90-70257
ISBN No. 87762-617-0

TABLE OF CONTENTS

Introduction vii

2.1 The Application of Demulsification Chemicals in Recycling, Recovery, and Disposal of Oily Wastes 1
 Sam Delchad, Ph.D.

2.2 Solvent Extraction of Organic Pollutants from Wastewaters 9
 Roger R. Argus, Gregory R. Swanson

2.3 Air Stripping Technology 19
 William D. Byers

2.4 Vacuum Extraction: Effective Cleanup of Soils and Groundwater 33
 James J. Malot, P.E.

2.5 Catalytic Hydrodechlorination 45
 Bijan F. Hagh, David T. Allen

2.6 Detoxification of and Metal Value Recovery from Metal Finishing Sludge Materials 55
 L. G. Twidwell, D. R. Dahnke, S. F. McGrath

2.7 Using Alternating Current Coagulation to Treat and Recycle Wastewaters
 Containing Hazardous Substances 63
 P. E. Ryan, T. F. Stanczyk

2.8 EPP Process for Stabilization/Solidification of Contaminants 77
 S. L. Unger, H. R. Lubowitz

2.9 Electromembrane Process for Recovery of Lead from Contaminated Soils 87
 William F. Kemner, E. Radha Krishnan

2.10 Electrolytic Treatment of Waste Pickling Liquor 97
 Dr. David W. Scarooson, Ed Flemming

2.11 Demonstration Results for Three Innovative Technologies 101
 Howard M. Feintuch, Ph.D.

2.12 Vitrokele®, High-Efficiency, Selective, Metal-Chelating Adsorbents:
 Integrated Use with Electrolytic Metal Recovery for Closed-Cycle
 Elimination of Metal Wastes 111
 Bruce E. Holbein, Ph.D.

2.13 Recovery and Disposal of Nitrate Wastes 117
 John M. Napier

2.14 Minimization of Arsenic Wastes in the Semiconductor Industry 131
 Darryl W. Hertz

2.15 UV-Catalyzed Hydrogen Peroxide Chemical Oxidation of Organic Contaminants in Water 143
 D. G. Hager

2.16 Polysilicate Heavy Metals Mitigation Technology 155
 George J. Trezek

2.17 Electrochemical Oxidation of Refractory Organics **165**
    Brian G. Dixon, Myles A. Walsh, R. Scott Morris
2.18 PACT® Systems for Industrial Wastewater Treatment **177**
    John A. Meidl
2.19 Destruction of Cyanides in Electroplating Wastewaters Using Wet Air Oxidation **193**
    H. Paul Warner
2.20 In Place Treatment of Contaminated Soil at Superfund Sites: A Review **199**
    M. Roulier, J. Ryan, J. Houthoofd, H. Pahren, F. Custer
2.21 Control of Air Emissions from Soil Venting Systems **205**
    F. A. M. Buck, Craig A. Smith
2.22 Landfill Leachate Control Treatment **211**
    Trevor P. Castor
2.23 Innovative Practices for Treating Waste Streams Containing Heavy Metals:
    A Waste Minimization Approach **221**
    Douglas W. Grosse
2.24 Solidification/Stabilization Techniques: Promising Treatment Technologies for
    Remediating Superfund Sites **237**
    Joseph DeFranco

INTRODUCTION

Treatment is defined in the Resource Recovery and Conservation Act as, "Any method, technique, or process, including neutralization designed to change the physical, chemical, or biological characer composition of any hazardous waste so as to neutralize it, or render it non-hazardous or less hazardous or to recover it, make it safer to transport, store, or dispose of, or amenable for recovery, storage, or volume reduction." Treatment processes are typically classified as either thermal, physical/chemical, or biological, reflecting in general, those principles most applicable to the subject technology or process.

As the United States has moved toward requiring more restrictive regulations for treating and disposing of hazardous wastes, there has been an increase in the number of new or innovative hazardous waste treatment technologies either under development or in use. In this volume are contained chapters devoted to covering 24 such new processes that are generating interest in the hazardous waste management field for treating a variety of hazardous waste streams using physical or chemical principles. The reader is encouraged to contact the authors at the addresses listed for further discussions of the processes.

This volume is the second in a three-part series. The first volume covers innovative thermal processes. The third volume covers biological processes. We would appreciate any suggestions by our readers regarding processes that they feel should be included in future volumes of the series.

The Application of Demulsification Chemicals in Recycling, Recovery, and Disposal of Oily Wastes

Sam Delchad, Ph.D.[1]

ABSTRACT

This paper deals with the application of demulsification technology to the processing of oily wastes. A series of specialized chemicals have been tailor-made for the complex, everchanging oil-water separation problems facing generators, reclaimers, and service companies.

The unique features of emulsion systems and the mode of action of demulsifiers are briefly outlined. The following practical aspects will be dealt with in greater detail: a) types of waste streams successfully treated, b) preferred methods of chemical addition, and c) recommended procedures for chemical selection and monitoring.

Examples are selected from field case histories to emphasize the versatility, simplicity, and cost-effectiveness of demulsifiers.

The handling and volume reduction of oily sludge is highlighted as a special problem facing many companies. Chemical treatment is presented as a viable and proven technology for dealing with many types of oil field, refinery, and industrial sludges.

INTRODUCTION

Although chemical demulsifiers have been widely used in the oil field for over fifty years, their use in industrial and remedial applications has only recently achieved prominence.

Increasing disposal costs and restrictive regulations have become critical, rather than tolerated, factors in the closely-monitored oil reclamation industry. Moreover, the proliferation of environmental services involved in hazardous waste remediation and resource conservation has emphasized the economic and environmental advantages of chemical treatment.

For generators of oily waste, be it contaminated oil, oily water, or sludge, and for recyclers/reclaimers, the correct application of demulsifiers has proven to be both simple and cost-effective.

This paper offers a wide-angle view of the role played by this innovative technology in the processing of oily wastes, especially oily sludges.

DEMULSIFICATION TECHNOLOGY

Nature of Emulsions

Those of us involved in the handling of oil and water have come to realize that the old saying, oil and water do not mix, can be far from the truth.

Water is sometimes so finely dispersed in oil (or oil in water) that no amount of heating, settling, pulling, or waiting will cause them to separate. Such mixtures of liquids, which are intrinsically different, yet do not separate under normal physical conditions, are known as *emulsions*.

In the oil industry, emulsions are endemic, being generated in crude oil production and refining, in petrochemical processes, or being artificially (intentionally) produced for cooling and lubrication purposes. Recycling and reclamation plants, as well as companies involved in environmental services, are faced with the most complex and challenging emulsions.

It would be beneficial to explain how and why emulsions come into existence, for this is the key to devising the solution. When oil and water come into contact under "quiet" conditions, they will layer according to their specific gravities, normally oil above water. If you agitate this mixture vigorously, it will take a few minutes to separate. Now, if you add a

---

[1]Emulsions Control Inc., 829 Hoover Avenue, National City, CA 92050

Table 1. Factors Affecting Emulsion Stability

1. Temperature
2. Concentration of emulsifiers
3. Specific gravity differential
4. "Age" of the emulsion
5. Contamination with "bulk" impurities

small amount of a surfactant (e.g., detergent) and then agitate, a new phase is formed, which may not separate readily.

This is what happens: mechanical energy introduced into the mixture will shear the oil–water surface, creating small droplets that become dispersed in the oil phase. Since all systems eventually drift towards a lower energy state, these droplets, moving around as a result of natural convection motion, will collapse against each other (coalesce) and form successively larger droplets, eventually settling out.

However, surfactants, whose physiochemical characteristics are such that they will gravitate towards the oil–water interface, will provide a molecular coating around each water droplet. The molecules of this thin film are polarized in such a manner that the droplets will repel each other and are maintained in a dispersed state. Naturally, the smaller the droplets, the greater the surface area and thus the tendency to form a stable dispersion.

These particular surfactants are called *emulsifiers*. They may occur naturally as asphaltenes, naphthenic acid, or clay particles, or they may be man-made, for example detergents and dispersants.

An emulsion of water in oil has oil as the continuous phase and is known as a *regular* emulsion. A *reverse* emulsion has water as the continuous phase and consists of oil in water or oily solids in water. Usually, but not always, the predominant phase determines the type of emulsion.

In the case of oily sludges, the emulsion is further stabilized by the presence of oil-coated solids, which acquire a gravity close to that of water and thus have no incentive to separate.

Factors Affecting Emulsion Stability

Some of the variables that are known to affect the stability of an emulsion are shown in Table 1.

The presence of significant levels of solid particles is the most noticeable feature of oily sludge, as seen in Figure 1. The particle size and the stability of the oil coating impart to the sludges their high viscosity and thixotropic properties.

These solids may be inorganic, such as silicates, clay particles, iron sulfide, carbon, metal fines, metal oxides, spent catalysts, etc., or organic, such as paraffins, asphaltenes, resins, protein solids, etc.

Solids may range in size from submicron particles to large sand granules, although it is the finer particles that pose the challenge for effective separation.

The water content may vary from zero, as in some tank bottoms, up to 60% or 70%, as in refinery API basin sludge.

In appearance, oily sludges may be watery, low-viscosity oily, high-viscosity oily, granular, gelatinous, lumpy, pasty, or semisolid. Much can be learned about oily sludges from their physical properties. It is more difficult to evaluate them based on their chemical composition.

The Need for Demulsifiers

Handling of oil/water emulsions is governed by two primary considerations: the desired quality of the separated liquids and the economics of treatment.

Since an emulsion is, by definition, an unnatural

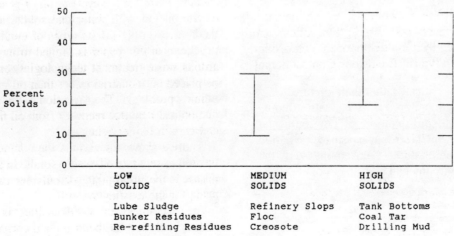

FIGURE 1. Typical sludge compositions.

Table 2. Factors Affecting Demulsification

1. Temperature/viscosity
2. Mixing of the demulsifier
3. Solids and other trace contaminants
4. Settling time
5. Type of treating equipment
6. Demulsifier type and concentration

state for an oil/water mixture, it must then be brought back to equilibrium with a specially designed demulsifier. Whereas other methods, such as heating, use of separators, centrifuges, etc., resort to brute force for pulling the two phases apart and end up concentrating the emulsion; demulsifiers, added in trace amounts, just balance out the stabilizing forces in the shortest possible time, and permit natural settling.

Of course, a mechanical system used after chemical demulsification can be very beneficial in terms of improved separation.

Several factors affect the extent and quality of demulsification. Some of these are outlined in Table 2.

The Selection of Demulsifiers

Given the complexity and diversity of emulsions, an emulsion breaker is expected to perform many tasks (Table 3). In addition, a demulsifier should be formulated with built-in ability to handle minor, and occasionally significant, fluctuations in the emulsion stream. This is particularly helpful for those recyclers who are frustrated by the constant variations in the oily wastes, and whose present physicochemical processes are unable to accommodate those changes.

Based on the requirements listed in Table 3, it is reasonable to assume that no single-component chemical can perform all those functions effectively. The various surface phenomena, which occur in three different phases (oil, water, and solids), can only be addressed by a multicomponent demulsifier.

For example, a typical Emulsions Control demulsifier consists of two, three, or four components blended in a specific order and in carefully controlled proportions. Thus, various demulsifiers containing different building blocks will perform differently in a given system. Occasionally, a new product has to be tailor-made for a particular application.

Evaluation of Demulsifiers

Tests for selecting demulsifiers are comparative, empirical, and do not lend themselves to complex instrumentation. Theoretical relationships, equations, factors and rate constants, Reynolds numbers and interfacial tension help very little in pinpointing the optimum demulsifier type or concentration.

Bottle and jar tests remain the only meaningful tests for the selection of a demulsifier and correlation with actual field conditions. Although those tests are fairly simple, the results obtained can be very meaningful. In carrying out demulsification tests, the operator is aided by a wealth of information:

a. Advances in unique formulations for demulsifiers
b. Development of lab procedures and techniques that are more informative and more consistent
c. Large body of field data correlated with lab results
d. Better understanding of emulsions and separation problems

THE APPLICATION OF DEMULSIFIERS

Versatility of Demulsifiers

Experience has shown that emulsions are generally associated with the presence of oil. At virtually every stage in the life cycle of oil, it is susceptible to contamination with water and solids, and, thus, to the formation of various types of emulsions. Since the scope of this paper is limited to innovative hazardous waste treatment technologies, emphasis will be placed on industrial rather than oil production or refining problems. The exception to this is the remediation and resource recovery from oil field slop pits and certain refinery sludges.

Figure 2 shows the various ways demulsifiers may be used to separate oil–water–solids. In almost every instance, the appropriate demulsifier can be tailor-made for the existing system.

The usefulness of a demulsifier is completely realized only if it has been applied correctly. Mixing of the chemical with the emulsion is of paramount

Table 3. Desirable Features of Demulsifiers

1. Have suitable solubility properties
2. Result in rapid coalescence and water break
3. Water-wet solid particles
4. Leave minimum unresolved interface
5. Yield oil-free water
6. Tolerate small fluctuations in treatment rate, temperature, BS&W, pH, etc.

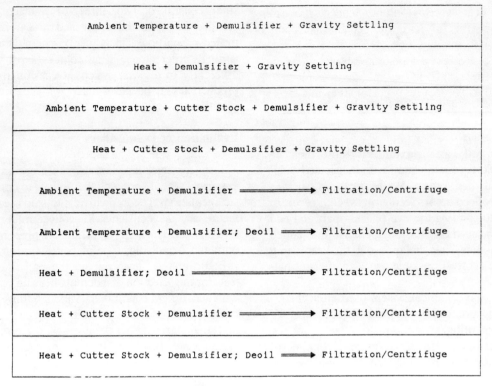

FIGURE 2. Demulsification options.

importance. This is discussed in more detail in Appendix 1.

Emulsion Problems in Industry

Some of the waste streams that have been successfully treated with demulsifiers are outlined in Table 4, which matches demulsification solutions to specific separation problems in various industries. The list is merely intended to present examples of treatable waste streams and is by no means complete.

Treatment of Oily Sludges

It is the author's opinion that true emulsions (chemically stabilized) are rarely destabilized by us-

Table 4. Emulsion Problems in Industry

Industry	Problem	Solution
Formulators	• Water entrainment in oil • Lube plant pipeline flushings • Off-space finished product • Spent coolant at users plant	• Demulsibility additives • Demulsifier for high-water lube • Dehazer • Deoiling with reverse demulsifier
Metal-Working	• Automotive plant lube • Aluminum plant coolant • General plant effluent • Steel mills/railroad • Aluminum can cleaners (synthetic cutting fluids)	• Demulsification and recycling • Deoiling with reverse demulsifier • Water clarification • Waste oil recycling; oily sludge reduction • Tramp oil removal
Recyclers/reclaimers	• Hydraulic oil recyclers • Waste crankcase oil • Industrial oily wastes • Spent coolants	• Dehydration prior to distillation • Demulsification for fuel oil • Demulsification for re-refining • Deoiling with reverse demulsifier

ing mechanical processes alone. Indeed, processes such as centrifugation are only capable of isolating liquids and solids after oil–water and oil–solids separation. Belt presses and plate-and-frame filters require oil-free solids to avoid blinding the filter medium, to minimize the need for filter aid, and to produce an acceptable filter cake. Polyelectrolytes, which are water soluble, work by tightly binding together oil droplets and solid particulates but do not offer the required destabilization prior to mechanical handling. In many cases, demulsification with an oil-soluble, solids-dispersing chemical, may be followed by a small concentration of polymer to condition fine solids for better capture.

Other physical methods that may be used in conjunction with demulsifier treatment are heat and diluents. Heat reduces the viscosity of oily sludges, expediting handling, mixing, and separation. Additionally, heat improves dewatering characteristics and makes demulsifiers less dependent on sludge variability and subsequent mechanical shearing [1]. Dilution with a light hydrocarbon serves a similar purpose. This process may be practical if heat is not desirable or available, if the diluents are inexpensive, and if the recovered oil need not meet rigorous specifications. Certain types of oily sludge may benefit dramatically from a combination demulsifier and levels of diluent as low as 5–10%.

A comprehensive treatment process, including typical data, is shown in Figure 3. This is not intended to be limiting or biased, but merely an aid for optimum unit selection.

Advantages of Demulsifier Use

Some of the advantages offered by demulsification chemicals have been alluded to earlier. Many of the objections expressed by past users of these chemicals have been addressed by a handful of chemical companies specializing in demulsification chemicals. These companies are now able to offer a more reliable and systematic approach to separation problems. The following are some of the advantages of demulsifiers.

1. *Simplicity:* Once a chemical has been selected for a given separation problem, it can be readily adapted to an existing system. The user of the chemical will have simple, clear instructions for how and when to use the chemical to achieve predetermined desired results.

2. *Cost-Effectiveness:* A chemical demulsification program may be implemented with little or no initial investment in process systems. As the type and

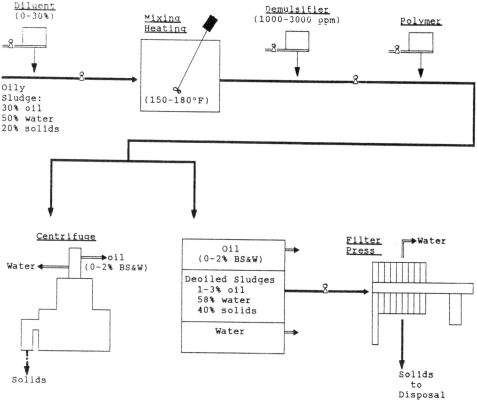

FIGURE 3. Comprehensive treatment system for oily sludge.

Table 5. Sludge Handling Methods: A Qualitative Comparison

	Cost	Environ. problems	Recovery	Volume reduction	Complexity	Treatment time	Supervision
Landfarming Landfill	+?	−	−	−	+	+	+
Incineration	−	−	−	+	−	+	+
Biological	+	+	−	−	+	−	+
Centrifuge	−	+	+	+	−	+	−
Filtration	−	+	−	+?	−	+	−
Solvent extraction	−	+	+	+	−	+	−
Chemical demulsification	+	+	+	−	+	+	+

+ Advantage
− No advantage

concentration of the demulsifier are optimized, ongoing treatment costs are usually small, heating requirements minimized, and failed treatments virtually eliminated.

3. *Resource Recovery:* This is probably the most obvious advantage of demulsification over conventional methods of handling oily wastes. Since a demulsifier is by design a means of phase separation, the recovery of a valuable resource such as oil and an environmentally critical stream such as water is the minimum that is achieved. The chemical in no way affects the quality and usefulness of the recovered streams.

4. *Volume Reduction*: In addition to removing clean oil and clean water from a hazardous waste stream, the volume of the hazardous residual is significantly reduced. It is therefore a double gain to convert a disposal liability into a usuable resource. This is particularly relevant in view of the current federal and state laws governing resource recovery, source reduction, and disposal sites, just to name a few.

5. *Enhancement of Mechanical Process:* As mentioned earlier, a chemical emulsion can only be destabilized chemically. In order for a mechanical system, such as a centrifuge or a filter-press, to function effectively, distinct phases have to be released from the emulsion to be separated. Recent successes in applying demulsifier/dispersant blends to the centrifuge or filter have demonstrated the completeness of chemical/mechanical systems.

A dramatic example, which highlights the advantages of demulsification technology over more conventional processes, is the area of oily sludge handling. Table 5 shows a qualitative comparison among various methods.

RECENT SUCCESSES WITH DEMULSIFIERS

Waste Oil Treatment

Background

Two waste oil recyclers were faced with the problem of economically dehydrating waste oil to produce a marketable product. Reduced efficiency was attributed to the inability to heat all incoming streams, and the high cost of boiling off high concentrations of water. Emulsions Control's demulsifier, RECOVEROL* ECO 19, was tested by both companies, and it afforded them the flexibility of treating various streams utilizing existing process units.

Treatment

Using a single-step chemical process with or without heat, high-quality oil was recovered, and clear water separated below a concentrated layer of solids. One reclaimer processed 20,000-gallon batches of low-water crankcase oil for one and one-half cents per gallon. Overnight settling at 160°F, or two to three days at ambient temperature, yielded clean oil (0.2–1% BS&W). The other reclaimer processed 5000-gallon batches of tightly emulsified lube bottoms for less than one cent per gallon. Overnight settling at 150°F gave complete oil recovery.

Demulsifier/Centrifuge Combination

Background

An industrial waste oil recycler added a centrifuge

*Trademark of Emulsions Control Inc.

to their system to speed up the rate of separation and improve oil quality. The demulsifier used in the system was not always successful, even at high concentrations. In addition, sulfuric acid was needed, and separation often required several days. Emulsions Control was brought in and conducted tests on representative samples of the oily emulsions under a variety of conditions. RECOVEROL* ECO 6N50 offered consistent removal of water and solids to less than 0.5% BS&W at 150°F.

Treatment

The customer reported rapid separation and consistently good results when treating 8000-gallon batches with eight gallons of ECO 6N50, followed by centrifugation after brief contact. Not only were chemical costs reduced to one-fourth of what they had been, but the entire operation was simplified and became less labor intensive. The elimination of acid alleviated safety problems associated with the handling of sulfuric acid. Furthermore, pH and odor problems were no longer a concern.

Spent Coolant

Background

A waste treatment facility was unable to get consistent clarification of oily wastewater, especially soluble oil. Wide fluctuations in surfactant concentration, oil content, pH, and soluble salts made it difficult for any one chemical to produce sewerable water. After extensive testing on dozens of incoming streams, Emulsions Control developed a demulsifier, ECA* 4FC, which successfully treated all types of oily water. Added at 500–2500 ppm and followed by caustic addition to pH of 9–10, ECA 4FC resulted in rapid settling of oily floc from clear water.

Treatment

The customer routinely treated mixtures of oily water by adding ECA 4FC during transfer from the holding tanks to settling tanks, injected upstream of the transfer pump. Downstream of the pump, sodium hydroxide (50% strength) was added to give a final pH of 9.5. The treated water was settled in 10,000-gallon tanks. The floc compacted to 5–10% by volume overnight and was removed for further dewatering. The clear water was tested prior to discharge to the sewer (typical oil content: 25–100 ppm). The typical cost for treatment was around one to two and one-half cents per gallon.

Refinery Sludge

Background

A major West Coast refinery was faced with a deadline for cleaning up an API separator. The separator had accumulated 20,000 barrels of API sludge and DAF skimmings, which limited viability of conventional treatment. The mixture resisted separation with a centrifuge provided by a contractor, even at high polymer concentrations. Emulsions Control analyzed samples composited from several locations in the basin and found that RECOVEROL* ECO 3NH, mixed with 10–15% jet fuel (or other light material), released oil-free solids and recovered clean oil within a few hours.

Treatment

The sludge was pumped from the separator basin into a series of Baker tanks, from which it was transported by vacuum trucks and transferred to a settling tank. During the transfer, jet fuel and ECO 3NH were added at a T-junction upstream of the transfer pump. The treated mixture separated into three layers within a few days. Oil recovery was in excess of 90% and oil quality greater than 95.5%. Overall savings to the refinery compared to off-site disposal were around $300,000.

APPENDIX 1

The Importance of Demulsifier Mixing

Introduction

The mixing of chemical demulsifiers with an emulsion is not a simple blending operation. In this case it is definitely not true that more mixing is better. It is a parameter that is difficult to quantify and one that becomes more critical as the volume of the emulsion increases.

RECOVEROL* demulsifiers and ECA* Water Clarifiers are surface-active chemicals that work on the particulate level. They neutralize the effects of emulsifiers by coagulating solids and water droplets, thereby allowing the solids and water to separate under the influence of gravity.

Once the demulsifier binds itself to a portion of the emulsion, it will not readily distribute to other portions. Thus, some of the emulsion will be overtreated and some undertreated.

Unacceptable Mixing

1. Pouring the chemical over the top of the tank and expecting downward movement
2. Pouring the chemical into an empty tank and then filling it with waste stream

Unreliable Mixing

Air Mixing

The chemical is poured on top of the emulsion in a holding tank while compressed air is introduced at the bottom.

Problems:
- Air pushes upwards through a layer of loosely stabilized emulsion in the presence of fine solids. This causes remixing with a cleaner phase at the top.
- Severe agitation may shear the water/oil interface and cause re-emulsification.
- Cold air mixing for an extended period may cool the heated tank too rapidly for effective separation.

Pump Circulation

The emulsion is circulated from top to bottom or bottom to top for at least one turn-around while the chemical is poured in at the top of the tank.

Problems: In a large tank, especially where both suction and inlet points are on the same side, this is what happens:
- A significant volume of the emulsion will receive little or no chemical.
- The chemical concentration based on the total volume may overtreat the actual volume circulated. This may worsen the problem.
- Continued agitation may re-emulsify phases which have already separated.
- A poor choice of pump, such as a centrifugal pump, may actually make the results worse than the initial problem.

The ECI Method

ECI has promoted the concept of TRANSFER-INJECTION for demulsifier mixing and has been successful in treating oilfield, refinery, and industrial waste streams using this method.

Procedure:
- Preheat emulsion if necessary and pump from a holding tank to a receiving tank with a positive displacement pump.
- Inject RECOVEROL demulsifiers continuously into the suction or discharge side of the transfer pump by means of a chemical proportioning pump.
- Once transferred, allow the treated emulsion to settle, maintaining heat, if necessary.

Benefits:
- UNIFORM DISTRIBUTION of the chemical throughout the entire volume of oil.
- IDEAL AGITATION for the chemical. Short/vigorous mixing followed by longer, gentle mixing.
- EASY OPERATION. Chemical pump can be switched on at the same time as the transfer pump. This permits intermittent treatment, batch splitting, and closer monitoring.
- MINIMAL DISTURBANCE in both tanks, which allows natural settling and the prevention of re-emulsification.
- SIMPLE AND RELIABLE means of monitoring chemical action. The operator can determine the separation in the tank by sampling the treated emulsion. This allows optimization of the chemical concentration during the transfer, rather than waiting until the end of the mixing period when it may be too late.

REFERENCE

1. Shabi, F. A. and M. J. Hansbury. "Aspects of Sludge Treatment," *Effluent and Water Treatment Journal*, pp. 343–351 (July 1976).

Solvent Extraction of Organic Pollutants from Wastewaters

Roger R. Argus,[1] Gregory R. Swanson, Ph.D.[1]

INTRODUCTION

Countercurrent, liquid–liquid solvent extraction of toxic organic pollutants from wastewater streams is a promising treatment technology that is currently the focus of research and development activities at the S-CUBED Division of Maxwell Laboratories, Inc. in San Diego, California. The development of solvent extraction processes for specific wastewater treatment or pretreatment applications involves: 1) selecting an appropriate solvent (or solvent mixture), 2) identifying the general extraction process flow scheme, including provisions for solvent recycle and raffinate stripping, if applicable, and 3) optimizing the process design and operating parameters to achieve the desired pollutant removal efficiency within acceptable spent solvent specifications and cost-effectiveness considerations. Each of these engineering activities is interrelated, as evidenced by the information presented in the balance of this chapter.

Two significant process designs have been developed as a result of the solvent extraction technology development efforts at S-CUBED: 1) Solvent EXtraction of Organic Pollutants (SEXOP), and 2) solvent EXtraction of Pollutants for Energy Recovery/Thermal Destruction (EXPERT). The patented SEXOP process [1] involves the removal of toxic organic pollutants from wastewater in a countercurrent extractor utilizing a volatile, water-immiscible solvent. This process incorporates recycle of the extraction solvent and has been designed with an emphasis on energy efficiency. The EXPERT process, for which patent applications have been filed, also involves the removal of toxic organic pollutants from aqueous media in a countercurrent extraction system. However, the solvent extract is not recycled, but rather directly utilized for energy recovery. Both of these processes are described further in the Process Description section.

APPLICABILITY OF SOLVENT EXTRACTION PROCESSES

Investigations conducted to date indicate that solvent extraction technology is potentially applicable to a wide variety of toxic industrial wastewaters. Specific testing performed by S-CUBED has demonstrated the feasibility of solvent extraction for the cleanup of wastewaters containing various levels (low ppm to percent) of pesticides, halogenated aliphatic hydrocarbons, aromatic hydrocarbons, chlorinated aromatic hydrocarbons, ketones, phenols and chlorinated phenols, and nitroaromatic compounds.

Determining the technical applicability of solvent extraction to the removal of particular pollutants from wastewaters requires an evaluation of the potential extraction efficiency. In generalized solvent extraction process models, the extraction efficiency for organic pollutants is a function of the solvent/water equilibrium partition coefficient for each pollutant, the solvent to water flow ratio within the extractor, and the number of theoretical equilibrium contact stages in the extractor. Thus, evaluating the potential extraction efficiency involves determining or estimating the applicable partition coefficient and then modelling the solvent extractor or extraction system.

---

[1]S-CUBED, 3398 Carmel Mountain Road, San Diego, CA 92121-1095

Partition Coefficient Estimation

The equilibrium partition coefficient (Kp), sometimes referred to as the distribution coefficient, may be defined as:

$$Kp = y/x$$

where:

y = the concentration of pollutant in the solvent phase
x = the concentration of pollutant in the aqueous phase

This relationship, which is only valid for solvent and water streams that are in equilibrium with each other, leads to the definition of the theoretical equilibrium contact stage as one in which the exiting solvent and water streams are in equilibrium with each other.

Kp values for many solvent/pollutant/water systems have been determined through both theoretical and empirical means and are available in the literature. However, a more directly useful value of Kp for each solvent/pollutant/water system of concern can be attained through single-stage laboratory extraction testing with a simulated or actual wastewater sample. Table 1 presents Kp values for various solvent/pollutant/water systems, as determined through laboratory testing at S-CUBED.

Extractor Modeling

Figure 1 illustrates a series of N theoretical equilibrium contact stages corresponding to a countercurrent liquid–liquid extractor. A mathematical model of this system has been developed and is utilized by S-CUBED in the assessment of process applicability and in process optimization. This model relates extraction efficiency to the partition coefficient (Kp), to the solvent–water flow ratio, and to the number of theoretical stages. Figure 2 gives an example output of the countercurrent extraction model.

As can be seen through application of the model, a high partition coefficient (>100) is most conducive to the effective extraction of pollutants in a limited number of theoretical stages. However, efficient extraction of pollutants with relatively low partition coefficients (5–10) can still be achieved through the use of extractors incorporating a larger number of theoretical stages and through an increased solvent–water flow ratio in the extractor. For a given Kp, the model allows the definition of a cost-effective compromise between the solvent–water flow ratio and the number of theoretical stages.

For most extraction processes, solvent selection is based primarily on Kp for the particular wastewater pollutant. However, other criteria to be considered when selecting an extraction solvent are flammability, water solubility, boiling point (if the solvent is to be recycled), cost, and tendency to form a solvent/water emulsion.

Table 1. Partition Coefficients Determined in Laboratory Testing at S-CUBED

Pollutant	Solvent	Partition Coefficient (Kp)
Pesticides		
DDT	Hexane	320,000
	Pentane	270,000
	Isopropyl Ether	230,000
	Ether	960,000
	Butyl Chloride	370,000
Toxaphene	Hexane	74,000
	Pentane	39,000
	Isopropyl Ether	140,000
	Ether	180,000
	Butyl Chloride	160,000
Chlordane	Hexane	11,000
	Pentane	26,000
	Isopropyl Ether	84,000
	Ether	130,000
	Butyl Chloride	240,000
Norflurazon	Hexane	17
	Pentane	27
	Isopropyl Ether	61
	Ether	190
	Butyl Chloride	200

Table 1. (continued)

Pollutant	Solvent	Partition Coefficient (Kp)
Diuron	Hexane	3.5
	Pentane	3.0
	Isopropyl Ether	53
	Ether	180
	Butyl Chloride	95
2,4-D	Hexane	0.24
	Pentane	9.2
	Isopropyl Ether	84
	Ether	66
	Butyl Chloride	5.7
Bromacil	Hexane	1.3
	Pentane	1.3
	Isopropyl Ether	11
	Ether	18
	Butyl Chloride	29
Glyphosate	Hexane	0.11
	Pentane	0
	Isopropyl Ether	0.048
	Ether	0.11
	Butyl Chloride	0.030
Halogenated Aliphatics		
Chloroform	Mixed Hydrocarbon*	20
1,2-Dichloroethane	Mixed Hydrocarbon*	>4
Carbon Tetrachloride	Mixed Hydrocarbon*	>15
1,2-Dichloropropane	Mixed Hydrocarbon*	>50
Trichloroethene	Mixed Hydrocarbon*	>35
Bromoform	Mixed Hydrocarbon*	>50
Tetrachloroethene	Mixed Hydrocarbon*	>10
1,1,2,2-Tetrachloroethene	Mixed Hydrocarbon*	>45
Halogenated Aromatics		
Chlorobenzene	Mixed Hydrocarbon*	>25
1,2-Dichlorobenzene	Mixed Hydrocarbon*	>1450
Ketones		
2-Butanone (MEK)	Mixed Hydrocarbon*	>10
4-Methyl-2-Pentanone (MIBK)	Mixed Hydrocarbon*	15
Phenols		
Phenol	Mixed Hydrocarbon*	30
2-Chlorophenol	Mixed Hydrocarbon*	97
2-Methylphenol (o-Cresol)	Mixed Hydrocarbon*	26
4-Chloro-3-methylphenol	Mixed Hydrocarbon*	79
4-Nitrophenol	Mixed Hydrocarbon*	>8000
4,6-Dinitro-2-Methylphenol	Mixed Hydrocarbon*	32
Pentachlorophenol	Mixed Hydrocarbon*	>1800
Misc. Aromatics		
Benzene	Mixed Hydrocarbon*	>35
Toluene	Mixed Hydrocarbon*	>25
Nitrobenzene	Mixed Hydrocarbon*	290
Naphthalene	Mixed Hydrocarbon*	575
2,4-Dinitrotoluene	Mixed Hydrocarbon*	555
Dimethyl Phthalate	Mixed Hydrocarbon*	104
Diethylphthalate	Mixed Hydrocarbon*	863
4-Nitroanaline	Mixed Hydrocarbon*	436

*Proprietary solvent.

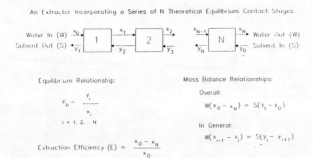

FIGURE 1. Basic model of a countercurrent solvent extractor incorporating N theoretical equilibrium contact stages.

PROCESS DESCRIPTION

The solvent extraction process may be designed to achieve treatment levels that permit reuse of the water within an industrial process or meet applicable pretreatment requirements for discharge to a conventional wastewater treatment facility. The countercurrent extractor design is tailored to the pollutant to be removed, the solvent to be used, and the duration of solvent contact needed to effect efficient extraction of the pollutant from the water. For many pollutants ($Kp > 100$), a single-stage countercurrent extractor suffices to substantially remove the organic pollutant from the water. However, for pollutants that are difficult to extract or are present in high concentrations, a multiple-stage extraction process is typically required.

The SEXOP Process

Figure 3 illustrates the SEXOP process flow diagram. In this process a stream of wastewater containing organic pollutants is fed to an extractor, where it is contacted with a countercurrent stream of high-volatility, water-immiscible solvent. Efficient extraction of pollutants from the water is assured through the use of a rotary disc contactor, or other extraction column incorporating intimate micro-scale mixing of the solvent and water, and an adequate number of theoretical stages.

The aqueous raffinate exiting the extractor may contain significant amounts of dissolved organic solvent, which may render the water unfit for reuse or discharge. Consequently, the raffinate stream is passed through a stripper, wherein residual solvent present in the raffinate is stripped by a stream of inert carrier gas. The treated water, now substantially free of both the pollutant and the solvent, is reused or discharged.

The solvent extract is passed to an extract stripper, wherein the extract is sparged by a carrier gas that enters the stripper toward its lower end and bubbles upward through the extract to vaporize the solvent. For efficient use of solvent, a minimum of 90% of the solvent entering the extract stripper is stripped by the carrier gas. The remaining solvent, containing a heavy concentration of the pollutants, is withdrawn from the bottom of the extract stripper and collected. Depending on the economic value of the pollutants, the final solvent extract containing the pollutants may either be recovered or incinerated.

The vaporized solvent streams from both the extract and raffinate strippers are condensed and recycled to the extractor.

The methods and apparatus of the SEXOP process are intended to minimize the use of external energy. The process achieves this energy efficiency principally through the use of a low-boiling solvent, preferably with a low heat of vaporization. Low-boiling solvents may be efficiently vaporized by sparging with a gas at ambient or slightly elevated temperatures, whereas the vaporization of high-boiling solvents by sparging at low temperatures is inefficient and impractical.

The EXPERT Process

Figure 4 conceptually illustrates the general EXPERT process. In this process, a stream of waste-

Definitions:

- Kp: Equilibrium partition coefficient
- W/S: Water–solvent flow ratio in extractor
- X_0: Initial pollutant concentration in wastewater
- X_f: Desired pollutant concentration in wastewater
- N: Theoretical equilibrium contact stage number
- X_N: Pollutant concentration in wastewater exiting stage N
- Y_N: Pollutant concentration in solvent exiting stage N

$$X_0 = 1000 \text{ ppm}$$
$$Kp = 20$$
$$W/S = 10$$
$$X_f = 1 \text{ ppm}$$

N	Y_N ppm	X_N ppm	Overall Extraction Efficiency
0	0	1000	—
1	9990	499.50	50.05%
2	4985	249.25	75.08%
3	2482.50	124.13	87.59%
4	1231.25	61.56	93.84%
5	605.63	30.28	96.97%
6	292.81	14.64	98.54%
7	136.41	6.82	99.32%
8	58.20	2.91	99.71%
9	19.10	.96	99.90%

FIGURE 2. Countercurrent solvent extraction model output.

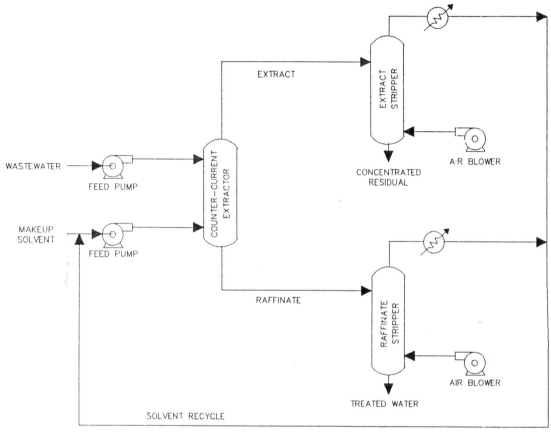

FIGURE 3. SEXOP process flow diagram.

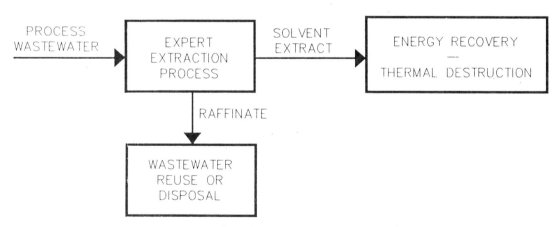

FIGURE 4. General concept of the EXPERT process.

water containing organic pollutants is fed to an extractor, where it is contacted by a countercurrent stream of water-immiscible solvent that extracts the pollutants from the water. As the EXPERT process is designed to incorporate once-through extraction without solvent recycle, the solvent used in this process is specifically selected for its amenability to energy recovery or thermal destruction operations, low cost, and high partition coefficient for the pollutants to be extracted. The raffinate, having greatly reduced concentrations of toxic constituents as a result of the extraction treatment, is intended for reuse or discharge to a POTW, an on-site biological treatment system, or a land disposal scenario in accordance with regulatory pretreatment requirements established by federal, state, or local regulatory agencies.

ADVANTAGES OVER CONVENTIONAL TREATMENT TECHNOLOGIES

The recently promulgated federal restrictions on the land disposal of hazardous wastes mandate that effective treatment technologies must be identified and implemented for wastewaters containing high levels of toxic organic compounds. For wastewaters that are currently land disposed because they are not amenable to conventional treatment technologies, including low-cost biological treatment processes, solvent extraction can provide a highly cost-effective method for satisfying regulatory treatment or pretreatment requirements. Additionally, the integration of solvent extraction with energy-generating processes (steam-generating boilers, industrial furnaces, etc.) may form part of an effective hazardous waste minimization program.

The primary advantage of solvent extraction over other potentially applicable treatment technologies is that solvent extraction isolates the toxic pollutants in an organic liquid matrix that is suitable for use as a boiler, furnace, or kiln fuel. The use of the solvent extract as a boiler/furnace fuel or the incineration of the extract, where required, provides for the efficient thermal destruction of extracted organic pollutants. Additionally, the types of solvent extraction processes described herein are operated at ambient pressure and temperature, thereby limiting the potential for secondary air emissions or treatment residuals. Because the ultimate destruction of toxic wastewater pollutants is assured, the environmental and regulatory acceptability of the solvent extraction treatment process is higher than that of competing technologies, which do not achieve these goals.

Another advantage of solvent extraction is the flexibility inherent in the process. With appropriate solvent selection, solvent extraction is applicable to a wide variety of both volatile and semivolatile organic pollutants. With proper extractor design and solvent/water ratio selection, high removal efficiencies can be achieved for essentially any applicable pollutant. Also, the system design may be tailored to achieve solvent extract characteristics that are most amenable to the subsequent use of the extract as a fuel.

Finally, cost evaluations indicate that, for many applications, S-CUBED's solvent extraction processes can be economically competitive with conventional treatment processes. The primary economic advantage of the SEXOP process is that, due to the selection of a volatile extraction solvent, the process is highly energy efficient. With the selection of a low-cost solvent, the EXPERT process is cost-effective in that extraction occurs in a single-pass flow scheme in which solvent recycle is not employed. Also, the pollutant-laden solvent from the EXPERT extraction process may be used as fuel in on-site energy recovery operations.

STATE OF DEVELOPMENT

S-CUBED's SEXOP process is currently entering the pilot stage of development. Under the California Department of Health Services Waste Reduction Grant Program, a preliminary design has been prepared for the construction of a mobile pilot-scale SEXOP system.

The EXPERT process is in the laboratory phase of development, in which bench-scale and modeling studies are being performed to confirm the initial feasibility study results. S-CUBED anticipates developing a pilot-scale EXPERT system in the 1989/90 timeframe.

S-CUBED is also initiating a marketing effort to license its patented solvent extraction technologies for direct use or for further development for specific industrial process applications.

Additionally, laboratory and bench-scale process studies are continuing at S-CUBED in order to determine partition coefficients for a variety of solvent/pollutant/water systems.

PROCESS OPERATIONS AND TESTING RESULTS

The Bench-Scale Process

A bench-scale solvent extraction (SEXOP) process was designed and fabricated by S-CUBED in 1983, and experimental studies were conducted to test its capability. The SEXOP bench-scale system incorporates a two-phase, countercurrent flow rotary disc extraction column and a packed column for stripping solvent from the solvent extract.

The rotary disc contactor (RDC) can be rotated using a variable-speed motor at speeds up to 1800 revolutions per minute. The extractor column is constructed from 316 stainless steel segments and is approximately 122 cm (4 ft.) high and 7.6 cm (3 in.) in diameter. Three of the segments are fitted with glass viewports to permit observation during the course of the experimental runs. Figure 5 shows the assembled RDC mounted in a walk-in hood.

Testing Results

The bench-scale RDC has been utilized in a series of test runs in order to evaluate solvents and solvent–water flow ratios for the extraction of a variety of organic pollutants. To date, testing of the SEXOP process has focused on pesticide-contaminated

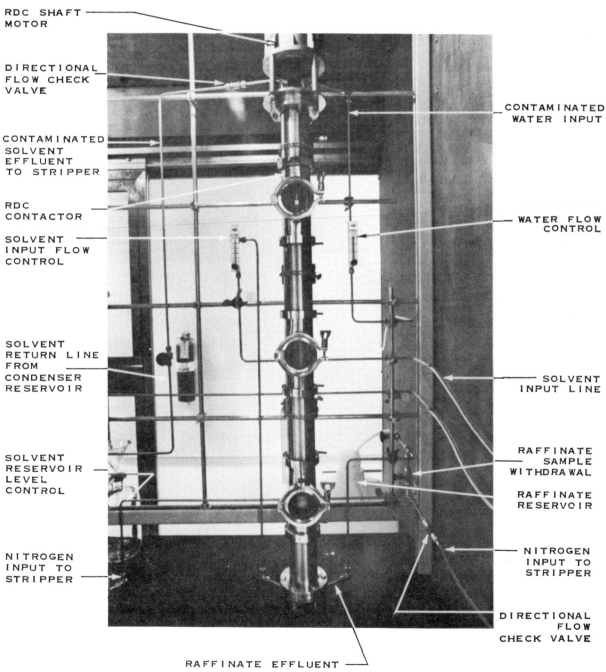

FIGURE 5. Rotary disc contactor with attendant plumbing detail.

Table 2. Results of Bench-Scale RDC Unit Testing Performed to Date

Wastewater Description/Contaminants	Contaminant Concentration	Extraction Solvent	Solvent/Water Flow Ratio	Number of Test Runs Performed	Extraction Efficiency	Ref.
Synthetic DDT-Contaminated Wastewater						
DDT	0.16 ppm	Isopropyl Ether	0.1	1	83%	4
		Hexane	0.1	2	>99.99%	
Real-World DDT-Contaminated Wastewater						
DDT	3.8 ppm	Hexane	0.1	5	90–>99%	4
DDE	25.6 ppm				90–>99%	
DDD	5.2 ppm				90–>99%	
Chlorinated Benzenes	139 ppm[1]				90–>99%	
Synthetic Bromacil-Contaminated Wastewater						
Bromacil	431 ppm	Hexane	0.5	1	30.2%	5
Real-World Bromacil-Contaminated Wastewater						
Bromacil	150 ppm[1]	Hexane	0.5	1	0.3%	5
		Butyl Chloride	0.5	2	90.2%	5
Real-World 2,4-D Plant Wastewater (pH = 2)						
2,4-D		Butyl Chloride	0.5	1	95.6%	5
			0.25	2	88.8%	5
Rocky Mountain Arsenal Basin F Wastewater-Dilute[2]		Hexane	0.1	2	77%[3]	2
Rocky Mountain Arsenal Basin F Wastewater[2]		Hexane	0.1	4	67%[3]	
			0.5	4	81%[3]	3

[1] Estimate based on information presented in the referenced report.
[2] A wide variety of organic contaminants were present in this wastewater.
[3] An average extraction efficiency for all organic compounds present in the wastewater.

Table 3. Number of Theoretical Equilibrium Stages Required to Achieve Extraction Efficiencies for Various Partition Coefficients and Solvent–Water Flow Ratios

Partition Coefficient	Solvent/Water Flow Ratio	No. of Theoretical Stages Required to Achieve Extraction Efficiency**			
		90%	99%	99.9%	99.99%
$K_p = 2$	1.0	4	6	9	13
$K_p = 5$	1.0	2	3	5	6
	0.5	3	5	7	10
	0.25	5	14	*	*
$K_p = 10$	1.0	1	2	3	4
	0.5	2	3	5	6
	0.25	3	5	7	10
	0.15	4	9	15	17
$K_p = 20$	1.0	1	2	3	4
	0.5	1	2	3	4
	0.25	2	3	5	6
	0.10	3	6	9	13
	0.07	4	11	17	*
$K_p = 50$	1.0	1	2	2	3
	0.5	1	2	3	3
	0.25	1	2	3	4
	0.10	2	3	5	6
	0.05	3	5	7	10
	0.025	5	14	*	*

* >20 Theoretical Equilibrium Stages Required.
** An initial pollutant concentration of 1000 ppm is assumed for purposes of this example.

wastewaters. Table 2 summarizes the results of these tests.

Modeling Results

Results of bench-scale testing are utilized to refine and apply the mathematical model described in the Applicability of Solvent Extraction Processes section to the specific application under consideration. Thus, the results of a few simple bench-scale tests provide input to the model and extend the ability to assess the effectiveness of solvent extraction over a wide range of design conditions and operating parameters.

Table 3 presents a generalized output of the model and relates extraction efficiency to the number of theoretical equilibrium contact stages, to the solvent–water flow ratio, and to the partition coefficient. However, the model is most useful when applied to a specific solvent/pollutant/water system, as was illustrated previously by Figure 2.

Test Conclusions

As evidenced by the bench-scale testing results and the process modeling results presented in Tables 2 and 3, respectively, solvent extraction can be very effective in extracting organic pollutants from wastewater. For many pollutants, extraction efficiencies of >99% can be achieved by using the appropriate solvent and optimum process design and operating conditions.

S-CUBED CONTACTS

For additional technical information on solvent extraction processes or licensing information with regard to either the SEXOP or EXPERT processes, contact Dr. Gregory R. Swanson, S-CUBED's Chemical Engineering Program Manager responsible for directing solvent extraction technology development activities, or Mr. Roger Argus, the project engineer for S-CUBED's solvent extraction technology development activities. Dr. Swanson and Mr. Argus may be contacted directly at S-CUBED, 3398 Carmel Mountain Road, San Diego, CA 92121-1095, (619)453-0060.

REFERENCES

1. Burns, E. A. and B. N. Colby, U.S. Patent 4,518,502 (May 21, 1985).
2. Reynolds, S. L., "A Preliminary Study of the Applicability of SEXOP for the Removal of Contaminants from Basin F Wastewater," S-CUBED, April 14, 1982 (private communication).
3. Hiler, G. V. and S. L. Reynolds. "Engineering Optimization of SEXOP for the Removal of Contaminants from Basin F Wastewater," S-CUBED Report No. SSS-R-82-5657, Department of the Army Contract No. DAAA05-82-M-0685 (July 1982).
4. Reynolds, S. L. "Extraction of Pesticides from Process Streams Using High Volatility Solvents, A Feasibility Study," NTIS: PB83-209767, EPA Report No. EPA600/2-83-0404 (May 1983).
5. Hiler, G. V. and S. D. Cameron. "Liquid–Liquid Extraction of Trace Level Pesticides from Process Streams, Draft," S-CUBED Report No. SSS-R-83-6328, EPA Contract No. 68-02-3629, Task 13 (August 1983).
6. "Step III Grant Application for Construction of a Mobile System for the Extraction of Organic Pollutants from Process Streams," S-CUBED Report No. SSS-P-87-8241107 (January 28, 1987).
7. Reynolds, S. L. and T. Kruger. "Step II Grant for the Design of a Mobile System for the Extraction of Organic Pollutants from Process Streams," Grant No. 85-00166, Draft Final Report, S-CUBED Project No. 41021, S-CUBED Report No. SSS-R-87-8560 (March 1987).
8. "Solvent Extraction of Organic Pollutants from Wastewater Sources, A Process Overview," S-CUBED, 1987 (private communication).

2.3

Air Stripping Technology

William D. Byers[1]

INTRODUCTION

Volatile organic compounds (VOCs) in groundwater are a widespread problem both in the United States and abroad. The problem has been widely reported by both popular and professional journals, and has been examined in the United States by public and private researchers. For example, a U.S. Environmental Protection Agency survey reported that 22 percent of 466 randomly sampled utilities produced drinking water with detectable levels of VOCs [1]. Such findings have prompted increased attention from regulators in a variety of environmental areas. This attention has included setting standards for drinking water and for industrial pretreatment.

Increased Regulation

For drinking water, maximum contaminant levels (MCLs) and maximum contaminant level goals (MCLGs) have been finalized for eight VOCs [2]; limits for more contaminants are in the regulatory pipeline.

Industrial pretreatment standards are being expressed, in some situations, as limits on individual VOCs of local concern. In other situations, the standards are expressed as limits on total toxic organics, which may in some cases include all of the compounds on the EPA's Hazardous Substances List. With this environment, the situations requiring removal of VOCs from water and wastewater are increasing considerably.

Removal Methods

Reports of the problem prompt an obvious question: how to remove VOCs from water? A wide variety of treatment processes have been studied [3,4]. Most include air stripping or adsorption on activated carbon or synthetic resins. Destructive techniques (such as use of UV light with an oxidant, or wet air oxidation) are sometimes considered; but, because of their cost, they tend to have limited application. Biological treatment can be effective; however, for many VOCs, removal is primarily caused by air stripping during the aeration process. Of these alternatives, air stripping has generally been shown to be the most cost-effective treatment for removing VOCs from water and wastewater.

WHAT IS AIR STRIPPING?

Air stripping exploits the natural tendency of VOCs to volatilize from water into air by allowing intimate contact between these two phases. This contact can be accomplished in a variety of physical arrangements, including bubbling air through a pool of water, spraying water into a stream of air, whirling air and water together in a mechanical device, and forcing air to flow through a packed bed wetted by a stream of water. Important features of an efficient air stripping system include a high surface area of contact between the two phases and a reasonable amount of turbulence to bring VOCs to the water surface where transfer occurs.

One of the most commonly used devices for achieving interphase mass transfer is the packed tower. A packed tower consists of a tower enclosure partially filled with a bed of small pieces of material

---

[1]CH2M Hill, 2300 N.W. Walnut Blvd., P.O. Box 428, Corvallis, OR 97339

(packing). A schematic of a packed-tower air stripper is shown in Figure 1.

In countercurrent flow, which is the most common flow arrangement, air is introduced to the tower below the packed bed and water is distributed over the top of the bed. Thus, the "cleanest" air contacts the cleanest water at the bottom of the packed bed, providing an opportunity for the treated water to reach very low levels of contamination.

This paper focuses on the design of countercurrent packed-tower air strippers for the removal of VOCs from water. It discusses the hardware typically included in packed towers, describes a methodology for sizing packed towers, and presents design and performance data, as reported for stripping towers operated at fourteen sites.

AIR STRIPPER HARDWARE

Figure 2 shows a cut-away view of a typical packed-tower air stripper. A typical tower includes a tower shell, packing support, packing, bed limiter, liquid distributor, and mist eliminator (optional). Excellent articles have been published on the subject of packed tower internals [5,6]. The following is a brief summary of the items important in an air stripper.

The tower shell typically provides a reservoir for water in the bottom of the tower and may be equipped with an internal seal (or an external trap in the outlet piping) to prevent air from escaping through the water outlet nozzle. The shell also provides a plenum below the packing, which allows the inlet air to distribute itself evenly throughout the tower cross section before traveling upward through the packing.

A packing support is required at the bottom of the packed bed. Packing supports should have a high open area, with openings small enough to prevent the packing from slipping through. The most common types are the vapor injection type and the subway grating type.

Virtually any material that will cause intimate contact between the air and the water can be used as tower packing. Material as simple as redwood slats have been used. At the other end of the spectrum are a variety of highly engineered structured packings.

Probably the most common packings for VOC stripping have been the random dumped packings. These include such nonproprietary packings as pall-type rings and saddles available from a variety of suppliers. Random packings also include proprietary packing materials such as Jaeger Tripacks, Glitsch Cascade Minirings, and Ceilcote Tellerettes. The proprietary packing materials tend to be more efficient and have lower pressure drop than the commodity packings, but have a higher initial cost.

A bed limiter is a plastic or stainless netting mounted on a frame attached to the tower wall. The purpose of the bed limiter is to prevent the entrainment of packing into the liquid distributor or out of the top of the tower.

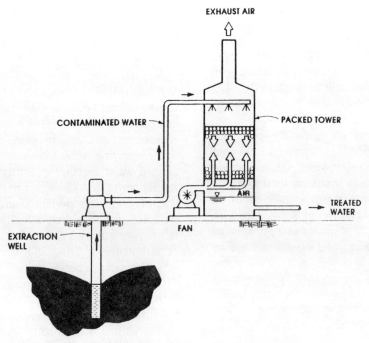

FIGURE 1. Packed tower air stripping.

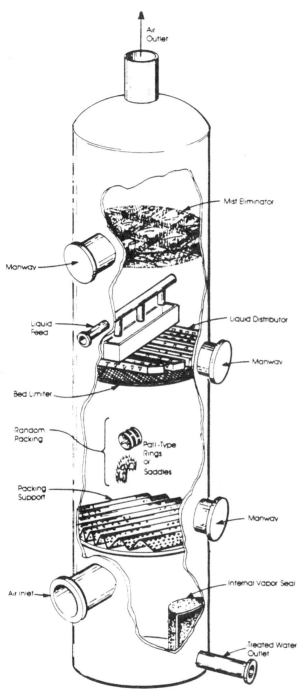

FIGURE 2. Packed tower internals.

Other than the packing, the liquid distributor may be the most important device in the tower. Many instances of poor tower performance have been traced to poor liquid distribution in the tower rather than improper sizing of the packed bed. A good liquid distributor will apply streams of water to the top surface of the packed bed rather than fine droplets or sprays. Examples of good liquid distribution devices include weir-trough distributors, orifice-riser distributors, and orifice-pipe distributors. Spray headers are not recommended.

Good liquid distribution is most important at low hydraulic loadings. At loadings above 10 gpm/ft^2, a tower may tolerate some minor maldistribution without noticeable impact on performance. However, at loadings below 5 gpm/ft^2, the need for even liquid distribution is compelling. In general, it is good practice to use well-engineered distribution devices. If just 2 percent of the liquid runs down the wall of the tower without significant contact with the air, the column will do no better than 98 percent removal, no matter how deep the packed bed may be.

A mist eliminator is an option in an air stripping tower. It can be used to reduce the amount of entrained mist that leaves the tower to form an unsightly plume at the air outlet. Two types are available for general use: Chevron-type and mesh pads. Occasionally, a bed of packing is used as a mist eliminator.

PRESSURE DROP

The pressure drop of the air flowing through a packed bed is one of the major cost items in operating an air stripper. The pressure drop that will occur in a packed bed can be predicted by a method developed by Eckert [7] as given in Reference [8]. The method uses the gas flow rate, the air-to-water ratio, air and water physical properties, and a packing parameter called the packing factor to predict the pressure drop through each unit of bed depth. Tables of packing factors for a variety of packing shapes are given in References [8,9] and are available from the packing manufacturers.

MASS TRANSFER THEORY

A mass transfer theory that is useful in predicting the operation of air stripping towers is known as the two-resistance theory [8,10]. This theory says that when two phases are in contact with each other, an equilibrium will be established at the interface between the two phases. In air stripping, a VOC in water will tend to migrate through the water toward that interface because the VOC is more concentrated in the bulk aqueous phase than it is at the interface. Similarly, the VOC will tend to migrate away from the interface and into the bulk air phase because the VOC is more concentrated at the interface than it is in the bulk air phase.

This model is represented in Figure 3. Note that the concentrations at the interface are not the same but are in equilibrium with each other. The phases

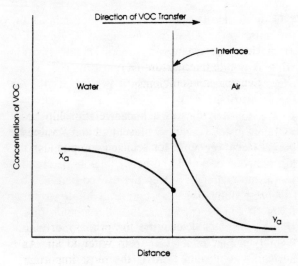

FIGURE 3. The two-resistance concept.

on both sides of the interface present a resistance to transfer of VOCs, and this resistance causes a concentration gradient to occur on each side of the interface. These gradients provide the driving force for mass transfer through each of the phases.

This model is a microview of a process that is repeated randomly throughout an air stripping tower, as liquid spreads over the surface of one piece of the packing and then moves on to another. As such, the model provides a useful view of the forces at work, allows identification of the variables that affect the rate of mass transfer, and ultimately leads to general techniques for predicting the performance of air stripping towers.

VARIABLES AFFECTING AIR STRIPPING

Application of the model suggests three types of variables that will affect the performance of air stripping towers:

- equilibrium properties, which influence the air-phase and water-phase concentrations at the interface
- diffusion properties, which influence the resistance to VOC transfer that is inherent in the two phases
- general system properties (such as flow, temperature, and system geometry), which affect the dynamics of the stripping process

Equilibrium Properties

The equilibrium distribution of a VOC between air and water is the primary variable influencing the suitability of air stripping for removing that VOC from water. This distribution is typically represented by a ratio of vapor-phase concentration to liquid-phase concentration at equilibrium. As suggested by the two-resistance theory, this will be the ratio of concentrations at the interface between the air and water. The higher this ratio, the higher will be the driving force for transfer of the VOC from water into air, and the more effective air stripping will be as a treatment process.

For many of the sparingly soluble, or hydrophobic, contaminants, the equilibrium distribution ratio does not vary significantly over a wide range of concentrations. For such compounds, the equilibrium ratio is said to follow Henry's Law, explained below:

$$P_a = H_a x_a$$

where:

P_a = Partial pressure of component a
H_a = Henry's law constant for component a
x_a = Concentration of component a in the liquid phase

The partial pressure is typically expressed in atmospheres. Liquid-phase concentrations are commonly expressed as mole fractions or as moles per unit volume (e.g., moles/cubic meter). When the liquid concentration is expressed in mole fractions, the Henry's constant has the units of atm/mole fraction and is commonly reported as simply atmospheres. When the liquid-phase concentration is expressed in moles/cubic meter, Henry's constant has the units of atm/(mole/m³), often expressed as atm-m³/mole.

Henry's constants are also expressed as ratios of concentrations in mg/L. This type of Henry's constant (mg/L in air/mg/L in water) is frequently referred to as the dimensionless Henry's constant. When looking up Henry's constants for a particular compound, it is important to note the dimensional system applied when the Henry's constant was reported.

Experimentally determined Henry's constants are available in the literature for a variety of environmentally significant VOCs [11–22]. Experimental techniques vary widely, and results of several different researchers should be reviewed before selecting a value to use.

When experimental values are not available, Henry's constants may be estimated for a particular VOC by taking the ratio of its vapor pressure to its water solubility [23]. This method is sometimes referred to as a method of "calculating" Henry's constant. It is important to remember that the technique is only as good as the solubility and vapor pressure data available and that it depends on the assumption that the Henry's constant is indeed constant for all concentrations up to the compound's solubility limit.

Table 1 presents a selection of experimental and calculated Henry's constants found in the literature. Note that these are selected values, not necessarily recommended values. Readers are urged to review the references and draw their own conclusions about the best values to use.

Effect of Temperature

Henry's constants are strongly influenced by temperature. A change of 10°C in temperature can cause a two- to three-fold change in Henry's constant. The relationship of the Henry's constant to temperature can be modeled by a van't Hoff-type relation in the form:

$$\ln H = A - B/T$$

where

H = Henry's constant
T = Absolute temperature (K)
A, B = Regression coefficients

This expression indicates a linear relationship between $\ln H$ and $1/T$. Table 2 presents A and B values reported in the literature for selected compounds.

Diffusion Properties

Henry's constant determines the primary driving force for transfer of a VOC from water to air. As such, this equilibrium ratio is the most important property of a particular VOC for determining if air stripping is a practical treatment approach.

Of secondary importance in determining the rate of VOC transfer in a stripping tower are the diffusion properties of the VOC in the air-water system. These

Table 1. Selected List of Henry's Constants

Compound Name	Molecular Weight	Solubility (mg/l)	Vapor Pressure (mm Hg)	Calculated	Measured	Ref.[a]
Bromoform	252.8	3,033	5.6	34	34	20
Chloromethane	50.5	5,350	4,275	521	490	21
Methylene Chloride	84.9	19,400	438	140	122	21
Chloroform	119.4	7,900	192	212	204	21
Carbon Tetrachloride	153.8	1,160	113	1,764	1,689	21
Chloroethane	64.5	5,710[b]	755[b]	628[b]	617	21
1,1-Dichloroethane	98.9	5,100	226	321	312	21
1,2-Dichloroethane	98.9	8,700	82	68	54	20
1,1,1-Trichloroethane	133.4	720	124	1,678	956	21
1,1,2-Trichloroethane	133.4	4,420	30	67	66[c]	20
1,1,2,2-Tetrachloroethane	167.8	3,000	6.5	27	26	20
1,1-Dichloroethylene	96.9	400	598	7,303	1,450	21
cis-1,2-Dichloroethylene	96.9	3,500	206	417	227	21
trans-1,2-Dichloroethylene	96.9	6,300	326	367	521	21
Trichloroethylene	131.4	1,000	74	713	532	21
Tetrachloroethylene	165.8	140	19	1,612	983	21
Benzene	78.0	1,780	95	306	302	20
Toluene	92.2	515	28	372	369	20
O-Xylene	106.2	175	6.6	293	274[c]	20
M-Xylene	106.2	162	8.3	395	384[c]	20
P-Xylene	106.2	185	8.8	368	389[c]	20
Ethylbenzene	106.2	152	9.5	486	468	20
Ethylene Dibromide	187.9	1,696	2.0	18	18[c]	20
Naphthalene	128.2	34	0.1	22	27	20
Acenaphthene	154.2	4	0.02	13	8	20
Chlorobenzene	112.6	472	11.9	207	210	20
Acetone	58.1	2.3E+06	200	0.38	NA	22
Methyl Ethyl Ketone	72.0	353,000	90	1.34	NA	22
Methyl Isobutyl Ketone	100.0	15,000	36	17.3	NA	22

[a]Reference shows source of measured Henry's constant. All other data from Reference 20, except ketones (from Reference 22).
[b]Data taken at 20°C.
[c]Recommended value from Reference 20 (not experimental).
Note: H (atm/mole fraction) at 25°C unless noted.

Table 2. Temperature Regressions for Henry's Constant

$$\ln H = A - B/T$$

Compound	A	B	r^2
Tetrachloroethylene	23.38	4918	0.996
Trichloroethylene	22.30	4780	0.996
1,1-Dichloroethylene	19.77	3279	0.994
cis-1,2-Dichloroethylene	19.40	4192	0.979
trans-1,2-Dichloroethylene	20.27	4182	0.994
1,1,1-Trichloroethane	20.70	4133	0.995
1,1-Dichloroethane	19.56	4128	0.994
Chloroethane	16.90	3120	1.000
Carbon tetrachloride	22.22	4411	0.995
Chloroform	20.77	4612	0.996
Dichloromethane	17.58	3817	0.951
Chloromethane	20.28	4215	0.990

Note: Based on studies from approximately 10 to 35°C. H, atm/mole frac.; T, K (adapted from Reference 21).

diffusivities are represented by diffusion coefficients for each of the phases.

Diffusion Coefficients

Diffusion coefficients indicate the rate at which one compound will migrate by molecular diffusion through a given phase or media under the influence of a concentration driving force. In general, the higher the value of the diffusion coefficient, the faster will be the rate of mass transfer or migration. In the case of air stripping, the diffusion coefficients of a specific VOC in air and in water are important in determining the rate at which that VOC will transfer between the two phases.

Diffusion coefficients are best determined experimentally. However, they are difficult to measure, and experimental data, therefore, is scarce. In the absence of experimental data, a variety of prediction techniques can be used to estimate diffusion coefficients.

For estimating diffusion coefficients of VOCs in air, the Wilke-Lee method and the Fuller, Schettler, and Giddings (FSG) method are the most highly recommended [23]. Of these, the FSG method is much easier to use than other methods and is similar in overall accuracy. The FSG method uses temperature, molecular weights, and molar volumes to estimate the diffusion coefficient. Most diffusion coefficients of VOCs in air are on the order of 0.1 cm²/sec at ambient temperatures.

For aqueous-phase diffusion coefficients, the prediction method of Hayduk and Laudie is recommended [23]. This method uses the viscosity of water and the molar volume of the VOC to estimate the diffusion coefficient. Diffusion coefficients of VOCs in water are generally on the order of 10^{-5} cm²/sec at ambient temperature.

System Properties

The equilibrium and diffusion properties of the materials to be treated are not directly under the control of the designer of an air stripping tower. Instead, the designer must vary other properties of the system to achieve the desired removal efficiency. Such general system properties include temperature, flow rates, and system geometry.

Temperature

The performance of an air stripping tower is extremely sensitive to the temperature at which it operates. A temperature change of 10°C can cause the Henry's constant of a VOC to change by a factor of 2 or 3. Although of lesser importance, diffusion coefficients also increase with temperature. Awareness of these strong temperature effects is important to the designer in two ways:

- It is essential that the designer be realistic or even conservative in assigning the temperature used as a basis of design. An air stripping tower will be severely under-designed if the design basis assumes a higher temperature than will be experienced in the actual system.
- Elevating the temperature can be considered by the designer as a means of enhancing the performance of a stripping tower. This is particularly true of the less volatile VOCs, for which stripping is difficult to achieve at ambient temperatures.

Flow Rates

The flow rates of each of the two phases (air and water) are variables that will affect the sizing of the air stripping tower.

Water flow rate is typically determined by other factors; however, if discretion is allowed, it is not unusual to take impact on tower size into consideration when determining the water flow to be treated.

Air flow rate is one of the more important discretionary variables available to the designer of a packed tower. Typically expressed as a ratio of air flow to water flow, air flow rates may range from as low as 10 to as high as 300 ft³ of air/ft³ of water. Increasing air flow improves removal efficiency but also increases blower horsepower, and thus operating costs. A tradeoff exists between increased capital cost for a deeper packed bed (or for more efficient packing material) vs. the cost of operating a blower of higher horsepower.

System Geometry

The primary system geometry variables are the tower cross sectional area (or diameter for circular cross section), the depth of the packed bed, and the size and shape of the packing material.

The tower cross-sectional area is usually determined by the hydraulics of the tower. Typically, air stripping towers are sized to provide hydraulic loading rates in the range of 5 to 30 gpm per square foot. Generally, the lower the hydraulic loading rate (larger tower cross section), the higher the removal efficiency of the tower and the lower the pressure drop through the tower. However, increased tower cross section requires additional capital cost and, as such, becomes an input into the economic tradeoffs involved in optimizing tower size. Recent literature indicates 20–25 gpm/ft² as a cost-effective hydraulic loading rate for trichloroethylene [24,25], one of the more commonly encountered VOCs.

The required depth (or height) of a packed bed is largely determined by the properties of the VOCs and the required removal efficiencies. The more difficult the removal task, the deeper will be the required packed bed. Lack of bed depth can be compensated for, to some degree, by increasing the air flow rates.

SIZING METHODOLOGY

The preceding discussion shows that a substantial number of variables affect the performance on an air stripping tower. In fact, there are literally an infinite number of tower arrangements that will achieve a desired VOC removal function. Tower sizing is therefore a process of optimization, i.e., of finding the combination of variables that will perform the desired function at the lowest overall cost.

The following concepts and equations are used to predict the performance of an air stripping tower. The concepts are followed by a suggested computational sequence that can be used to arrive at an optimal air stripping tower size.

Stripping Factor

The stripping factor is a concept of fundamental significance in sizing air stripping towers. The stripping factor for a given system is as follows:

$$R = (H/P_t)(G/L)$$

where:

R = Stripping factor
H = Henry's constant (atm/mole fraction)
P_t = Total system pressure (atm)
G = Gas (air) flow rate (moles/hr)
L = Liquid (water) flow rate (moles/hr)

or:

$$R = H_c \text{ (air/water)}$$

where:

H_c = "Dimensionless" Henry's constant (mg/L air/mg/L water)
Air/Water = Air-to-water flow ratio (ft³/ft³)

The stripping factor is truly dimensionless. Either of the above equations will give the same value for the stripping factor.

It can be shown by material balance that a stripping factor of 1.0 is the minimum stripping factor that could, in theory, achieve complete removal of a specific VOC [8]. Further, it can be shown that a stripping factor of 0.3 will only allow a maximum of 30 percent removal of a specific VOC. Thus, for a VOC with a dimensionless Henry's constant of 0.2, an air-to-water ratio of at least 5 will be required to approach complete removal.

A stripping factor of 1.0 is the minimum value for approaching complete removal. However, a packed tower would need to be infinitely tall to achieve complete removal at a stripping factor of 1.0. Therefore, in actual practice, air-to-water ratios are commonly selected to give stripping factors greater than about 5 to avoid a requirement for an exceptionally tall tower.

It is also important to note that a stripping factor is subject to the "law of diminishing returns" (results decrease as quantity increases). A stripping factor of 5 is much improved over a stripping factor of 1. Increasing the stripping factor from 5 to 20 will have a noticeable effect. Beyond a stripping factor of 20, it is difficult to see much improvement from an increase in the stripping factor.

Transfer Units

Another important concept in sizing air strippers is that of a transfer unit. A transfer unit is an artificial construction of a section of a packed bed. The difficulty involved in achieving a specified removal efficiency is indicated by the number of transfer units (*NTU*) required to perform the removal. The relative efficiency of a packed bed is indicated by the height of transfer units (*HTU*) within the bed.

Techniques are available for predicting the *NTU* required to achieve a specified removal efficiency. Techniques are also available for predicting the *HTU* in a packed bed at specified operating conditions. The product of these two parameters predicts the height (or depth) of packed bed that will be required to achieve the desired removal efficiency:

$$Z = NTU \times HTU$$

where:

Z = Required packed bed depth (ft)
NTU = Required number of transfer units
HTU = Height of a transfer unit (ft)

Number of Transfer Units

The required number of transfer units (*NTU*) depends upon the difficulty of the removal the tower is expected to achieve. Mathematically, the required *NTU* can be calculated by the following [8]:

$$NTU = \frac{R}{R-1} \ln \left[\frac{(x_{in}/x_{out})(R-1) + 1}{R} \right]$$

where:

NTU = Number of transfer units
x_{in} = Concentration in water entering tower
x_{out} = Concentration in water leaving tower
R = Stripping factor

The proper use of this equation requires that the stripping factor remain constant throughout the entire packed bed. This, in turn, requires that both the gas/liquid (*G/L*) and the Henry's constant do not vary through the tower. These conditions will generally be met if the tower is operating at ambient temperature, where there is a minimum of evaporation to change the *G/L* ratio and a minimum of temperature change to affect the Henry's constant. At elevated temperatures, it is unlikely that these conditions would be met.

Height of a Transfer Unit

The height of a transfer unit is independent of the degree of removal desired from the tower. It depends instead on the resistance to mass transfer that is inherent in the system.

Of the several empirical correlations available for predicting the transfer unit height the method of Onda et al. [9] seems the consensus favorite of authors on the subject [10,15–19, 26–28]. The correlation predicts well the effects of packing type, liquid rate, and VOC properties. There appears to be a problem with the correlation's ability to predict the effect of air flow rate, particularly for the less volatile of the VOCs, but the problem is usually minor [29].

The equations involved in the Onda model are tedious to use by hand but are easily programmed on a handheld calculator or a personal or mainframe computer. Most designers who use the model more than a few times will adapt the model to computing devices they have available. The model's equations are not repeated here but can be found in several of the referenced literature sources.

In general, the height of a transfer unit will decrease if the:

- hydraulic loading rate is lower
- air flow rate is higher
- packing surface area is higher
- VOC diffusion coefficients are higher

Suggested Computational Sequence

A number of articles have been written describing methods for optimizing air stripping tower size by using computerized optimization search programs [24–26]. For designers without access to such computing tools, a reasonable tower size can be determined by a trial and error process, as described below:

1. Select a packing with good operating experience.
2. Select a tower diameter (cross-sectional area) that will result in a hydraulic loading rate of 10–30 gpm/ft^2 for VOCs of high volatility such as TCE. For less volatile organic compounds, a lower hydraulic loading rate may be appropriate to accommodate a higher air flow rate.
3. Select an air flow rate that will give a stripping factor of 5–10 for the VOC with the lowest Henry's constant.
4. Determine the required number of transfer units from: a) the stripping factor and b) the required removal efficiency of the VOC with the lowest Henry's constant.
5. Determine the height of a transfer unit using the Onda model.
6. Determine the required packed depth.
7. Check the removal efficiency for any other VOCs present to be sure their respective removal effi-

ciencies will be achieved. If not, adjust the packed depth accordingly.
8. Check the pressure drop through the tower and compute the required blower horsepower.

Optimizing the tower size involves repeating the above sequence for a number of air flow rates, tower sizes, and packing selections, comparing the capital and operating costs of the various cases to determine the most cost-effective system design. Plotting the trends from one case to another can assist in arriving at an optimal tower configuration.

AIR STRIPPER OPERATING EXPERIENCE

Table 3 presents relevant data concerning fourteen different installations where air stripping has been used to remove VOCs from water. Many entries in the table are self-explanatory. References are identified for readers who might want additional information. Highlighted below are some aspects of the cases that are of general interest and may not be fully represented in the table.

Operation and Maintenance

The most common problem associated with operation and maintenance of packed towers is plugging of the packed bed. There are two common mechanisms for this to occur in VOC stripping: biological fouling and inorganic scaling.

Biological fouling occurred at the Wurtsmith Air Force Base stripper installation. The fouling was thought to be caused by iron bacteria, due to the relatively high iron content of the water, but such bacteria were not identified in the tower. The original 5/8" packing plugged was removed from the tower and cleaned, then replaced in the column. When the tower plugged again, the packing was replaced with 1" packing. Constant chlorination of the tower feed was instituted to control growth.

A temporary air stripper installed at the site listed in Table 3 as Confidential Site 4 plugged with inorganic scale after four months of operation. The scaling is thought to have resulted from some combination of iron precipitation, iron bacteria buildup, and/or complex coprecipitation of iron.

Acid washing was tried for tower cleaning but was not practical due to significant foaming. The tower packing was replaced with 3-1/2" Tripacks (same as original packing). An attempt to control deposits by continuous influent treatment with phosphates was not successful. Options for routine tower monitoring and cleaning are being investigated. Alternative tower packings will be considered if fouling cannot be controlled.

Fouling of the packed bed can often be avoided by providing a chemical recirculation pump and using it to routinely wash the tower with acid or hypochlorite solution. The wash solution can be introduced through the liquid distributor or through a separate nozzle bank located above the mist eliminator. A new tower should be washed every two to four weeks until an operating history is developed. If there is no evidence of fouling, the wash frequency may be reduced. The practice of waiting until the tower pressure drop increases before washing the tower results in a layer on the packing that is too thick for adequate removal with the packing in place. In the extreme, the partial plugging can result in dead zones in the tower that the wash solution cannot reach to effect a cleaning.

Packing selection may also have an effect on the tendency of the tower to plug. Selecting a large, open packing such as 3-1/2" pall-type rings may provide a tower that is more resistant to plugging than a tower with very small packing. The clearances between surfaces in the large rings is so great that it is difficult for growth or scale to bridge the gaps. Thus, for some applications, the large rings may be desired, even though their mass transfer efficiency is lower than that of most other packings.

The Verona Wellfield stripping tower uses 3-1/2" pall-type rings and has been rather routinely washed with a hypochlorite solution. The tower is located in Battle Creek, Michigan, within 100 miles of the Wurtsmith AFB system. One of the groundwater extraction wells in the Verona Wellfield system plugged solid with bacterial growth. In spite of this potential for biofouling, the Verona Wellfield stripping tower has operated for three years with no evidence of fouling. Inspection of the packing after the first year showed a thin, rust-colored layer coating the packing but no evidence of bridging or occluding the packing surface.

Air Emission Control

The Verona Wellfield air stripper was one of the first in the nation required to install air emission control equipment. A variety of emission control devices were considered for the system, including catalytic incineration, UV-oxidant destruction systems, and adsorption on activated carbon. Vapor-phase activated-carbon adsorption was selected as the most cost-effective treatment. Due to the low levels of VOC emissions expected, a nonregenerable carbon system was installed. Figure 4 is a schematic of the system.

Table 3. Air Stripper Sizing and Performance Histories

Case	Water Flow (gpm)	Temp. (°C)	No. of Towers	Tower Diam. (ft)	Packing Type	Packed Depth (ft)	Air:Water Ratio	Contaminant(s)	Influent Conc. (µg/l)	Effluent Conc. (µg/l)	Removal Efficiency[2] (%)	Stripper Cost ($1,000s)	Installed Cost ($1,000s)	Fan (hp)	Ref.
Ambient Temperature															
Port Malabar	1,000	amb.	1	9 x 9 square	structured polypropylene	12	70	TCE trans total	50 50 113	NA NA NA	>90p 99a	40	185	5	30
Tacoma Well 12A	3,500	amb.	5	12	1" plastic saddle	20	300	1,1,2,2 TCA	350	NA	89p 95a	450	900	60	31
Verona Wellfield	2,000	amb.	1	10	3-1/2" plastic p.r.	40	20	CHC	40	NA	>90	NA	926	NA	32
Wurtsmith AFB	300–600	amb.	2	5	5/8" plastic p.r.	27	10–25	TCE	500	NA	86–98ai 96–99.9as	NA	200	NA	33
Schofield Barracks	7,000	amb.	5	12	3-1/2" plastic t.p.	17.5	60	TCE	29–47	<0.5	97.2p >98.9a	360	3,950	10	30
Sydney Mine	175	amb.	1	4	3-1/2" plastic t.p.	24	200	aromatics MeCl other CHC's TCFM	31 503 41 71	ND ND ND ND	>99.8a >98.6a	NA	NA	NA	35
Savannah River	20 50 330	amb. amb. amb.	1 1 1	14 in. 20 in. 4.5	1" plastic p.r. 1" plastic p.r. 1" plastic p.r.	18 28 34	80 50 45	CHC CHC CHC	150,000 150,000 150,000	10 <1 <1	>99.9a >99.9a >99.9i	NA NA NA	NA NA NA	NA NA NA	36, 37, 38
Confidential Site 1	20	amb.	1	1.5	1" Nor-Pac	15	100	TCE	138	1.6	98.9	10	80	1.5	38
Confidential Site 2	100	amb.	2	3.25	structured polypropylene	13.5	420	aromatics	700	NA	>99.9a	44	101	3/tower	38
Confidential Site 3	60	amb.	1	38 in.	3-1/2" plastic t.p.	20	200	CHC benzene	549,230 34,000	NA NA	99.0p 99.6a	55	250	1	38
Confidential Site 4	250	amb.	1	4	3-1/2" plastic t.p.	20	50	PCE TCE trans	59–90 2.5–18 10	0.1–1.3 ND–0.3 NA	98.6–>99.8a 96–>98.3a 90p	72.4	72.4	5–10	38
Elevated Temperature															
Hydro Group	100	49–82	1	3.5	1-1/2" plastic p.r.	15	30–75	MEK	1,000	NA	61–99.9a	NA	NA	NA	39
McClellan AFB	800	65	1	8	NA	23	20	CHC MEK MIBK acetone	1,300–17,150 4,900–25,000 1,200–3,700 5,100–35,000	<0.5 45–000 45–130 100–6,300	>99.9a 97–99a 96a	NA	NA	NA	40
Gilson Road	300 300 300	71–81 11–81 60–01	1	4	structured flexipack	16	29–104 29–513 20–310	CHC THF 2-propanol	11,000 170 280	NA NA NA	>99.6a 33–>99.9a 50–87a	NA	NA	NA	41

[1] t.p. = tripacks; p.r. = pall-type rings. [2] Removal efficiency abbreviations: a = actual, p = predicted, i = individual tower, s = towers in series. NOTES: NA = not available; ND = not detected.

Key to contaminant abbreviations:
TCE   Trichloroethylene   TCA   Trichloroethane   MEK   Methyl ethyl ketone
PCE   Tetrachloroethylene   CHC   Chlorinated hydrocarbons   MIBK Methyl isobutyl ketone
MeCl Methylene chloride   TCFM Trichlorofluoromethane   aromatics Benzene, toluene, xylenes, ethylbenzene
trans trans-1,2-dichloroethylene THF Tetrahydrofuran

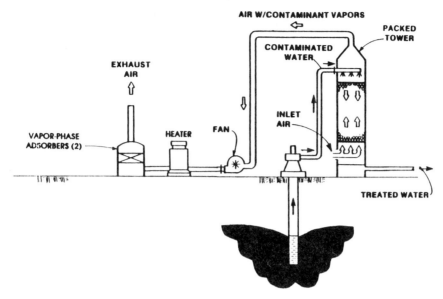

FIGURE 4. Packed tower with emission control (Verona Wellfield).

Actual experience from the project showed that adsorbing the VOCs on vapor-phase carbon used less carbon per pound of VOCs adsorbed than did an aqueous-phase adsorption system that was installed temporarily at the site. In fact, the vapor-phase carbon adsorbed about ten times as much VOC per pound of carbon as the aqueous-phase carbon [32].

Elevated Temperature Stripping

For those VOCs whose Henry's constant is too low to allow convenient stripping at ambient temperatures, elevated temperature air stripping is being used. Three cases in Table 3 are examples of elevated temperature air stripping. As shown, the removal efficiency of such hard-to-strip compounds as MEK can exceed 99 percent with elevated temperatures.

Elevated temperature air stripping does not necessarily have an exorbitant operating cost associated with heating the water. A typical full-scale application would include heat recovery equipment to preheat the feed by cooling the treated effluent. An efficient system may only require sufficient heat input to heat the feed water by 5 to 10°C after the feed is preheated by the treated effluent.

Analyzing and predicting the performance of an elevated temperature air stripper can be a more complex task than it is for an ambient temperature stripper. Depending upon the tower temperature and the air-to-water ratio, there may be a significant amount of evaporation. This evaporation will change the gas flow rate significantly, violating the simplifying assumption of constant G/L through the tower. Evaporation of water will also cool the water as it flows through the tower. This cooling will cause the Henry's constants of the VOCs to decrease as the water flows down through the packed bed, causing even more error in the simplified analysis that applies at ambient temperatures.

More rigorous tools are available for analyzing nonisothermal packed towers, but more vapor–liquid equilibrium data are required before these tools will be widely useful. In the meantime, designers are cautioned to stay very close to their pilot-scale design parameters when scaling up elevated temperature air strippers.

SUMMARY

- VOC contamination in water is receiving increased regulatory pressure and more streams will require treatment in the future.
- Air stripping is the treatment of choice for many applications of VOC removal.
- Packed-tower air stripping is an efficient process whose performance can be predicted by a relatively simple mathematical model.
- Packed-tower air stripping has been used in a variety of applications with reasonable success. O&M problems can be overcome by reasonable care in design and by a routine program of maintenance. Air emissions can be controlled when required.
- Elevated temperatures can be used to strip VOCs that are not readily strippable at ambient temperatures, but more elegant computational

techniques are required for analysis and prediction of the stripper performance.

REFERENCES

1. Mayo, F. T., C. A. Fronk and R. M. Clark. *Removal of Organic Contaminants from Groundwater; Status of EPA Drinking Water Research Program*, Water Engineering Research Laboratory, U.S. Environmental Protection Agency, Cincinnati, Ohio. Report number EPA/600/D-86/134. Reproduced by National Technical Information Service, U.S. Department of Commerce, Springfield, Virginia. Accession Number PB86-217130 (July 1986).
2. "National Primary Drinking Water Regulations; Synthetic Organic Chemicals; Monitoring for Unregulated Contaminants," *Federal Register*, Vol. 52, No. 130, July 8, 1987.
3. Ruggerio, D. "Removal of Organic Contaminants from the Drinking Water Supply at Glen Cove, NY," Municipal Environmental Research Laboratory, U.S. Environmental Protection Agency. Report Number EPA-600/2-84-029 (January 1984).
4. Love, O. T., Jr. and R. G. Eilers. "Treatment of Drinking Water Containing Tri-Chloroethylene and Related Industrial Solvents," *Journal AWWA* (August 1982).
5. Chen, G. K. "Packed Column Internals," *Chemical Engineering* (March 5, 1984).
6. Fadel, J. M. "Selecting Packed-Column Auxiliaries," *Chemical Engineering* (January 23, 1984).
7. Eckert, J. "Design Techniques for Sizing Packed Towers," *Chemical Engineering Progress*, Vol. 57, No. 9 (1961).
8. Treybal, R. E. *Mass Transfer Operations*, Third edition. McGraw Hill Book Co. (1980).
9. Perry, R. H. and C. H. Chilton. *Chemical Engineer's Handbook, Fifth edition*. McGraw Hill Book Company (1973).
10. Roberts, P. V., G. D. Hopkins, C. Munz and A. H. Riojas. "Evaluating Two-Resistance Models for Air Stripping of Volatile Organic Contaminants in a Countercurrent, Packed Column," *Environmental Science and Technology*, Volume 19, No. 2 (1985).
11. Warner, H. P., J. M. Cohen and J. C. Ireland. *Determination of Henry's Law Constants of Selected Priority Pollutants*. Cincinnati, OH:Wastewater Research Division, Municipal Environmental Research Laboratory (April 1980).
12. Dilling, W. L. "Interphase Transfer Processes, II. Evaporation Rates of Chloro Methanes, Ethanes, Ethylenes, Propanes, and Propylenes from Dilute Aqueous Solutions. Comparisons with Theoretical Predictions," *Environmental Science and Technology*, Vol. 11, No. 4 (April 1977).
13. Munz, C. and P. V. Roberts. "Mass Transfer and Phase Equilibrium in a Bubble Column," *American Water Works Association 1982 Annual Conference Proceedings*.
14. Lincoff, A. H. and J. M. Gossett. "The Determination of Henry's Constant for Volatile Organics by Equilibrium Partitioning in Closed Systems," *Gas Transfer Water Surfaces*. Brutsaert and Jurka, eds. Reidel Publishing Company (1984).
15. Cummins, M. D. *Field Evaluation of Packed Column Air Stripping*. Riviera Beach, FL:USEQP Office of Drinking Water (1985).
16. Cummins, M.D. *Field Evaluation of Packed Column Air Stripping*. Glen Cove, NY:USEPA Office of Drinking Water (1982).
17. Cummins, M. D. *Field Evaluation of Trichloroethylene Removal by Packed Column Air Stripping*. Delavan, WI:USEPA Office of Drinking Water (1982).
18. Cummins, M. D. *Field Evaluation of Trichloroethylene Removal by Packed Column Air Stripping*. Wausau, WI:USEPA Office of Drinking Water (1982).
19. Cummins, M. D. *Field Evaluation of Packed Column Air Stripping*. Bastrop, LA:USEPA Office of Drinking Water (1985).
20. Mackay, D. and W. Y. Shiu. "A Critical Review of Henry's Law Constants for Chemicals of Environmental Interest," *Journal of Physical Chemistry Reference Data*, Vol. 10, No. 4 (1981).
21. Gossett, J. M. "Measurement of Henry's Law Constants for C1, and C2 Chlorinated Hydrocarbons," *Environmental Science Technology*, Vol. 21, No. 2 (1987).
22. Ehrenfeld, J. R. et al. *Controlling Volatile Emissions at Hazardous Waste Sites* (Noyes, 1986). Appendix C: RCRA Waste Categorization based on Aqueous Volatility (Henry's Constants), pp. 393–399.
23. Lyman, W. J., W. F. Reehl and D. H. Rosenblatt. *Chemical Property Estimation Methods*, McGraw-Hill Publishers, Table 15-4 (1982).
24. Hand, D. W., J. C. Crittenden, J. L. Gehin and B. W. Lykins, Jr. "Design and Evaluation of an Air-Stripping Tower for Removing VOCs from Groundwater," *Journal of AWWA* (September 1986).
25. Nirmalakhandan, N., Y. H. Lee and R. E. Speece. "Designing a Cost-Efficient Air-Stripping Process," *Journal AWWA* (January 1987).
26. Wallman, H. and M. D. Cummins. *Design Scale-up Suitability for Air-Stripping Columns*. Cincinnati, OH:Water Engineering Research Laboratory, U.S. Environmental Protection Agency. Report Number EPA/600/2-86/009. Reproduced by National Technical Information Service, U.S. Department of Commerce, Springfield, Virginia. Accession Number PB86-154176 (January 1986).
27. Morton, C. M., T. R. Card and W. D. Byers. *Treatment of Contaminated Groundwater by Air Stripping and Carbon Adsorption*, CH2M HILL, Bellevue, Washington. Presented at 1984 Annual Water Pollution Control Federation Conference, New Orleans, Louisiana.
28. Gossett, J. M., C. E. Cameron, B. P. Eckstrom, C. Goodman and A. H. Lincoff. *Mass Transfer Coefficients and Henry's Constants for Packed-Tower Air Stripping of Volatile Organics: Measurements and Correlations*. Tyndall AFB, FL:Engineering & Services Laboratory, U.S. Air Force Engineering and Services Center, Report No. ESL-TR-85-18 (1985).
29. Byers, W. D. and C. M. Morton. "Removing VOC from Groundwater; Pilot, Scaleup, and Operating Experience," *Environmental Progress*, Vol. 4, No. 2 (May 1985).

30. McIntyre, G., J. K. Cable and W. D. Byers. *Cost and Performance of Air Stripping for VOC Removal*, presented at the 1987 ASCE National Conference on Environmental Engineering, Lake Buena Vista, FL.

31. Rosain, R. *Design and Operation of a 3,500 gpm Air Stripping System for VOC Removal*, presented at the 45th Annual Meeting, International Water Conference, Pittsburgh, PA (October 1984).

32. Byers, W. D. *Control of Emissions from an Air Stripper Treating Contaminated Groundwater*, presented at the 1986 Summer National Meeting of the American Institute of Chemical Engineers, Boston, MA.

33. Gross, R. L. and S. G. TerMaath. Packed Tower Aeration Strips Trichloroethylene from Groundwater. *Environmental Progress*, Vol. 4, No. 2 (May 1985).

34. Marske, D. and R. Torres. *Air Stripping Towers Remove TCE Contamination from Schofield Barracks Well Water Supply*, presented at the 1987 Annual Hawaii Water Pollution Control Association Conference, Honolulu, Hawaii.

35. McIntyre, G. T., N. N. Hatch, Jr., S. R. Gleman and T. J. Peschman. "Design and Performance of a Groundwater Treatment System for Toxic Organics Removal," *Journal WPCF*, Vol. 58, No. 1 (January 1986).

36. Steele, J. L. *On-site Treatment of Volatile Organic Compounds in Groundwater*, presented at the Preconference Worskhop of the Fall Conference of the Water and Pollution Control Association of South Carolina (1983).

37. Steele, J. L. *Groundwater Pollution Control*, for presentation at the 1983 DuPont Environmental Forum.

38. Internal CH2M HILL records and files.

39. Sullivan, K. M., Frank Lenzo and T. Johnson. *Pilot Testing and Design of a High Temperature Air Stripping System for MEK Removal*, presented at the 40th Annual Purdue Industrial Waste Conference (May 14–16, 1985).

40. Blaney, B. L. and M. Branscome. *Air Strippers and Their Emissions Control at Superfund Sites*, for presentation at the 80th annual meeting of APCA, New York, NY (June 21–26, 1987).

41. Roy F. Weston, Inc. *Gilson Road Hazardous Waste Site Pilot Plant Treatment Study*, prepared for New Hampshire Water Supply and Pollution Control Commission, Concord, NH (May 1983).

2.4

Vacuum Extraction: Effective Cleanup of Soils and Groundwater

James J. Malot, P.E.[1]

INTRODUCTION

Vacuum extraction technology effectively removes volatile and semivolatile compounds from soils and groundwater. Vacuum extraction is typically implemented in-situ; however, treatment of excavated soils on-site using vacuum extraction technology is also effective. Groundwater is removed simultaneously from vacuum extraction wells to further enhance the recovery of contaminants and reduce the time frame for cleanup.

Cleanup of contaminated soils using vacuum extraction is often the most important aspect of rapid, cost-effective remediation at sites contaminated with volatile organics. A leaky tank or pipeline, surface spill, storage lagoon, landfill, or other release can quickly contaminate a large volume of subsurface soil and rock. Frequently, underground pollution problems go undetected for years and take many more years before remediation begins. Vacuum extraction removes contaminants directly from the source area, eliminating further migration. Implementation can be immediate, without risk of complicating future cleanup efforts.

Transport of volatile contaminants within the vadose zone is a four-phase phenomenon consisting of complex interactions between immiscible liquid-phase components, solutes in soil water, vapors in the air-phase, and immobilized components adsorbed to soil particles. However, migration of hydrocarbons from near-surface sources is largely controlled by the hydrogeologic setting; soil permeability, depth to groundwater, underlying stratigraphy, and hydraulic gradients are some of the parameters that affect the movement of hydrocarbons underground.

Vertical migration of liquid-phase contaminants through the vadose zone is usually rapid, leaving a trail of immobile residual contamination. Contaminants percolate downward to the capillary fringe, where the pore spaces in the soil gradually become water-saturated with depth. In the capillary fringe liquid-phase contaminants that are lighter than water tend to accumulate and spread out, forming a lenticular body floating on the water table. Dense nonaqueous-phase liquids (DNAPL) follow a similar pattern in the vadose zone and also tend to accumulate on the capillary fringe before building up enough pressure to penetrate the saturated zone [1].

Most spills and leaks leave behind more residual contaminants in the soils than ever enter the aquifer or that may be recovered as free-floating hydrocarbons. The amount of liquid-phase hydrocarbons filling the pore spaces in the vadose zone as residual contaminants can be 15% to 40% [2]. This means that a volume of fine-grained soils 40 ft. by 30 ft. by 30 ft. deep can hold 10,000 gallons of gasoline. Hydrocarbons floating on the aquifer spread out and often move slowly in the direction of groundwater flow. However, as the water table fluctuates, free product rises and falls with it, and a great volume of soil becomes filled with immobile, residual hydrocarbons.

Typically, a vertical "smear zone" of residual hydrocarbons is apparent within a few feet of the water table. This smear zone expands vertically with seasonal changes in the water table and contains the highest contaminant levels beyond the source area itself. As floating hydrocarbons migrate, the trail of residual hydrocarbons that are left behind is a greater

---

[1]Terra Vac, 356 Fortaleza St., P.O. Box 1591, San Juan, Puerto Rico 00903

and more lasting threat to groundwater quality than the floating free product itself.

Even if the source is small and liquid-phase contaminants are never detected on the water table, residual contaminants within the vadose zone will leach out, continuing to contaminate the aquifer for years. A significant flux of contaminant mass is transferred downward in a vapor phase. Vaporized from residual contaminants, VOC vapors, which are heavier than air, equilibrate with water in the unsaturated zone, spreading contaminants further. Percolating rainwater and flowing groundwater slowly dissolve floating contaminants and transport them great distances within the aquifer. As a result, cleanup of residual hydrocarbons is often the most important aspect of subsurface remediation.

Considering the natural flushing of residual gasoline from the vadose zone due to precipication by advective forces, Baehr and Corapcioglu [3] use a multiphase transport model to predict that benzene will be leached into the aquifer for about twenty years; other components would take several decades longer to leach out from the vadose zone. Stringent cleanup goals for groundwater quality cannot be achieved without direct remediation of all of the residual hydrocarbons within the soils.

The high incidence of soil contamination observed at leaky underground storage systems, industrial and Superfund sites requires that efficient and cost-effective solutions be found to achieve cleanup objectives. Experience with groundwater remediation has shown that where soils are contaminated and allowed to remain in place untreated, the cleanup process is costly, lengthy, and may never reduce contaminant levels to where the water resource can again be used for a potable supply.

Vacuum extraction is an in-situ cleanup process that removes volatile and semivolatile contaminants directly from the soils. Successful cleanup of residual contaminants in soils and groundwater using vacuum extraction has been demonstrated at numerous sites [4–7]. The mechanics of vacuum extraction are described along with conditions and results at several sites. A comparison of the advantages and disadvantages of vacuum extraction with some of the more common soil remediation techniques is summarized here. Results of in-situ soil treatment are presented with respect to stringent soil and groundwater cleanup objectives that are enforced in many states.

DELINEATION OF SOIL CONTAMINATION

Prior to full-scale remediation, an assessment of the subsurface conditions is required. Subsurface contaminants existing in four distinct plume types or phases must be delineated prior to the design and installation of a vacuum extraction system. These plumes may include:

1. Liquid-phase contaminants on the water table
2. Dissolved contaminants within the aquifer
3. Residual contaminants within the soil matrix
4. Vapor plume emanating from the other plumes

The initial focus of the subsurface investigation is often the delineation of the groundwater plume, or liquid-phase plume, using groundwater monitoring wells. The delineation of the residual contaminants within the vadose zone is often neglected, although the quantification of the extent and magnitude of contamination in the soils is critical to the effective remediation of the groundwater at the site.

The fourth type of plume, vapors, emanates from the other three "sources." Due to the nature of volatile contaminants, vapors are generated from dissolved components within the aquifer, floating hydrocarbons at the water table, and residual contaminants within the vadose zone above the groundwater. Driven by diffusion, in response to concentration gradients and advection caused by natural changes in subsurface pore pressures, vapors move away from subsurface source areas. A vacuum extraction system is designed to remove volatile contaminants directly from all four types of plumes, eliminating further contamination of the groundwater.

COMPARATIVE SOIL REMEDIATION METHODS

Numerous methods for treating soils contaminated with volatiles are available. Each method has limitations with respect to the contaminant, soil types, depth, time frame, and the residual concentration that can be achieved. Unfortunately, there is no single method that is applicable for all chemicals at all sites. For sites with a wide variety of chemicals present, combination with other technologies may be required to adequately treat the soil. Some of the more common approaches to remediate soils include: excavation and treatment, pump and treat, soil flushing, bioremediation, and vacuum extraction.

Perhaps the most commonly used method to remediate soil contamination is excavation and disposal, although transporting the removed soils to a suitable disposal location is expensive and merely transfers the contamination, rather than eliminating it. Removal is also disruptive and may be impractical due to physical conditions at the site, such as the locations of buildings, roadways, or utilities, and the proximity to residents. Studies have shown that the process of soil excavation can release 60 to 90% of the volatile contaminants into the atmosphere [8], even when engineering controls are in place. Accordingly, toxic volatiles rapidly released from the process of excavation will often violate air emissions regulations in several states or cause unnecessary health risks to workers and nearby residents [8–10]. One significant advantage of the vacuum extraction process is that soils and groundwater are treated in situ, without excavation.

However, if excavation is required, treatment of volatiles can be handled effectively by using a process known as "heap vacuum extraction." In this process, contaminated soils are placed on a horizontal vacuum system and covered. All of the volatiles remaining in the soil are extracted by the "ex-situ" vacuum system. Thus, volatiles in soils are contained and treated.

Advantages of vacuum extraction systems are that the cleanup of contaminated soil and groundwater is in-situ, rapid, and low cost. Vacuum extraction systems are not limited by depth to groundwater, with successful application demonstrated at sites with groundwater as deep as 300 feet and as shallow as 3 inches. In shallow groundwater applications "dual extraction," or simultaneous extraction of groundwater and vapors from the same well, is particularly effective.

Pumping and treating groundwater is usually conducted at all sites where groundwater quality is significantly impacted. However, pump and treat systems alone do not treat the soils and source areas directly. Initial capital costs may be low to moderate, but high operations and maintenance costs are required as contaminants continue to leach from the soils. Pumping and treating groundwater alone is very slow; it takes decades to restore aquifers to the level of common cleanup goals. Where free product is present, pump and treat systems can cause groundwater to become worse, as residual hydrocarbons are spread throughout the cone of depression of a pumping well, causing groundwater concentrations to rise significantly whenever pumps are turned off. However, if pumping and treating is combined with vacuum extraction, complete remediation of soils and groundwater can be achieved.

Vacuum extraction has been demonstrated to be effective in virtually all hydrogeologic settings: clays, silts, sands, gravel, alluvium, glacial till, and fractured rock. Vacuum extraction eliminates residual hydrocarbon within the unsaturated zone. Treatment of vapors is required at most sites, although risk assessment often indicates that dispersion is sufficient to render groundwater contaminants harmless. However, carbon adsorption and catalytic oxidation are effective if complete destruction of contaminants is an objective.

For multimedia contamination sites, those sites with numerous types of compounds (i.e., VOCs, PCB, pesticides, and metals), a phased approach is often required. In these cases it is prudent to remove VOCs first by using vacuum extraction so that other technologies can then be applied more cost-effectively and safely. For example, since both chemical treatment and soil incineration require excavation, the health risks associated with the excavation of VOC-contaminated soils are minimized if the majority of the VOCs are first removed in-situ by vacuum extraction. Many methods used to chemically stabilize metals are more effective after vacuum extraction has removed VOCs.

MECHANICS OF VACUUM EXTRACTION

The first step of vacuum extraction design is the delineation of the extent and magnitude of both the soil contamination and the liquid-phase hydrocarbons that may be floating on the water table. Wells are designed with a vacuum-tight seal near the surface and an extraction zone (screen) corresponding to the profile of subsurface contamination. Extraction systems may be vertical (wells) or horizontal (screens installed in trenches or horizontal borings). Horizontal systems are particularly effective in areas where the groundwater and the contamination are very shallow (i.e., less than 10 feet) and the removal of groundwater is to be minimized.

The spacing of vacuum extraction wells is critical to efficient remediation. Depending on the depth to groundwater and the soil type, the radius of influence of an extraction well can range from tens of feet to hundreds of feet. Soil permeability, porosity, moisture content, stratigraphy, and depth to groundwater are important factors in determining the radius of influence. Vapor flow models are often calibrated to site conditions to determine design parameters and sensitivity before pilot testing or full-scale cleanup is implemented.

Vacuum is applied to the vacuum extraction system via a manifold system. The relative vacuum at the wellhead is directly related to the radius of influence of the well and the cleanup rate that can be achieved. When vacuum is transmitted to the well, subsurface vacuum propagates laterally, volatilizing contaminants in place. Subsurface air and vapors migrate towards the vacuum extraction well in response to the negative pressure gradient around the well. Figure 1 illustrates the subsurface vacuum and flow regime developed by the vacuum extraction process.

As vapors and air are removed from the subsoils under vacuum, air is recharged from the ground surface or through recharge wells. Although conceptu-

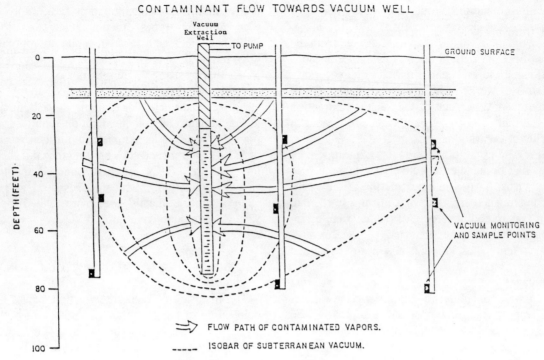

FIGURE 1. Subsurface vacuum profile and vapor flow.

ally appealing, the use of recharge wells has limited effectiveness for increasing the recovery rates from a vacuum extraction well except in certain types of multilayered hydrogeologic settings. Also, the use of injection wells has the disadvantages of dispersing contaminants further in the ground (causing more groundwater contamination) and driving contaminants, uncontrolled, into the ambient air.

Vacuum extraction technology is effective in treating soils containing virtually any chemical with a volatile character. All of the volatile priority pollutants and many of the semivolatiles have been successfully extracted with the vacuum process [11]. Gasoline, jet fuel, kerosene, and diesel have also been rapidly recovered from soils and from groundwater.

The basis for successful application of vacuum extraction to clean up soils and groundwater is Henry's Law. Simply stated, Henry's constant is the ratio of a contaminant's concentration in air divided by the concentration in water. Compounds with higher Henry's constants will clean up faster. Effective recovery of chemicals with Henry's constants greater than 0.001 (dimensionless) have been demonstrated [11].

If contaminants were released to the subsurface in a nonaqueous liquid phase, the partitioning to the vapor phase will follow Raoult's Law. This applies to the extraction of residual contaminants in the unsaturated zone and hydrocarbons floating on the water table. Cleanup of liquid-phase contaminants is faster using vacuum extraction than with conventional skimmer pumps or dual pump systems. For example, two feet of floating product has been recovered from a single well within forty-two days of vacuum extraction operation [12].

Where contaminants are within the saturated zone and groundwater is relatively shallow (i.e., less than 30 feet deep), a "dual extraction" approach is most effective. Dual extraction is a term used to describe the process of extracting groundwater and vapors under vacuum using the same well. In the simplest form, operating a submersible pump within a vacuum extraction well will lower the water table and increase the effective unsaturated zone in which the vacuum extraction process will vaporize contaminants. Figure 2 illustrates a typical dual extraction well.

Simultaneous extraction of groundwater and vapors under vacuum has several benefits that enhance the rate of groundwater cleanup. First, the rate of contaminant removal increases compared to groundwater extraction alone, since contaminants have two pathways for removal: aqueous phase and vapor. Even in areas where there have been no sources of soil contamination other than groundwater movement beneath the water table, the dual extraction process often yields the same mass flux (i.e., lbs/day) from the vapor phase as from the aqueous phase. In medium- to low-permeability aquifers the maximum rate at which groundwater can be extracted from a given well increases two- to three-fold. The net effect of these two phenomena can yield a six-fold increase in the overall contaminant removal rate, and hence, a six-fold reduction in the time required to reach cleanup objectives.

The treatment of vapors produced by the process is typically handled in one of three ways: dispersion, carbon adsorption, or thermal destruction. Other methods, such as condensation, biological degradation, ultraviolet oxidation, and others, have been applied, but only to a limited extent. Dispersion can render the vapors harmless, since quite often the health risks posed by resultant concentrations in air are much less than those posed by these vapors in groundwater or soils. Methods that destroy the contaminants are preferable; however, they do increase costs. Thermal destruction by incineration and catalytic oxidation is quite effective, especially for hydrocarbons. Chlorinated hydrocarbons can be effectively adsorbed by activated carbon. Regeneration of the carbon, on-site or off-site, is preferred to landfilling, so that chemicals extracted from carbon can either be recycled or destroyed separately.

SUMMARY OF APPLICATIONS

Vacuum extraction is applicable for cleanup of VOC contamination in soils and groundwater in virtually every hydrogeologic setting. Successful vacuum extraction systems have been operated in clays, silts, sands, gravel, fractured rock, and karst. Other hydrogeologic features, such as heterogeneities, mixed or layered materials (such as alluvium or glacial till), and shallow groundwater, are incorporated into the site-specific design. These critical design features and typical operating results are discussed for five sites. The characteristics of the sites are summarized in Table 1.

Verona Site

Verona Wellfield Superfund Site has several source areas, one of which is the former Raymond Road Thomas Solvents. This was a chemical storage and distribution facility that contained twenty-two underground storage tanks. Numerous chemicals have been identified on-site, including twenty-six volatile

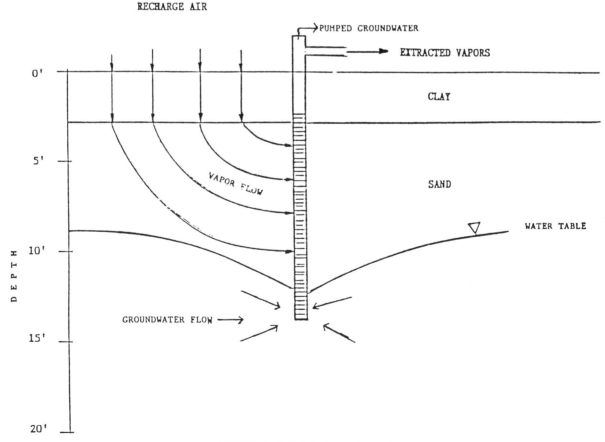

FIGURE 2. Typical dual extraction well.

Table 1.

Site	Soil Types	VOC	Depth to Groundwater	Other
Verona	Silt sand	PCE, TCE, 24 solvents	25 ft	Floating product Multiple sources
Barceloneta	Clay	Carbon tetrachloride	300	Beneath active tank farm
Tinkhams	Clay/sand layers	PCE & others	2	Dual extraction Bedrock at 13 ft
Belleview	Clay silty sand Karst	Gasoline	50	Perched water at 20 ft
Underground Tanks	Fill	Solvents	15	Monthly leak/spill monitoring

compounds monitored for the vacuum extraction cleanup. The predominant chemicals are PCE and TCE. Liquid-phase chemicals were observed floating in two of the groundwater extraction wells on-site. Groundwater was about 25 feet deep and soils consisted of silty sands.

Critical design considerations at this site were the radius of influence. Modeling indicated a radius of influence of about 60 feet, which was confirmed to be slightly higher in the pilot test. Existing underground tanks could not be safely removed without dispersing much of the contamination to the air, in violation of the local air regulations. Therefore, the impact of underground tanks on the subsurface vacuum and vapor flow regime had to be considered and quantified.

The remedial investigation had revealed that about 1700 pounds of VOCs were present at the site. The vapor treatment system was designed accordingly. However, during the initial testing of the vacuum extraction system, VOCs were extracted at a rate of 4400 lbs/day. Subsequent design considerations were made to adjust for the higher extraction rates.

Another consideration at this site was about 2 feet of floating hydrocarbons observed in one of the extraction wells. Extraction of this product by conventional means (bailing and skimming) was relatively futile, and could only remove hydrocarbons from immediately around the well, even with the groundwater depression pumps in operation. Utilization of the vacuum system to volatilize the hydrocarbons in place and remove them as vapors proved to be thorough and rapid. Floating hydrocarbons were eliminated within forty-two days of operations. Knowing that the vacuum extraction system's radius of influence extended more than 60 feet assured that the liquid-phase hydrocarbons had been completely removed over a large distance.

Barceloneta Site

Approximately 15,000 gallons of carbon tetrachloride leaked from an underground storage tank within an industrial tank farm. Beneath the tank, between 40 and 210 feet of clays and silts were present. Underlying the fine-grained soils was a karst limestone formation, highly permeable and riddled with solution channels. The top of the unconfined aquifer was within the limestone at a depth of 300 feet and served as a primary drinking water source. A subsurface investigation revealed extensive contamination in the unsaturated zone and widespread contamination of the aquifer. Roughly 4,400,000 cubic yards of soil and bedrock were estimated to be contaminated within the vadose zone.

A pilot test was conducted in the tank farm area where the highest concentrations were recorded. The vacuum extraction system was installed in the contaminated clayey soils to depths ranging from 75 to 180 feet. Approximately three weeks after a vacuum of 29.9 in. Hg. (357.19 Pa) had been applied to the wells, the subsurface vacuum had propagated 3 feet away. The subsurface vacuum continued to develop until it stabilized after about 90 days with a radius of influence of more than 10 feet. Propagation of the subterranean pressure gradient caused the flow rate of contaminants recovered by the vacuum system to increase. After the first three weeks, carbon tetrachloride was being extracted at a rate of 250 pounds per day. This illustrates a phenomenon frequently observed with low-permeability clay soils.

After the pilot test, the vacuum extraction system was expanded to treat the entire volume of contaminated soil and bedrock. When extracted vapor concentrations were lowered by 99.98% to levels below which equilibrium groundwater concentrations were less than drinking water standards, the system was shut off and soils in the area of highest previous contamination were tested. Results: carbon tetrachloride was nondetectable (less than 10 ppb) in all soil samples.

Tinkhams Garage Site

Tinkhams Garage Superfund Site is an example of successful application of dual vacuum extraction in low-permeability soils and shallow groundwater. Solvents, predominantly PCE, were released at the surface, contaminating humic clay soils. Contaminants eventually seeped into the bedrock aquifer below. Residual contaminants in the shallow soils were a continuing source of groundwater contamination, as the water table was only 2 feet below the ground surface. Since restoration of groundwater quality was the objective, cleanup of soils and shallow groundwater by dual vacuum extraction was considered the most cost-effective solution for the site [13].

On-site soil sampling and analysis indicated that most of the contaminants were held within the humic clay soils in the upper 2 feet. Also, the highest groundwater concentrations were located in the shallow sand aquifer, a direct source to the bedrock aquifer. Pilot testing of the dual vacuum extraction system demonstrated the effectiveness of VOC removal from the clayey humic soils, even though the test was conducted in subzero weather.

Submersible pumps installed in four vacuum extraction wells could produce a maximum of 2.5 gpm. When vacuum was applied to the wells, the flow rate increased to 6 gpm. The resultant radius of influence of both the vacuum system and the groundwater pumping system increased significantly. The overall removal rates of VOCs also increased as the saturated soils and shallow groundwater were cleaned up simultaneously. As a result, the time frame and cost for cleanup would be reduced dramatically compared to pumping and treating groundwater and excavating contaminated soils.

Full-scale implementation of the dual vacuum extraction system will incorporate treatment of vapors using activated carbon, pretreatment of extracted groundwater using spray aeration prior to discharge to a POTW. Based on data derived from the pilot test, full-scale cleanup to the 1-ppm goal is expected to take eight months.

Belleview Site

Three public supply wells had to be abandoned in Belleview, Florida due to leaks from a service station located approximately 600 feet upgradient from the city wellfield. Because groundwater quality was of primary concern, delineation and cleanup of the residual hydrocarbon was critical. After assessing remedial alternatives for the site, vacuum extraction was selected for demonstration, for without source control of the residual hydrocarbons, the pumping and treating of the groundwater would have been futile.

The surficial geology at the site consists of four generalized units. The surficial sands contain very little water. Where a shallow water table is encountered, it exists as perched groundwater of limited areal extent situated above relatively impermeable clays at depths of about 20 feet. Beneath the clay a relatively dry unsaturated zone of silty sands exists above variable weathered limestone.

A pilot test of the vacuum extraction process was initially conducted for three weeks but was later extended to four months to further the cleanup operations. Soil samples were analyzed from each of the four vacuum extraction well locations and three shallow borings in the tank pit area. The extent and magnitude of the subsurface contamination in vacuum extraction locations indicated high levels of residual hydrocarbons. The benzene concentrations observed in subsoils are summarized in Figure 3. Similar trends are apparent for other indicator parameters and total hydrocarbons.

At each of four test borings a vacuum extraction monitoring well was installed. Two wells were installed primarily for vacuum extraction of subsurface hydrocarbons from above and below the confining clays at 20 feet. At the other two boring locations, multilevel, dual purpose wells were installed. These wells were designed to monitor subsurface vacuum and extraction of hydrocarbons from three different hydrogeologic zones based on the contaminant profile developed from the on-site data. A profile of these extraction wells and the subsurface stratigraphy is shown in Figure 4.

Apparently, the clay layer at 17 to 20 feet has caused significant lateral migration of hydrocarbons from the source area beneath the leaky pipeline. Based on the magnitude of the original leak, an estimated 10,000 gallons, and the free product observed in downgradient groundwater monitoring wells, free product apparently migrated laterally across the top of the clay layer over much of the site where the clay exhibits low permeability. Since the clay layer is discontinuous at the monitoring well locations beyond the site boundaries, significant amounts of residual hydrocarbons remain in the soils above the clay as a lingering source of groundwater contamination.

Relative recovery rates of hydrocarbons from pilot

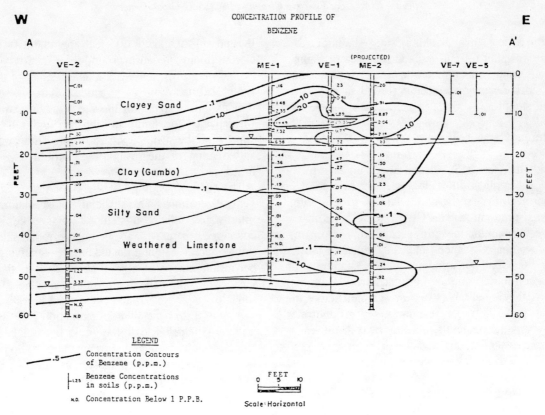

FIGURE 3. Benzene concentration contours in soil (ppm), Belleview, Florida.

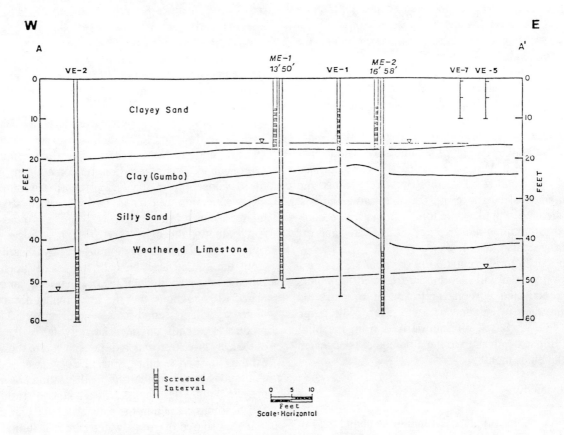

FIGURE 4. Subsurface profile and extraction well configuration, Belleview, Florida.

test wells ranged from 450 to 2000 pounds per day. During the pilot test and subsequent vacuum extraction operations, about 22,000 lbs of hydrocarbons were extracted from the subsoils at the site, corresponding to approximately 2750 gallons. As the subsoils were cleaned up by the vacuum process, the extracted vapor concentration and the hydrocarbon recovery rates delcined with time.

A comparison of the initial extracted vapor concentrations and the final vapor concentration is useful to evaluate the relative cleanup level that was achieved by the vacuum extraction process. Figure 5 indicates the relative cleanup achieved in the extraction well installed in the clayey sands. Similar results were observed in the other vacuum extraction wells.

The relative decline in extracted vapor concentrations is expected to be proportional to an aggregate soil concentration within the radius of influence of each vacuum extraction well. Essentially, the extracted vapors are in equilibrium with dissolved and adsorbed hydrocarbons within the soil matrix around the well. Overall, the relative degree of cleanup achieved by the vacuum extraction process during the demonstration period, based on initial and final wellhead concentrations of total hydrocarbons, ranged from 95.9 to 99.7%. Following the EPA test period, further reduction in concentrations were observed, ranging from 99.0 to 99.99% reductions.

Underground Tank

Early detection and cleanup of spills and leaks from underground storage facilities minimizes the impact on groundwater quality. Vacuum extraction systems can be designed to provide continuous or intermittent monitoring of storage tanks and immediate cleanup if a leak or spill is ever detected.

With one-time installation, extraction wells are constructed to assure a tight seal within the vadose zone. Wells are located beneath or adjacent to the underground tank, storage facility, spill area, or pipeline that is being monitored. The spacing of the extraction or monitoring wells is based on the radius of influence of the wells so that sufficient overlap exists.

Unlike simple vapor detection systems or tank inventory systems, the vacuum extraction system recovers a spill or leak of any volatile material within the storage facility. Based on the recovery rate and the type of release (leak or spill), approximate location and magnitude can be determined from on-site data. Cleanup operations can begin at the moment they are detected, avoiding costly groundwater

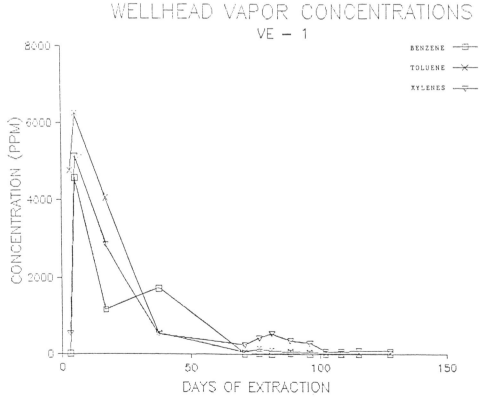

FIGURE 5. Cleanup rate at Belleview site, Florida.

assessment and cleanup. Installation, testing, and cleanup can all be conducted without interruption of the facility operations. Leak detection rates as low as 1 gram/hour have been achieved.

CLEANUP CRITERIA

Two criteria for evaluating cleanup effectiveness are often considered at sites:

1. Achieving prescribed residual soil concentration
2. Eliminating impact on groundwater quality

Evaluation of the concentration versus time data is used to estimate the time frame for cleanup. Depending on the local regulations, anticipated land and groundwater uses, type of contaminant, depth to groundwater, and other political constraints, cleanup levels for VOCs in soils can range from 5 ppb to 100 ppm. Although these levels can be achieved, it will invariably take longer to reach lower cleanup goals.

As a practical alternative to prescribed soil concentrations, the Florida Department of Environmental Regulations has defined excess soil contamination as 500 ppm of hydrocarbons in the headspace of soils. The concentration extracted from the wellhead may be considered essentially an aggregate "headspace" concentration of hydrocarbons in the soils around the well screen. Assuming that any disparity between static equilibrium established under a headspace test and extracted concentrations after the well has been allowed to requilibrate were negligible or at least quantifiable, the 500 ppm hydrocarbon response would represent the upper limit of a cleanup. However, lower values may be required at certain sites in order to protect groundwater resources. At the Belleview site mentioned above, cleanup to these levels was demonstrated within five months of vacuum extraction operations.

An alternative criterion for considering the cleanup goal would be based on a drinking water standard for indicator parameters, such as benzene, TCE, or some other compound, for which groundwater standards could be compared. Goals for cleanup of the vadose zone may be determined using the concentration of the indicator compound in the soil water just above the water table and Henry's Law to calculate the associated vapor concentration. Presumably, if the concentration of a contaminant in the soil water is at groundwater standards, the source of groundwater contamination has been eliminated. Since groundwater is the primary exposure pathway for contaminants in the soils, applying the groundwater standard to the soil water concentration just above the water provides a standard to which vacuum extraction rates are evaluated.

As an example, Florida has adopted a groundwater cleanup criterion for benzene of 1 ppb. Based on Henry's constant for benzene, the soil gas concentration in equilibrium with water at 1 μg/L would be about 0.18 μg/L or 0.05 ppm benzene in air. Assuming the extracted vapors from the wellheads are sufficiently close to equilibrium conditions, an area within the radius of influence of the well may be considered clean if the concentration of extracted benzene vapors is less than 0.05 ppm. This is analogous to a leaching model that is often applied at Superfund sites to determine acceptable maximum concentration levels (MCLs) for soils to prevent degradation of groundwater quality.

CONCLUSIONS

Cleanup of contaminated soils using vacuum extraction is often the most important aspect of rapid, cost-effective remediation at sites contaminated with volatile organics. Vacuum extraction technology effectively removes volatile and semivolatile compounds from soils and groundwater. Vacuum extraction is typically implemented in-situ; however, treatment of excavated soils on-site using vacuum extraction technology is also effective. The system can also be applied to monitor underground storage facilities. Dual vacuum extraction treats groundwater and soils simultaneously to further enhance the recovery of contaminants and reduce the time frame for groundwater cleanup. Stringent cleanup goals for soils and groundwater can be achieved using vacuum extraction.

REFERENCES

1. Wilson, J. L. and S. H. Conrad. "Is Physical Displacement of Residual Hydrocarbons a Realistic Possibility in Aquifer Restoration?" *Proceedings of Petroleum Hydrocarbons and Organic Chemicals in Groundwater*, Nov. 5–7, 1984, NWWA, Worthington, OH.

2. Schwille, F. *Dense Chlorinated Solvents in Porous and Fractured Media, Model Experiments*. Translated by J. F. Pankow. Chelsea, MI:Lewis Publishers (1988).

3. Baehr, A. and M. Y. Corapcioglu. "A Predictive Model for Pollution from Gasoline in Soils and Groundwater," *Proceedings of Petroleum Hydrocarbons and Organic Chemicals in Groundwater*, Nov. 5–7, 1984, NWWA, Worthington, OH.

4. Malot, J. J. and P. R. Wood. "Low Cost, Site Specific, Total Approach to Decontamination," *Environmental and Public Health Effects of Soils Contaminated with*

Petroleum Products. P. Kostecki, ed., Wiley & Assoc. (1988).

5. Applegate, J., J. K. Gentry and J. J. Malot. "Vacuum Extraction of Hydrocarbons from Subsurface Soils at a Gasoline Contamination Site." *Proceedings of "Superfund '87" Conference, November 1987, Washington, DC.*

6. Weston, USATHEMA.

7. Agrelot, J. C., J. J. Malot and M. J. Visser. "Vacuum: Defense for VOC Contamination," *Proceedings of the Fifth Annual Symposium on Groundwater Restoration, 1985, NWWA, Columbus, OH.*

8. "Public Health Risk Assessment of the Tyson's Site Remediation Plan," *ENVIRON*, 1987. Administrative Record of Tyson's Superfund Site.

9. Danko, Jo. CH2M HILL, Personal communication regarding air quality violations during excavation at Verona Superfund Site, 1987.

10. California Dept. Health Services Regulation Title 22. Regulations prohibiting excavation of soils contaminated with volatiles.

11. Terra Vac, Application for Demonstration of Vacuum Extraction Technology, Superfund Innovative Technologies Evaluation Program, 1986, EPA, Hazardous Waste Evaluation Laboratory, Cincinnati, OH.

12. Record of Decision, Tinkhams Garage, Cannons Superfund Sites, USEPA Region 1 Administor, 1988.

13. Terra Vac Monthly Report to CH2M HILL and EPA, November 1988, Raymond Road Thomas Solvents Site, Verona Wellfield Superfund Site, Battle Creek, MI.

2.5

Catalytic Hydrodechlorination

Bijan F. Hagh,[1] David T. Allen[1]

ABSTRACT

Chlorinated organics are among the most significant and widespread toxic materials in the environment. Because chlorinated organics are very difficult to destroy by incineration, new technologies must be developed for recycling or destroying these materials. A process for recycling chlorinated organics that has been under development at the UCLA Engineering Research Center for Hazardous Substances Control is catalytic hydrodechlorination (HDC). Catalytic HDC has potential applications in recycling of agricultural chemical wastes, in PCB and PCP remediation, and in oil recycling. All of these applications involve hydrogenating a carbon–chlorine bond. Therefore, this chapter will focus on the basic chemistry of HDC.

Our initial efforts in catalytic HDC of chlorinated organics were focused on identifying a suitable catalyst. A catalyst screening study in a batch autoclave reactor was performed in order to identify catalysts with a high dechlorination activity that would also be relatively unaffected by the wide variety of functionalities present in waste streams. Hydroprocessing catalysts used in heavy oil upgrading were logical candidates for robust dechlorination activity. Noble metal catalysts were also screened. Chlorobenzene and 1,2-dichlorobenzene were used as model compounds in the screening study. Presulfided NiMo supported on Al_2O_3 showed excellent dechlorination activity with both compounds. Reduced Pd supported on Al_2O_3 also showed good activity with chlorobenzene but was less successful with dichlorobenzene.

Based on these early results, presulfided NiMo supported on Al_2O_3 was used in all subsequent experiments. A fixed-bed, continuous-flow microreactor was constructed. Unlike the batch autoclave used in the catalyst screening, this reactor system allowed the determination of reaction mechanisms and reaction rates. These data on rates and mechanisms can be used to design and optimize a commercial reactor. The dechlorination reactions of chlorobenzene have been examined in the new reactor system, and the results indicate that the NiMo catalyst retains its remarkable activity over at least several days. Current work is focusing on the interactions between denitrogenation reactions and dechlorination reactions. These competing and interacting reaction pathways are particularly important in oil recycling and agricultural chemical recycling.

INTRODUCTION

This chapter will focus on the hydrodechlorination of chlorinated organics. This chemistry has applications in agricultural chemical manufacturing, oil recycling, and PCB remediation as described below.

Chemical Manufacturing

Chlorinated aromatics are widely used in pesticides and other agricultural chemicals. Often, only certain isomers of these chlorinated hydrocarbons have desirable properties. Since the synthesis processes employed are not completely stereospecific, the treatment and recycle of undesirable isomers pose an important waste minimization problem. Catalytic hydrodechlorination is being explored as a source reduction method, with the case study of chloropyridines in mind. Chlorinated pyridines are a major agricultural chemical. Only a few of the over 200 chloropyridine isomers are useful products, so in manufacturing chloropyridines, many streams are recycled internally. Still, a significant amount of these materials ends up in tars that must be incinerated. These materials are a logical candidate for hydrodechlorination. Preliminary results on the hydroprocessing reactions of chloropyridines will be presented later in this chapter.

Oil Recycling

Oil recycling is becoming a large industry. Recyclers typically separate lighter components from

[1]Department of Chemical Engineering, University of California, Los Angeles, CA 90024

waste oils using conventional distillation, then treat the residue using catalytic hydroprocessing. A difficult problem facing recyclers is catalyst deactivation due to the reaction of NH_3 and HCl, producing NH_4Cl. The NH_3 and HCl are products of denitrogenation and dechlorination reactions. Part of the data presented in this chapter will address the interaction of these reaction pathways.

PCB Remediation

One of the most significant applications of catalytic hydroprocessing is PCB remediation. Polychlorinated biphenyls, commonly called PCBs, are a group of chlorinated organic compounds in which one to ten chlorine atoms are attached to biphenyl. PCBs have extremely high thermal and chemical stability. They are relatively nonflammable and have excellent electrical insulating characteristics. The physical properties of PCBs make them highly useful in numerous commercial applications, including dielectric fluids in capacitors and transformers, heat transfer fluids, hydraulic fluids, lubricating and cutting oils. PCBs were commercially produced as complex mixtures beginning in 1929 and are not known to occur naturally. Commercial mixtures were synthesized by chlorination of biphenyl with chlorine gas over a metal chloride catalyst. The production of PCBs ceased in 1977 with a total worldwide production of PCBs estimated to be 2.4 billion pounds.

Animal studies with commercial mixtures of PCBs have shown a variety of chronic toxic effects, including birth defects, high offspring mortality rates, liver damage, and enlargement of the liver, kidneys, and heart (Ackerman and Scinto, 1983). PCB-contaminated cooking oil caused a total of 1,291 "Yusho" patients in 1968 in western Japan. Clinical tests done on the patients showed low birth weights, chloracne, and liver damage. The realization of the widespread distribution of PCBs in the environment and growing knowledge of their hazards led Monsanto, the world's largest producer of PCBs, to cease all production in 1977. Growing evidence of the problem of PCB contamination prompted a public outcry that culminated in the United States in 1976 with the regulation of PCBs by the Toxic Substances Control Act (TSCA). The TSCA banned the manufacture and use of PCBs in other than a totally enclosed manner by January of 1979, and a complete ban on manufacture, processing, and distribution in commerce of PCBs by July of 1979. Section 6e of TSCA required the Environmental Protection Agency (EPA) to regulate the disposal of PCBs already in use. Large amounts of PCB-containing products such as transformer oils and capacitors are in service. These devices are subject to or are currently awaiting disposal.

Disposal of PCB-contaminated products has generally been limited to incineration, although some chemical degradation processes (e.g., the Goodyear process) are permitted in the United States. Some of the problems associated with PCB incineration are listed below.

- The chlorine atoms are effective flame retardants. They tend to quench the hydrogen free radicals, which propagate the reactions in the incinerator. Therefore, it is difficult to achieve complete combustion.
- Incinerators have high operating costs due to high operating temperatures (1200°C–1600°C).
- The combustion stack gases present a severe corrosion problem at the high operating temperatures.
- The problem of greatest concern is the generation of products of incomplete combustion, which are formed in trace quantities. When PCBs are burned, dioxins and chlorinated dibenzofurans (CDBF) can be formed. Other materials contained in the stack effluents include HCl, CO, CO_2, NO_x, and O_2. Removing these materials from the stack gases can be very difficult and expensive.
- Currently there are only three licensed incinerator units in the United States (Texas, North Carolina, and Tennessee); therefore, PCB wastes from the rest of the country must be transported to those sites. As a result there are backlogs at the sites, and the transportation of the wastes can be very dangerous and expensive.

A number of chemical methods have been proposed for the treatment of PCB-contaminated materials. The best known of these include a sodium naphthalide method (Goodyear's) and a sodium–amine method (PCBX).

Goodyear Sodium Naphthalide Process

Goodyear Tire and Rubber Co. developed a chemical treatment process for PCB transformer and heat transfer fluids. Major steps in the process are shown in Figure 1. Preparation of the reagent begins with the dispersion of molten sodium in hot oil to produce highly reactive, finely dispersed droplets of liquid sodium. Agitation is continued while the mixture is rapidly cooled to room temperature, causing the

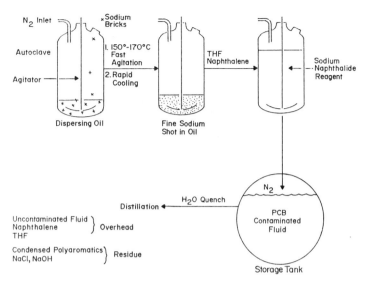

FIGURE 1. Goodyear process.

sodium droplets to solidify into fine, bright spheres. A solution of naphthalene and tetrahydrofuran is then added to the mixture to form sodium naphthalide. Additional stirring for 1–4 hours results in a homogeneous solution. The sodium naphthalide breaks the carbon–chlorine chemical bonds to produce sodium chloride and hydrogen ions. The reaction takes less than five minutes at room temperature, and a minimum ratio of 50 moles of naphthalide to 100 moles of chlorine is required to remove 98% of the PCBs from a standard heat transfer fluid containing 82 ppm of PCBs.

A number of concerns have hampered the widespread use of this process. Extreme caution is necessary with reactions involving metallic sodium. Hydrogen is rapidly generated when sodium makes contact with water. During the water-quench step, water must be added in small amounts over a long period to minimize possible hydrogen release. Nitrogen, or a similar inert gas, is used as a blanket to prevent the formation of potentially explosive hydrogen/oxygen mixtures. Furthermore, EPA has classified naphthalene as a restricted compound. The costs of this process are reported to be about $2.75/gal to reduce PCB concentrations from 500 ppm to zero (Fradkin and Barisa, 1982).

Sunohio PCBX Process

PCBX is a chemical destruction process. The chemical reactions are still proprietary information, but they probably involve organic sodium compounds in an amine solvent. The chlorine substituents in PCBs are converted to sodium chloride, while the organic portion of the PCB molecule, the biphenyl nucleus, is converted to polymeric solids, which are insoluble in water.

The various parts of a mobile unit are illustrated in Figure 2. The mobile processing unit is permitted to decontaminate 600 gal/hr of transformer oils containing up to 2600 ppm of PCBs. The EPA requires that the residual PCB concentration be under 2 ppm.

Since the PCBX process reduces the PCB concentration in cycles, the economics of the process are less favorable than Goodyear's. For example, three cycles are required to decrease the PCB concentration to 0.6% of the original contamination, e.g., from 100 ppm to 6 ppm. This process also involves the use of sodium in a potentially toxic solvent, so it possesses many of the same drawbacks as the Goodyear process. At present, costs are estimated at $3/gal if the transformer oil is reusable, and $7/gal if new oil is needed (Fradkin and Barisa, 1982).

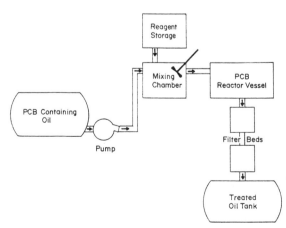

FIGURE 2. PCBX process.

Catalytic Hydrodechlorination as an Alternative Disposal Method

Catalytic hydrodechlorination can be used as an alternative disposal method for PCBs. The advantages of catalytic hydrodechlorination of PCBs over current incineration practices are listed below.

- Catalytic dechlorination can be effective at a much lower operating temperature than incineration (350°C vs. 1200°C to 1600°C).
- The biphenyl product would have value as a fuel additive and could be recovered rather than burned.
- The effluent from the catalytic dechlorination unit could be monitored much more closely than the effluent from an incinerator.
- The formation of dioxin, CDBF, NO_x, CO, and CO_2 are virtually eliminated. The mechanisms for the formation of dioxin from combustion reactions are poorly understood, but it is obvious that a source of oxygen is necessary. Catalytic hydrodechlorination requires no supply of oxygen; therefore, dioxin is unlikely to be produced.
- The greatest advantage of catalytic hydrodechlorination is its capability of being moved to or located at the site where the PCB wastes are being stored.

The operating costs of catalytic dechlorination appear to be lower than any of the present disposal technologies. During the past year, UOP has begun marketing the proprietary HDC technology for treating PCB-laden transformer fluids (Kalnes and James, 1988). Their process is shown in Figure 3. The remainder of this chapter will describe the work done in our laboratory over the past three years on HDC chemistry. The work is presented as three steps.

1. Various catalysts were screened for hydrodechlorination activity using chlorobenzene and dichlorobenzene as model compounds.
2. The most promising catalyst from the screening study was studied in a continuous-flow microreactor. Catalyst lifetimes, reaction mechanisms, reaction rates, and the effects of various contaminants were studied.
3. Using the experimental data obtained above, the suitability of HDC for processing a variety of wastes will be discussed.

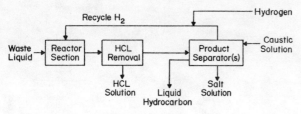

FIGURE 3. Process flow diagram for hydrodechlorination (Kalnes and James, 1988).

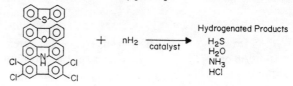

FIGURE 4. Process chemistry for hydrodechlorination.

PROCESS CHEMISTRY

Some of the reactions involved in hydrodechlorination of waste streams are shown in Figure 4. Note that the reactions frequently involve the cleavage of an aromatic carbon–chlorine bond. The goals of our work on process chemistry are to identify active catalysts, to quantify reaction rates, and to elucidate reaction mechanisms. Our progress to date is summarized in this section on Process Chemistry.

Catalyst Screening Study

The goal of the catalyst screening study was to identify catalysts with a high dechlorination activity that would also be relatively unaffected by the wide variety of functionalities present in toxic waste streams. Hydroprocessing catalysts used in heavy oil upgrading were logical candidates for robust dechlorination activity, since the carbon–chlorine bond energy is within the range of bond energies susceptible to hydrotreatment (Hendrickson et al., 1970). Noble

Table 1. Catalysts Selected for Screening

Catalyst	Composition	Pretreatment
CoMo on γAl_2O_3	(3.1% Co, 7.5% Mo)	None
CoMo on γAl_2O_3	(3.1% Co, 7.5% Mo)	10% H_2S in H_2 at 350°C, >1 hour
NiMo on γAl_2O_3	(2.0% Ni, 7.0% Mo)	None
NiMo on γAl_2O_3	(2.0% Ni, 7.0% Mo)	10% H_2S in H_2 at 350°C, >1 hour
NiW on γAl_2O_3, HR354	(2.7% Ni, 18.3% W)	3% H_2S in H_2 at 350°C, >2 hours
CoMo on γAl_2O_3, HR306	(2.4% Co, 9.3% Mo)	3% H_2S in H_2 at 350°C, >2 hours
NiMo on γAl_2O_3, HR348	(2.2% Ni, 10.75% Mo)	3% H_2S in H_2 at 350°C, >2 hours
Pd on γAl_2O_3	(0.5% Pd)	None
Pd on γAl_2O_3	(0.5% Pd)	H_2 at 400°C, >4 hours

metal and nickel catalysts have also been reported to have good hydrodechlorination activity (Dini et al., 1975; LaPierre et al., 1978a, 1978b, Coq et al., 1985, 1986).

Chlorobenzene and 1,2-dichlorobenzene were used as model compounds due to their simplicity, their availability, and their presence as contaminants in many waste streams. The catalysts used for the screening study were NiMo, CoMo, NiW, and Pd supported on Al_2O_3. The catalysts used in this study and the pretreatment applied are listed in Table 1. The screening study was performed in a batch autoclave reactor. Complete experimental details are available from the authors.

The results of chlorobenzene hydrodechlorination experiments are summarized in Table 2. The catalysts with the highest chlorobenzene conversion were then used to dechlorinate 1,2-dichlorobenzene. These experiments were carried out to determine the steric effects of the chlorine atoms. The results of the experiments with 1,2-dichlorobenzene are summarized in Table 3. The total absence of chlorocyclohexane and cyclohexane among the reaction products clearly indicated that chlorine atoms are removed before any hydrogenation of the aromatic ring occurs.

The catalyst screening study can be summarized as follows. First, presulfided NiMo supported on Al_2O_3 showed excellent dechlorination activity with both chlorobenzene and 1,2-dichlorobenzene. Reduced Pd supported on Al_2O_3 also showed good activity with chlorobenzene but was less successful with dichlorobenzene. In addition, the pretreated catalysts generally showed better dechlorination than untreated catalysts.

Reaction Rates and Mechanisms

The data collected in the batch autoclave reactor provided a reasonable basis for selecting a hydrodechlorination catalyst; however, the data were inadequate for designing a commercial process. Therefore, a high-pressure, continuous-flow microreactor system was constructed to provide the mechanistic and rate data crucial for process design. This reactor system can be used for catalyst lifetime studies and can be operated at differential conversions. Operating under differential conditions (see, for example,

Table 2. Dechlorination Activity of Screened Catalysts

Catalyst	Reaction Temperature (°C)	Reaction Time (hr)	H_2/Chlorobenzene Molar Ratio	Chlorobenzene to Catalyst Weight Ratio	Chlorobenzene Conversion (%)
Presulfided NiMo on γAl_2O_3	350	1	2.60	1.0	99.2
Reduced Pd on γAl_2O_3	352	1	2.60	1.0	99.6
Untreated NiMo on γAl_2O_3	350	1	2.60	1.0	23.8
Untreated Pd on γAl_2O_3	352	1	2.60	1.0	95.1
Presulfided HR306	352	1	2.60	1.0	84.5
Presulfided HR348	352	1	2.60	1.0	98.7

Table 3. Dechlorination Activity and Selectivity

Catalyst	Reaction Temperature (°C)	H_2/Dichlorobenzene Molar Ratio	Dichlorobenzene Conversion (%)	Benzene/Chlorobenzene Selectivity (%)
Presulfided NiMo on γAl_2O_3	350	2.9	96.92	97.62
Untreated NiMo on γAl_2O_3	350	2.9	72.13	72.02
Presulfided CoMo on γAl_2O_3	350	2.9	16.69	9.99
Reduced Pd on γAl_2O_3	350	2.9	96.90	61.50

In all runs, a reaction time of 1 hour and a dichlorobenzene/catalyst ratio of 1.0 mL/g were used.

Levenspiel, 1972) is essential for determining rate expressions, which in turn determine the optimum reaction conditions.

The microflow differential reactor system shown in Figure 5 consists of a one-liter stirred autoclave (rated for 2000 psi at 230°C), a SSI 100 LC pump controlled by a SSI 210 Guardian (rated for 0.05–9.99 cm³/min at 6000 psi), a Lindberg 55122 Moldatherm hinged tube furnace (rated for 1100°C) controlled by an Omega CN9111 temperature controller, a tubular fixed-bed reactor (1/4 inch diameter Hastelloy-C tubing), a 250-microliter sampling valve (Hastelloy-C), and two 500-ml pressure vessels (rated for 2000 psi at 232°C). The stirred autoclave is used to saturate the feed to the reactor with hydrogen, assuring that both the chlorinated organic and the molecular hydrogen are in the same phase. The feed to the reactor typically contains 0.01–0.05 weight fraction of reactant (e.g., chlorobenzene) in hexadecane saturated with dry, oxygen-free hydrogen at a controlled temperature and pressure. This saturated mixture is then pumped at a controlled rate over a heated reactor bed. The bed consists of a catalyst section, preceded and followed by inert alundum sections. The temperature of the catalytic section of the reactor was controlled within ±1°C, as measured by a thermocouple inserted directly into the catalyst bed. The product stream is sampled by using a high-pressure sampling valve. Depending on the degree of corrosivity, the product stream is stored in either a 500-ml

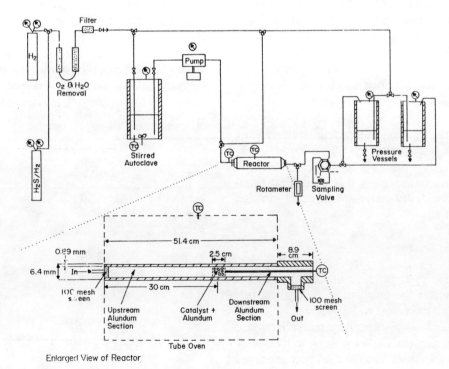

FIGURE 5. Hydrodechlorination microreactor.

stainless steel vessel or a 500-ml Hastelloy-C vessel. These vessels are used to maintain back pressure in the system and are periodically drained to maintain a constant reactor pressure. Products are quantified using gas chromatography; gas chromatography/mass spectroscopy is used to identify products. It must be emphasized that these reaction conditions are designed to elucidate process chemistry. They are *not* designed to achieve complete dechlorination or to represent actual waste mixtures. Demonstration of HDC on actual waste streams has been done, however, and is reported by Kalnes and James (1988).

In any study of catalytic rates and mechanisms, precautions must be taken to ensure that the data actually represent the chemistry occurring on the catalyst surface. Thus, a number of preliminary experiments were performed to address mass transfer effects, the catalytic properties of the reactor walls and packing materials, and the catalyst stability. These experiments are briefly described below.

Effects of Reactor Walls, Packing Materials, and Carrier Solvent

The reactor walls and the thermocouple, which measured temperature in the reaction zone, both contain traces of Ni and thus could serve as catalysts. To determine whether these materials were catalyzing dechlorination, a 0.0216 weight fraction chlorobenzene solution in *n*-hexadecane was pumped through a reactor filled with only the alundum packing material. At a flow rate of 1.6 cm³/min and at a reactor temperature of 350°C, less than 0.1% of the chlorobenzene was dechlorinated. At the same conditions, a reactor containing 0.04 g of catalyst dechlorinates 30% of the chlorobenzene fed. Thus, the reactor and packing materials do not play a significant role in the dechlorination reactions.

A second concern was that the solvent *n*-hexadecane might react, interfering with product analysis. At the most severe reaction conditions used in this work, less than 2% of the *n*-hexadecane underwent cracking reactions, and none of the products interfered with the analysis of the dechlorination reaction products.

Mass Transfer Effects

The rate of a catalytic reaction can depend not only on the rate of the reaction on the catalyst surface, but also on the rate at which reactant molecules arrive at the surface. To accurately measure surface reaction rates, the reactor must be able to deliver reactant molecules to the catalyst surface much faster than the surface reaction can consume them. Two

Table 4. Intraparticle Mass Transfer Effects

Catalyst Particle Size (μm)	Rate of Chlorobenzene Disappearance (mole CB/g cat-min) ×10⁴ᵃ	
	300°C	350°C
<250	4.2 ± 1.4	20.5 ± 3.0
250–355	4.3 ± 1.0	17.0 ± 4.1
355–470	3.1 ± 2.5	14.8 ± 4.1

ᵃError bands are based on 95% confidence limits.

mass transfer processes can be important. The first, transport of reactants to the catalyst pellet surface, was examined theoretically in this work and was found to have no effect. The second, transport of reactants within the catalyst pores, was examined experimentally by varying the size of catalyst pellets. The results, shown in Table 4, indicate that there were no intra-particle mass transfer effects.

Catalyst Stability

The catalyst activity remained relatively constant throughout the course of the experiments. Figure 6 shows hydrodechlorination rate as a function of time on stream. Because H₂S is frequently used in hydroprocessing to maintain catalyst activity, experiments were run both with and without H₂S. The presence of H₂S did not significantly affect catalyst lifetimes but did lower reaction rates slightly.

Dechlorination Rate Studies

The rate of the hydrodechlorination reaction will depend on the concentration of the reactants, on the reaction temperature, and, possibly, on the concentrations of the products. The rate data shown in Table

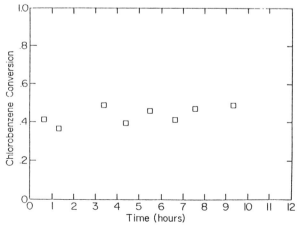

FIGURE 6. Catalyst activity over a 12-hour dechlorination run.

5 can be used to provide a preliminary assessment of the hydrodechlorination rate. The rate appears to be linearly proportional to the chlorobenzene concentration, and the temperature dependence of the rate is consistent with an activation energy of 10 kcal/gmol. These data point to either mass transfer or reactant adsorption as the rate-limiting step for dechlorination. Since experiments described earlier in this section seem to rule out mass transfer effects, we tentatively conclude that adsorption of reactants onto the catalyst controls the rate of surface reaction.

Hydroprocessing of Chloropyridines

The hydroprocessing reactions of chloropyridines were examined to assess the ability of HDC to treat waste streams containing both nitrogen and chlorine atoms. The presence of both nitrogen and chlorine in a waste stream may present problems, since the hydroprocessing products NH_3 and HCl can combine to form NH_4Cl. The ammonium chloride can sublime, plugging process lines and fouling the catalyst.

The hydroprocessing reactions of 3-chloropyridine were examined. Preliminary results indicate that the dechlorination reaction is much faster for chloropyridine than for chlorobenzene. Some of the pyridine resulting from the dechlorination reacts further. Plugging of the reactor is not a problem at reactor conditions; however, as the product stream cools down after leaving the reactor, precipitation and plugging become a serious problem. Additional work is planned to further investigate these preliminary findings.

Summary of Process Chemistry

- At moderate conversions, the dechlorination reaction is very selective, with very little hydrogenation of the aromatic ring. At the high conversions necessary for waste processing, some ring hydrogenation does occur, as reported by Kalnes and James (1988).
- The rate-limiting step in the catalytic hydrodechlorination of chlorobenzene appears to be adsorption of chlorobenzene onto the catalyst site. This process has an apparent activation energy of 10 kcal/mol.
- Hydrodechlorination over a NiMo catalyst at 300–350°C is a very fast reaction. Although adsorption appears to be the rate-limiting reaction step, in a commercial reactor, mass transfer rates will probably control the overall dechlorination rate.
- Dechlorination of waste streams containing organic nitrogen compounds can be very difficult. Reaction products from simultaneous dechlorination and denitrogenation can plug process lines.
- All of the experiments described in this chapter have used liquid-phase reactants at high pressure. The dechlorination reactions can proceed at low pressure if the hydrogen to chlorine molar ratio remains greater than 1 (La Pierre et al., 1978). The role of the high pressure is to ensure that enough hydrogen reaches the catalyst.

Table 5. Dechlorination Rate Data

Chlorobenzene Conc. (mole/L)[a]	Hydrogen Conc. (mole/L)[a]	H_2S Conc. (mole/L)[a]	Rate of Chlorobenzene disappearance (mole CB/g cat-min) $\times 10^{4}$[b]
\multicolumn{4}{c}{Experiments at 300°C}			
0.09	0.31	0	4.2 ± 1.4
0.18	0.24	0	7.1 ± 1.8
0.05	0.26	0	1.80 ± 0.08
\multicolumn{4}{c}{Experiments at 325°C}			
0.19	0.24	0	6.6 ± 0.4
0.05	0.25	0	3.1 ± 1.2
0.10	0.29	0.02	10.8 ± 1.8
\multicolumn{4}{c}{Experiments at 350°C}			
0.10	0.25	0	20.5 ± 3.0
0.18	0.23	0	11.9 ± 2.4
0.04	0.24	0	5.9 ± 0.3
0.09	0.25	0.02	17.3 ± 2.0

[a] Concentrations are calculated at reactor conditions.
[b] Error bands are based on 95% confidence limits.

Future Work

Because of the wide applicability of catalytic hydroprocessing, future work will be directed toward determining relative rates of dechlorination for a wide variety of chlorinated organics. These data will help define the limits of HDC applicability. Further, our preliminary work has indicated that the presence of some functionalities, such as nitrogen groups, makes dechlorination impractical. Identification of troublesome functionalities will also be a major focus of our continuing work.

CONCLUSIONS

Catalytic hydrodechlorination (HDC) is a potential treatment method for a variety of waste streams, including PCBs, agricultural chemicals, and waste oils. Although some applications are still in the conceptual phase, the technology has been demonstrated for PCB remediation and a few petrochemical waste streams (Kalnes and James, 1988). The goals of this chapter have been to present the basic process chemistry and to discuss the implications of the chemistry in the design of commercial-scale units. The main conclusions to be drawn from the work to date are that dechlorination is a viable process for many waste streams; however, the presence of certain combinations of compounds makes dechlorination impractical.

ACKNOWLEDGEMENT

This work was supported by the UCLA/NSF Engineering Research Center for Hazardous Substances Control (NSF grant CDR 86 22184).

REFERENCES

Ackerman, D. G. and L. L. Scinto. "Destruction and Disposal of PCBs by Thermal and Non-Thermal Methods," Noyes Data Corp., New York (1983).

Coq, B., G. Ferrat and F. Figueras. "Gas Phase Conversion of Chlorobenzene over Supported Rhodium Catalysts," *React. Kinet. Catal. Lett.*, 27(1):157–161 (1985).

Coq, B., G. Ferrat and F. Figueras. "Conversion of Chlorobenzene over Palladium and Rhodium Catalysts of Widely Varying Dispersion," *J. Catalysis*, 101:434–445 (1986).

Dini, P., J. C. Bart and N. Giordano. "Properties of Polyamide-Based Catalysts. Part I. Hydrodehalogenation of Chlorobenzene," *J. Chem. Soc. Perkin*, II:1479–1482 (1975).

Fradkin, L. and S. Barisa. "Technologies for Treatment, Reuse, and Disposal of Polychlorinated Biphenyl Wastes," Argonne National Lab, Illinois (Jan. 1982).

Hendrickson, J. B., D. J. Gram and G. S. Hammond. *Organic Chemistry*, New York:McGraw-Hill (1970).

Kalnes, T. N. and R. B. James. "Hydrogenation and Recycle of Organic Waste Streams," *Environmental Progress*, 7:185–191 (1988).

LaPierre, R. B., D. Wu, W. L. Kranich and A. H. Weiss. "Hydrodechlorination of 1,1-Bis(*p*-chlorophenyl)-2,2-dichloroethylene (p,p'DDE) in the Vapor Phase," *J. Catalysis*, 52:59–71 (1978a).

LaPierre, R. B., D. Wu, W. L. Kranich and A. H. Weiss. *J. Catalysis*, 52:230–238 (1978b).

Levenspiel, O. *Chemical Engineering Reaction*, 2nd Ed. New York:Wiley (1972).

Detoxification of and Metal Value Recovery from Metal Finishing Sludge Materials

L. G. Twidwell,[1] D. R. Dahnke,[2] S. F. McGrath[3]

INTRODUCTION

Metal-bearing hydroxide sludge material is generated by the metal finishing and electrochemical machining industry in the United States. These wastes are classified as hazardous materials and have traditionally been disposed of in hazardous landfill sites. Long-term maintenance of such sites is required, and metal values are lost unnecessarily. If metals are recovered from these sludges, it will alleviate or reduce the disposal problem and provide for conservation of energy and metal resources. The treatment of hydroxide sludge materials for metal value recovery will produce several beneficial results, i.e., economic benefits from the metal values recovered will help offset the cost of recovery/treatment; nonrenewable resource metals will be recycled for use by society; and there will be significantly less hazardous material to be disposed of in landfills. Detoxification will have resulted because the metal content will have been removed to such an extent that the waste products successfully pass the EPA TCLP toxic characterization test, or, at least, the quantity of material that has to be disposed of in hazardous waste sites will have been drastically reduced.

The experimental results from three major studies [1–3] are summarized in this chapter. A methodology to treat metal-bearing sludges by hydrometallurgical techniques is presented. The methodology emphasis is directed toward the application of known and industrially used hydrometallurgical technology, e.g., simple precipitation, solvent extraction, and cementation unit operations.

Leach solutions produced from multicomponent mixed metal electroplating and electrochemical machining sludge materials contain a mixture of metal species not normally encountered in most hydrometallurgical systems, e.g., iron, chromium, copper, zinc, and nickel. The demonstration that presently used industrial technology can be transferred to this new material is an important step forward in waste management.

Most large-scale generators of sludge material produce a multicomponent product. These sludge materials require that a multistage treatment sequence be used to economically recover the metal values. Most small-scale generators produce such a small quantity of waste material that they cannot afford to treat the material. Therefore, the most appropriate method of managing metal finishing and electrochemical machining sludge material from all sources is treatment in a centralized treatment facility.

WASTE STREAMS

The primary sources of hydroxide sludge materials are the metal finishing and the electrochemical machining industries. The treatment of rinsewaters and the disposal of spent electroplating electrolytes by the metal finishing industry, as well as the treatment of contaminated electrolytes by the electrochemical machining industry, result in the production of these hydroxide sludge materials.

The electroplating process creates large volumes of rinsewaters that contain relatively low concentra-

---
[1]Metallurgy–Mineral Processing Engineering Department, Montana College of Mineral Science and Technology, West Park Street, Butte, Montana 59701
[2]Kaiser Aluminum Company, Trentwood Works, Spokane, Washington 99215
[3]Montana Enviromet, Inc., Butte, Montana 59701

tions of metal ions. These metal ions are removed from the rinsewaters prior to disposal or recycle by solution neutralization. The solution pH is raised to a level at which metal hydroxides precipitate from the solution. Spent electrolyte solutions are often disposed of by addition to the rinsewater treatment system. The metal hydroxide solids produced by the precipitation process are separated from the solution by filtration. These solids are usually disposed of by transport to and placement in hazardous waste storage sites.

The electrochemical machining (ECM) process is an electrochemical process that is used to produce intricate designs in superalloy materials. The machining is accomplished by controlled electrochemical dissolution of the part material. This dissolution requires an electrolyte with special properties, including a relatively high solution pH. The solution pH, usually approximately 8-9, is high enough that, as the part material is dissolved, it precipitates as a metal hydroxide. The metal hydroxide solids are separated from the solution by filtration. These solids, especially if they are produced from superalloy material, contain high concentrations of nickel, chromium and sometimes cobalt, niobium, rhenium, and titanium. These solids are usually disposed of in hazardous waste sites, but because of their high value, some are reprocessed.

PROCESS DESCRIPTION

Recovery of metal values from multicomponent metal hydroxide sludge materials has been effectively demonstrated by Twidwell and Dahnke [1-3]. The process to accomplish the selective separations is based on the transfer of known hydrometallurgy technology and on the development of new precipitation technology. The technology presented in this chapter is documented in current technical literature and is readily available to the reader [1-12]. Conventional hydrometallurgical unit operations have been shown to be appropriate for selective recovery of copper, zinc, cadmium, and nickel. New technology has been developed for the effective and economical separation of iron from chromium, i.e., via a phosphate precipitation process (described later).

The results presented in this chapter are based on sequential studies of a number of flowsheets, first on a bench scale and then on a small pilot scale. The study was conducted in three phases: investigation as to whether known industrially used hydrometallurgical unit operations could be applied to complex multicomponent sludge leach solutions [1]; evaluation of the phosphate precipitation process for simplifying the separation of iron and chromium from divalent cations [2]; and testing selected unit operations on a small pilot scale [3].

The detailed experimental procedures used to develop the treatment process are presented elsewhere [3]. However, a typical study treatment sequence would be as follows:

- Each sludge was subjected to a standard optimized sulfuric acid leach developed at Montana Tech [3], e.g., one-half hour, 40–55°C, acid concentration of one gram acid per gram of solid in the sludge, an initial solid/liquid ratio of 0.8, and an agitation rate sufficient to suspend all particles in the solution phase. Residual solids were removed from the solution phase by vacuum filtration for the bench-scale tests and by a pilot-scale filter press for the large-scale tests.

- The leach solution was then treated for selective metal recovery. If the solution contained copper, it was removed by solvent extraction using LIX 622; separatory funnels were used as the contactors for both loading and stripping experiments; counter-current flow, one-gallon mixer-settlers were used for the large-scale test work. The raffinate from the solvent extraction was then treated by mild oxidizing conditions to ensure that the iron was completely in the ferric form. Precipitation experiments were then performed on the mixed metal solution for iron and subsequently for chromium separation and removal.

- After the iron and chromium trivalent cations were removed, the solution was further treated for recovery of the divalent metals by techniques, such as cadmium removal by cementation, zinc recovery by solvent extraction, and nickel recovery by precipitation.

The bench-scale test work was followed by large-scale tests. The large-scale test work was performed in 200-liter polypropylene reaction vessels equipped with heaters and agitators, one-gallon mixer-settler solvent extraction racks, and in a pilot-scale Ingersoll Rand filter press. The experimental conditions for the tests were based on the results of the optimized bench-scale studies.

STATE OF DEVELOPMENT

Currently, the treatment of metal hydroxide sludge materials is not extensively practiced. Small quantities of metal hydroxide sludge materials are disposed of by adding them to smelter charge mixtures and shipping them to foreign countries, but most of these sludges are disposed of in hazardous waste sites. Several chemical and metallurgical companies are currently (1988) investigating the feasibility and economics of metal value recovery from such materials.

The present state of development of the proposed treatment sequence is that the process has been studied extensively through bench-scale and small-batch pilot-scale test work.

DISCUSSION AND SUMMARY OF RESULTS

An extremely large data base has been generated at Montana Tech for the treatment of a wide variety of sludge materials. Nine master of science theses have been completed that are directly related to this project: Laney has investigated the effectiveness of removing copper from multicomponent sludge leach solutions [4]; Dahnke has summarized research on zinc and iron solvent extraction, and his work produced the initial data upon which the phosphate process is based [2]; Arthur investigated the application of the phosphate process to chloride-bearing sludge leach solutions [5]; Konda investigated iron–zinc separations from high zinc solutions in a sulfate solvent matrix [6]; McGrath studied the speciation of chromium in phosphate-bearing solutions and the kinetics of chromium phosphate precipitation [7]; Rapkoch studied the phosphate precipitation of trivalent cations from the ammonia/ammonium system [8]; Nordwick conducted studies on the rate of conversion of ferric phosphate to ferric hydroxide [9]; Quinn investigated the conversion of chromium phosphate to other more marketable products by soda ash fusion [10]; and Donelon investigated the application of the phosphate process to stainless steel pickle liquors [11]. These theses are presently available in published form from Montana College of Mineral Science and Technology.

Because of the problem of summarizing and representing such a large amount of data in the brief space available for this chapter, the approach used in this presentation is to summarize the results and conclusions without including a great deal of detail. The authors are, therefore, relying on the reader to solicit copies of individual theses of interest.

Mixed metal sludges often contain a variety of potentially valuable elements, e.g., chromium, nickel, cobalt; less valuable elements but still of commercial importance, e.g., copper, zinc, cadmium; and impurity elements of no commercial value, but elements which must be removed because they would otherwise contaminate the recovered valuable product metals or salts, e.g., iron and calcium. The concentrations in which each element is present in a sludge solid are extremely variable but the concentrations are often at levels high enough to provide commercial interest. Example sludge compositions are presented in Table 1.

The discussion of results will be presented by following a typical sequence of unit operations applicable to a sludge material containing copper, iron, chromium, cadmium, zinc, nickel, and calcium. The treatment of sludges containing these and other elements, e.g., titanium, niobium, and cobalt, has also been investigated by Twidwell and Dahnke, and their results are presented elsewhere [3,12]. The unit operations to be discussed in this chapter include sludge dissolution by leaching; solid/liquid separation; solvent extraction of copper; precipitation of iron and chromium as phosphates, and their conversion to marketable products; cadmium recovery by cementation on zinc; zinc recovery by solvent extraction; and nickel recovery by precipitation. The flowsheet for the treatment sequence is presented in Figure 1.

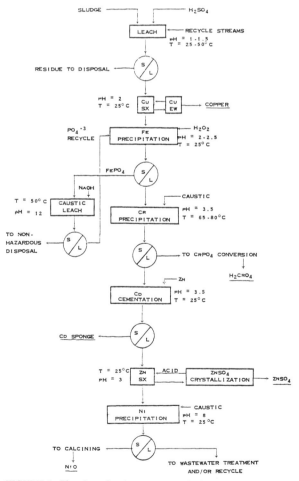

FIGURE 1. Flowsheet for the treatment of mixed metal hydroxide sludge materials.

Table 1. Example Metal Hydroxide Sludge Compositions [1,11]

Sludge Composition in Solids, %								Moisture, %
Cu	Fe	Cr	Cd	Zn	Ni	Al	Ca	
8.06	18.21	1.20	0.74	11.73	5.59	2.92	1.00	76.8
1.15	0.30	22.70	—	0.02	1.16	0.50	1.31	82.4
4.00	15.20	4.90	0.16	10.53	3.89	2.62	0.98	74.2
—	11.34	4.88	—	—	14.88	0.03	0.68	64.7

Leach

Sulfuric acid leaching is effective and efficient in redissolving metal values [1–3,12]. The dissolution is rapid and without control problems. Conditions can be specified to achieve greater than 90% extraction of all metal constituents. Conditions were described previously in the Process Description Section. Using the standard procedure results in a solution at a pH of approximately 1.5 and containing up to 80 grams per liter dissolved metals. Most of the sludge is redissolved, i.e., only about 10–15 % of the initial solids remain as a residue. Practically all the calcium present remains with the residue as calcium sulfate (the solution calcium concentration is less than 0.5 gram per liter). This residue may be considered hazardous, as is the starting electroplating sludge, and may have to be disposed of in the same manner as the starting sludge material. If copper is present, then the residue must be separated from the solution prior to copper extraction. If copper is not present, then the next unit operation would be the precipitation of ferric phosphate prior to the removal of solids. This technique ensures a simple effective solid/liquid separation because of the filtering properties of the ferric phosphate. Also, the residue–ferric phosphate mixture is considered nonhazardous according to the USEPA Toxicity Characteristic Leach Procedure [12].

Solvent Extraction of Copper

Bench- and large-scale test work have shown that copper can be effectively and selectively removed from leach solutions containing appreciable concentrations of iron, chromium, zinc, and nickel [1,3,4]. Copper extractions of better than 90% from solutions (pH of 1–1.5) containing about 3 grams per liter copper resulted in final raffinate solutions containing less than 30 mg/liter copper (three stages of extraction, two stages of stripping). Extraction of other metal ions from the leach solution by the organic reagent was below analytical detection limits using Inductively Coupled Plasma (ICP) analysis. Degradation of the organic extractant does not appear to be important for the conditions tested. Deterioration of the organic extractant was not significant for solution conditions of high metal content, high ionic strength, the presence of phosphate, low pH, and mild (to 55°C) temperatures. Tests were conducted for over 220 load/strip cycles.

Precipitation of Ferric Phosphate

Hydroxide sludge acid leach solutions contain a mixture of divalent and trivalent metal cation species. Phosphate precipitation is an ideal way to remove trivalent metal ions from divalent metal ions. The precipitations exhibit several important features:

- *Selectivity* in acid solution is excellent.
- *Solid/Liquid Separation* is excellent because of the precipitate morphology.
- *Conversion* of the phosphate precipitate to other products is possible.

Selectivity for removing trivalent cations from divalent cations can be predicted by use of solubility diagrams; an example for the separation of iron from nickel is presented in Figure 2. Such diagrams have been used extensively during the development of the phosphate process. In general, it can be stated that

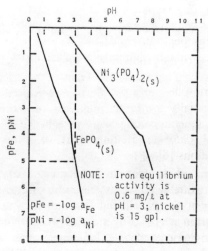

FIGURE 2. Selective iron removal from nickel by phosphate precipitation.

trivalent cation phosphates, under acid conditions, are much less soluble than are divalent cation phosphates. Therefore, selective separations of trivalent cations from divalent cations are possible.

Effective solid/liquid separation is an important feature of the phosphate process. The morphology of the precipitated solids is small, dense, spherical particles. This is in contrast to hydroxide particulates that precipitate as high surface area solids. The phosphate particle morphology is therefore less amenable to surface adsorption and contamination by other dissolved species.

An important feature of the phosphate process is that metal phosphates can be *converted to more marketable products* with the regeneration of phosphate reagent for recycle. Iron phosphate can be converted to ferric hydroxide for disposal by a simple elevated caustic leach [10].

Bench-scale test work showed and large-scale test work confirmed that phosphate precipitation is an effective and very selective way of removing iron from an acidic mixed metal solution (Figures 3 and 4) [1–5,8,11]. Some of the important features of the precipitation process are:

- Precipitated particle morphology is spherites and agglomerites of spherites (Figure 5). Because of the morphology, the filterability of the precipitated ferric phosphate is excellent, i.e., the filterability is about one hundred times better than that of precipitated ferric hydroxide.
- Precipitation is rapid and is essentially complete at room temperature in a residence time of 0.5–1.0 hour. Precipitation to less than 20 mg/liter can be accomplished from a room temperature solution at a pH of 2.5 without chromium, nickel, zinc, or cadmium coprecipitating (Figures 3 and 6). In fact, iron can be stripped from a solution containing 30 grams per liter iron and 150 grams per liter zinc without appreciable zinc contamination [6].
- Ferric phosphate precipitation–pH curves are essentially the same regardless of the media and aqueous metal species present, i.e., sulfate [3,6], chloride [5], nitrate [5], or ammonium solutions [8].
- Ferric phosphate can be converted to ferric hydroxide with the regeneration of phosphate reagent for recycle by leaching in a caustic solution at a pH of 12 at 50°C [9]. The precipitated ferric phosphate and the converted ferric hydroxide product both pass the TCLP test and, therefore, can be disposed of as nonhazardous solids [3].

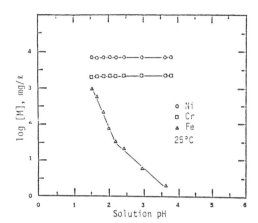

FIGURE 3. Selective removal of iron from chromium and nickel in metal hydroxide sludge leach solutions.

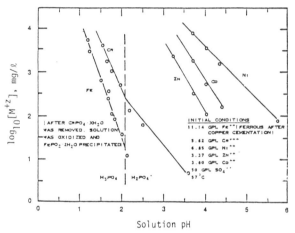

FIGURE 4. Selective removal of chromium and iron from divalent cations in metal hydroxide sludge leach solutions.

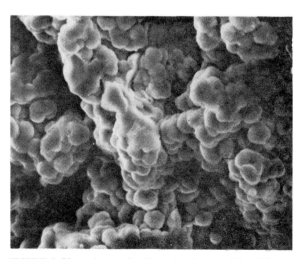

FIGURE 5. Photomicrograph of iron phosphate precipitated from a mixed metal aqueous leach solution (2000×).

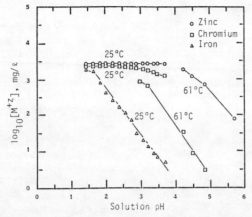

FIGURE 6. Selective precipitation of iron, chromium, and zinc from a mixed metal aqueous leach solution.

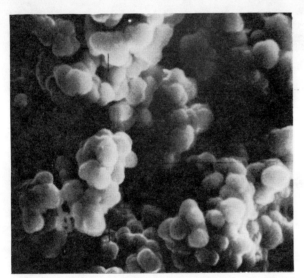

FIGURE 7. Photomicrograph of chromium phosphate precipitated from a mixed metal aqueous leach solution (2000×).

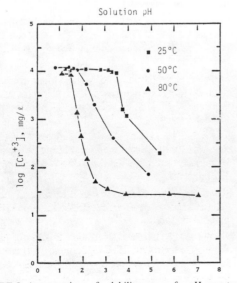

FIGURE 8. A comparison of solubility curves for pH scan tests on chromium solutions at 25°C, 50°C, and 80°C.

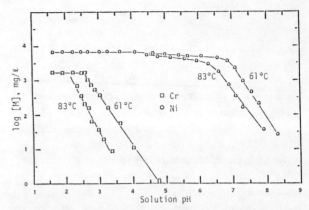

FIGURE 9. Selective separation of chromium and nickel from a mixed metal aqueous leach solution.

Precipitation of Chromium Phosphate

Bench- and large-scale test work showed that chromium can be very effectively stripped selectively from divalent cations by phosphate precipitation. Important features of the precipitation process include:

- Precipitated particle morphology is spherites. Therefore, filterability is excellent (Figure 7).
- Chromium phosphate precipitation is very temperature–time dependent. Room temperature precipitation does not begin until the solution pH is raised to at least 3.3, but precipitation is essentially complete at a pH of 3 at 80°C (Figure 8). The influence of time on the precipitation solubility curve is very important because it allows for good iron–chromium separations based on kinetics and temperature differences.
- Selective separation of chromium from nickel is excellent, even up to temperatures of 80°C (Figure 9).
- Chromium phosphate has a limited marketability. It has been demonstrated that it can be effectively converted to more marketable feedstocks, such as sodium chromate or chromic acid, by a soda ash fusion process [10].

Cadmium Cementation

Cadmium can be selectively separated from the solution by cementation on zinc. This is a widely practiced industrial unit operation [14]. The cementating reagent is zinc. Cadmium is precipitated onto the zinc surfaces, and the product is recovered from solution by filtration. This is an exchange process, and zinc ions enter the solution as the cadmium plates out. The zinc is subsequently recovered by the following solvent extraction unit operation.

Solvent Extraction of Zinc

At this point in the treatment sequence only zinc and nickel remain in solution. Large-scale test work has demonstrated that zinc can be very effectively stripped from solution by solvent extraction using DEHPA as the extractant [1,2,4]. Zinc extractions of better than 98% from solutions containing up to approximately 6 grams per liter were achieved without contamination of nickel. Deterioration of the organic solvent extraction reagent for extended load/strip cycling is not a problem [4].

Any calcium present in the leach solution will also be extracted by the DEHPA extractant. The calcium is stripped from the DEHPA by the strip acid (sulfuric acid) and precipitates as calcium sulfate solid. This solid can be removed from the strip circuit by filtration [4].

Precipitation of Nickel

Nickel can be stripped from solution by precipitation as nickel phosphate, nickel hydroxide, nickel sulfide, or nickel carbonate, or it may be recovered from solution by solvent extraction and electrowinning. Large-scale test work at Montana Tech has been directed toward recovery by sulfide precipitation. Nickel removals to below six mg/liter are achievable.

Bench-scale tests have shown and large-scale tests have confirmed that metal hydroxide sludge materials can be effectively treated for metal value recovery. The products that result are of sufficient purity to serve as feedstock for commercial uses, or they can be used for conversion to other, more marketable products. The treatment process consists of known hydrometallurgical unit operations, and simple, readily available equipment can be utilized, i.e., solvent extraction system vessels, simple stirred reactors, thickeners, and filtration devices.

Economic estimates have been presented by Twidwell and Dahnke [1,3,12] that suggest that the treatment of metal hydroxide sludge materials could be accomplished on a scale of 30–50 tons/day at a substantial profit.

CONTACTS FOR MORE INFORMATION

The contact person for more technical and economic information is L. G. Twidwell. Address: Processing Engineering Department, Montana College of Mineral Science and Technology, West Park Street, Butte, Montana 59701. Phone number (406)496-4208.

REFERENCES

1. Twidwell, L. G. "Metal Value Recovery from Hydroxide Sludges, Final Report," E.P.A. Hazardous Waste Research Laboratory, Cincinnati, OH, EPA 600/285/128, NTIS PB-86157294/As, 491 pp. (1984).
2. Dahnke, D. R. "Removal of Iron from Acidic Aqueous Solutions," M.S. Thesis, Montana College of Mineral Science and Technology, Butte, Montana, 359 pp. (1985).
3. Twidwell, L. G. and D. R. Dahnke. "Metal Value Recovery from Metal Hydroxide Sludges: Removal of Iron and Recovery of Chromium," E.P.A. Hazardous Waste Research Laboratory, Cincinnati, OH, NTIS PB-88176078, 202 pp. (1988).
4. Laney, D. G. "The Application of Solvent Extraction to Complex Metal-Bearing Solutions," M.S. Thesis, Montana College of Mineral Science and Technology, Butte, Montana, 140 pp. (1984).
5. Arthur, B. "Treatment of Iron, Chromium, and Nickel Aqueous Chloride Acidic Solutions by Phosphate Precipitation," M.S. Thesis, Montana College of Mineral Science and Technology, Butte, Montana, 300 pp. (1987).
6. Konda, E. "Study of Ferric Phosphate Precipitation as a Means of Iron Removal from Zinc-Bearing Acidic Aqueous Solutions," M.S. Thesis, Montana College of Mineral Science and Technology, Butte, Montana, 412 pp. (1986).
7. McGrath, S. F. "Rate of Chromium Precipitation from Phosphate Solutions," M.S. Thesis, Montana College of Mineral Science and Technology, Butte, Montana (1988).
8. Rapkoch, J. M. "Recovery of Metal Values from Metal Hydroxide Wastes: the Use of Ammonium Hydroxide and Ammonia Gas as Neutralizing Agents," M.S. Thesis, Montana College of Mineral Science and Technology, Butte, Montana, 152 pp. (1988).
9. Nordwick, S. "Conversion of Ferric Phosphate to Ferric Hydroxide," M.S. Thesis, Montana College of Mineral Science and Technology, Butte, Montana, 250 pp. (1987).
10. Quinn, J. "Conversion of Chromium Phosphate by Sodium Carbonate Fusion," M.S. Thesis, Montana College of Mineral Science and Technology, Butte, Montana (1988).
11. Donelon, D. "Recovery of Metal Values from Spent Stainless Steel Pickling Liquors," M.S. Thesis, Montana College of Mineral Science and Technology, Butte, Montana (1989).
12. Twidwell, L. G. and D. R. Dahnke. "Metal Value Recovery from Alloy Chemical Milling Waste: Phase II," EPA Contract No. 68-02-4432, Washington, D.C., 189 pp. (1987).
13. EPA, Toxicity Characteristic Leaching Procedure (TCLP), *Federal Register*, 51(51):40643–40653 (Friday, November 7, 1986).
14. Nomura, E. et al. "Modernization Process of Mitsui's Kamioka Electrolytic Zinc Operation," in *Hydrometallurgy Research, Development and Plant Practice*, K. Osseo-Asare and J. D. Miller, eds. Warrendale, PA, Met. Soc. of AIME, pp. 955–970 (1983).

2.7

Using Alternating Current Coagulation to Treat and Recycle Wastewaters Containing Hazardous Substances

P. E. Ryan,[1] T. F. Stanczyk[2]

INTRODUCTION

The separation of solids from liquids and the removal of soluble oils and metals from solution are vital to implementing effective hazardous waste management and waste minimization practices. This chapter describes a technology that can be readily integrated with existing processes and/or treatment methods in a cost-effective manner to achieve these objectives. The technology is known as alternating current electrocoagulation (ACEC). The A.C. electrocoagulator causes coagulation and separation without using chemical aids or polymers. It has no moving parts, is easy to maintain, and, because of residual effectiveness and short retention times, only a portion of the contaminated solution may require treatment. Thus, pumping and space requirements are minimized.

This technology has been used to separate suspended solids and heavy metals from mine drilling ponds [1] and to improve the performance of various coal cleaning processes [2]. ACEC has improved the filter dewatering rate of coal fines by 35 to 50% [2], has separated creosote from water [3], and has permanently broken a stable emulsion of rolling coolant oil while reducing the dissolved oil and grease by more than 99% [3]. It has the potential to provide a cost-effective alternative to hazardous waste management and waste minimization issues when used alone or in combination with conventional techniques.

Currently, wastewater treatment technologies are required to remove suspended as well as soluble pollutants in order to achieve mandated discharge criteria. As a result, strategies emphasizing the reduction of pollutants entering the wastewater have been adopted in an attempt to achieve these higher effluent quality levels. Appropriately, the emphasis has been focused on the separation and removal of fine and ultrafine solid products, which are typically found in industrial washwaters. These solid products are usually suspended, emulsified, and/or partially solubilized in aqueous media. They can be of a type and quantity to be deemed hazardous or to have toxic impact. Most of these waste washwaters require chemical addition to enhance solid agglomeration and settling before conventional mechanical dewatering systems can be employed. Unfortunately, the addition of these chemicals adds to the volume of waste generated [4].

As an alternative to chemical conditioning and flocculation, recent developments indicate that liquid–liquid and solid–liquid phase separations can be achieved using ACEC. The electrocoagulator facilitates the flocculation and settling of fine solids and emulsified oils without the use of chemical aids. It can be easily integrated with conventional process and waste control systems to achieve waste reduction goals, solid product recovery, and water purification.

This chapter discusses the theory and practice of ACEC. Operating variables are reviewed, and potential advantages and benefits are highlighted. The current stage of development is discussed in the context of applications dealing with water purification, waste minimization, and site remediation.

APPLICATIONS

Coal, pigments, pharmaceutical solids, ceramics, carbon, clays, metallic powders, and ores are among

---
[1]Electro-Pure Systems, Inc., 10 Hazelwood Drive, Suite 106, Amherst, NY 14150
[2]Recra Environmental, Inc., 10 Hazelwood Drive, Suite 106, Amherst, NY 14150

the categories of products that are wasted as suspended solids in aqueous-based wash solutions. Phase separation and recovery of these solids by conventional dewatering systems is often difficult and costly.

The ACEC facilitates agglomeration and settling of fine and ultrafine particles in aqueous suspensions, making recovery feasible. It also enhances the phase separation of emulsified oils and the removal of soluble pollutants, thereby cleaning the water and improving the performance of downstream conventional wastewater control systems.

General

Table 1 lists some industries that generate process waters and wastewaters containing suspended and emulsified solids applicable to separation by AC electrocoagulation.

The electrocoagulator can stand alone as a process unit; however, its effectiveness is generally recognized as an integrated treatment unit. Table 2 summarizes some of the conventional control systems that can benefit by integrating AC electrocoagulation.

Coal, lignin, pharmaceutical solids, graphite, carbon, clays, metallic powders, ores, pigments, ceramics, and organic chemicals are some of the solid products that, when subjected to water wash operations, produce suspensions of finely divided colloidal matter in aqueous media. The resulting wastewaters are generally difficult to phase separate, and the entrained suspended solids contribute to the loadings of inorganic and organic pollutants that are soluble in the aqueous matrix.

Table 1. Industries Generating Process Waters and Wastewaters Containing Suspended and Emulsified Solids Applicable to Separation by AC Electrocoagulation

Pulp and Paper
Petroleum Refining
Ceramic Industries
Coal Processing
Photographic Processing
Plastic Industries
Surface/Coating Industries
Agricultural
Food and Food By-Products
Rubber Industries
Electronic Industries
Metal Finishing/Plating
Pharmaceutical
Intermediate Dyes
Inorganic Chemicals
Chlor-Alkali Industries

Table 2. Control Systems That Can Benefit from the Integration of AC Electrocoagulation

In-Plant Water Purification
- Closed Loop Water Purification
- Pretreatment of Influent Water
- Cooling Tower Pretreatment
- Product Rinsing
- Equipment Maintenance/Washing

Air Abatement
- Removal of Particulates from Scrubber Solutions

Wastewater Treatment
- Segregate and Separate Solids Prior to Equalization
- Enhance Flocculation/Settling of Solids Produced After Chemical Treatment
- Pretreat High Water Content Suspensions Prior to Dewatering
- Separate Ultrafine Solids in Filtrate Prior to Treatment or Discharge
- Polish Treatments of Bio- and Carbon Treatment Effluents
- Enhance Preferential Separation of Soluble Inorganic/Organic Contaminants
- Pretreatment of Surface Run-Off

Thermal Treatment
- Improve Fuel Specifications by Phase Separating Organic Solids and/or Metals and Corrosive Salts from Aqueous Solutions

Electrocoagulation may be applied to four general categories of hazardous wastes:

- *Particulate suspensions*—Examples of hazardous particulate suspensions include certain pigments, metallic fines, and rinsing processes with particulate sizes ranging from molecular dimensions to one-fourth inch.
- *Oily emulsions*—Breaking oily emulsions and separating a portion of the oils can enhance the performance of conventional treatment systems with minimal residuals production.
- *Dissolved inorganics*—Dissolved inorganic waste streams include residuals from the electroplating, metal finishing, etching, and acid pickling processes.
- *Dissolved organics*—Examples include natural dissolved organic (humic) materials, synthetic organics, paint stripping solvents, and parts cleaning waste streams.

Industrial

ACEC offers a method to reduce the generation of waste by removing contaminants from process streams as they are created. The water can be reused. This technology can also be employed as a treatment step to segregate and concentrate constituents for reclamation or disposal. The volume of wastes leaving the generator's site will be dramatically reduced. A summary of general industrial applications follows:

- waste minimization through solids removal
- solids recovery
- water purification
- enhanced sludge thickening and dewatering
- separation of soluble oils from sludge and sediments
- reduction of soluble inorganic and organic pollutant concentrations

Waste Minimization Management

Some states [5] have established a hierarchy of preferred practices for the management of hazardous wastes. Generally, the reduction or elimination of hazardous waste generation is the first priority; followed by recovery, reuse, and recycle to the maximum extent practical; and finally, detoxification, treatment, destruction, and land disposal of residuals. ACEC has the potential to reduce the quantity of waste generated (eliminate chemical residuals), effect constituent recovery and reuse, and promote the reuse of water. Residuals remaining for thermal destruction or land disposal will be minimized.

Various studies [5,6,7] have determined that more than 95% of the hazardous wastes generated are aqueous in nature and are managed on the generator's site. Thus, a technology that addresses aqueous waste issues will serve the major part of the problem. These studies go on to observe that approximately 80% of these wastes are due to "routine contamination of process streams" [7], and that these contaminated streams could be substantially reduced through source reduction and on-site recycling. Thus, a technology that promotes contaminant removal and cleaning of process water for reuse will help meet the objectives mandated by various state laws.

Waste minimization management practices applicable to the four general categories of wastes cited earlier are discussed as follows.

Particulate Suspensions

Wastewaters containing suspended solids represent one of the largest waste volume segments. The presence of ultrafine particles in process waters can adversely affect product quality in applications such as electronics. Processes that generate these particulate suspensions could realize improvements in productivity and product recovery by separating solids prior to conventional wastewater treatment. Many of these solids (e.g., pigments, intermediate metallic fines, flux) comprise the pollutant loadings and may cause the waste streams to be classified as hazardous. These wastewaters, whether hazardous or nonhazardous, result in sludges requiring land disposal. The nonhazardous wastewaters represent potential sources of "future" hazardous and/or "specialty" wastes due to possible revisions in regulations that may require proper management of treatment residues.

Oily Emulsions and Soluble Oils

Wastewaters containing soluble oils contribute not only to the hazardous nature of sludge products, but also to the soluble contaminant loadings in water. The volumes associated with industrial sources are significant and become staggering when surface drainage is incorporated in the totals. Recent decisions [8] to list oils as hazardous could magnify the waste management problems throughout the country. The potential presence of toxic organics and heavy metals in the oil fraction will result in hazardous characteristics. Oils are already precluded from land disposal. The soluble oils present in wastewater treatment sludges can adversely impact conventional dewatering systems and chemical stabilization processes. The resulting solids are generally greater (30-60% by weight) and characteristically more toxic than they would be if the oil were absent. Unfortunately, many of the highly stable oil emulsions are not easily broken by conventional approaches, which utilize heat, chemical addition [4], or membrane separation. Further, the presence of soluble oils in the wastewaters and the resulting wastewater treatment sludges inhibit the recovery of metals that would, in turn, reduce the generation of solids residuals.

Dissolved Inorganic Solutions

Wastewaters containing soluble inorganic constituents are generated by metal plating, metal finishing, electronic and pigment industries, among others. Historically, the commercial landfills have buried large volumes of industrial wastewater treatment sludges. These sludges contain a multitude of contaminants generally categorized as heavy metals and inorganic salts (e.g., chlorides, sulfates, phosphates, and fluorides). The volume of wastewater and sludge is expected to increase significantly, as many industries convert organic solvent cleaning operations to aqueous-based media. Many of those converting to aqueous cleaners are finding that new treatment strategies are required to minimize sludge by-products and excess pollutant loadings in regulated discharges. The application of ACEC will enhance waste minimization by providing in-line processing for wastewaters containing low concentrations of metallics and salts, which generally require chemical addition for contaminant phase separation. Besides reducing unnecessary sludge volumes, this approach offers a solution to many of the generators pursuing aqueous-based cleaners.

Dissolved Organic Solutions

Theoretical observations and treatability studies [9] addressing the oxidation of methanol indicate that ACEC can also be used to oxidize dissolved organics. Additional research and empirical evaluations are required to assess the true potential of this application.

Superfund Site Remediation

A number of conventional and innovative technologies are being used alone or in combination with other treatment and/or remedial activities to reduce environmental risks and threats to human health. Complete systems are needed to control the potential multimedia transfer and fate of pollutants while reducing the volume of off-site waste shipments. ACEC may be used as an element in these complete systems to cost-effectively achieve EPA's goals for Superfund site remediation. The technology's flexibility and simplicity are two unique features that will undoubtedly benefit remediation and cleanup endeavors.

The applicability of ACEC to various waste matrices may benefit those in-situ remedial technologies being adversely impacted by waste properties and/or constituents. Some of these proposed beneficial applications are summarized as follows:

- Electrocoagulation may treat quenchwaters from mobile incineration systems containing ultrafine fume particulates that are not feasible to separate using conventional techniques. Besides reducing the potential pollutant loadings attributable to these particulates, there is the real potential for cleaning the water sufficiently to allow reuse.
- Electrocoagulation can separate emulsified oils from ground- and surface-waters. The completeness of this separation may be sufficient to allow the purified water to be used on-site as makeup water for slurry walls, soil flushing, and quenchwater.
- The ACEC unit is capable of removing suspended, as well as emulsified, oils and metals from waste residues, which would otherwise adversely impact the treatment performance of stabilization/solidification technologies.
- Conventional mobile dewatering performance may be improved, while reducing reliance on chemical aids and, thus, reducing the production of sludge by-products.
- Inorganic and organic fines suspended in waste matrices can be separated, thus, minimizing the need for conventional polish filtration.
- Organic and inorganic pollutants may be removed from sediments and high-water-content sludges by introducing an aqueous-based or petroleum-based fluid to preferentially extract pollutants in an immiscible phase that may then be readily separated by ACEC.

BACKGROUND, THEORY, DESCRIPTION AND OPERATION

Background

Solid particles and colloidal matter in wastewater may be influenced by electrical forces. Particulates in the form of colloidals maintain their stability due to repulsive electrostatic forces. Suspended inert colloidal matter, however, is susceptible to precipitation through redistribution or neutralization of charges.

Several studies [10,11,12,13] suggest that most solid particles suspended in aqueous media carry electrical charges on their surface. In suspensions with particles larger than atomic or molecular dimensions, the particles will tend to separate under gravitational or buoyancy force unless they are stabilized by electrical repulsion or by other forces. Such forces can prevent aggregation into larger particle masses, or flocs, which are more prone to settling. These surface charges may exist as an ionic double layer or as a neutralized electric dipole, as conceptually depicted in Figure 1.

These surface charges may originate in several ways:

- The particle crystal lattice may contain a charge resulting from lattice imperfections or substitutions. This charge is balanced by compensating ions at the surface (zeolites, montmorillonite, and other clay minerals).
- Solids may contain ionizable groups.
- Specific soluble ions may be adsorbed by surface complexes or by compounds formed on the surface.

Electrical double-layer theory dates from

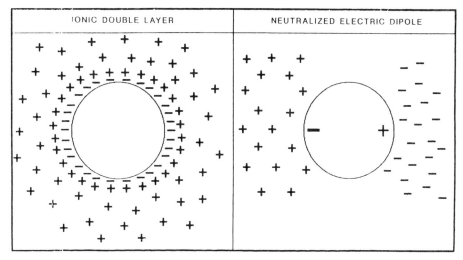

FIGURE 1. Surface charge distributions.

Helmholtz's [16] work in 1879. At about the same time, Shulze [16] showed that certain (lyophobic) colloidal sols were rendered unstable and flocculated by the addition of electrolytes, especially those of multivalent ions. In 1892, Linder and Picton [31] established that particles in a colloidal sol migrate under the influence of an electric field, indicating that they are electrically charged, the sign of which may be determined by the direction of migration. The Shulze–Handy Rule [16] indicates that flocculation is controlled by the valency of the ion of opposite charge to that of the colloidal particles. Hardy and Burton [32] demonstrated that the stability of lyophobic sols was related to particle electrophoretic mobility in an electric field.

When placed in a direct current (DC) electric field, colloidal particles migrate toward the charged surface. During this migration, a shear plane defines the boundary between the bulk solution and the "electrical double layer" that moves with the particle. The rate of migration in a given electric field may be used to calculate the Zeta potential. The oldest relation between measured electrophoretic mobility was derived by von Helmholtz [16], improved by von Smoluchowski [33], and extensively studied by others who proposed corrections for hydrodynamic and electrical factors.

The application of DC electrophoresis for the dewatering of aqueous suspensions has been investigated in detail by the U.S. Bureau of Mines [15]. However, there is little or no evidence of using AC power for electrophoresis and dewatering.

The stability of a colloid suspension is usually explained using the famous DLVO theory, as described in Reference [16]. It is now established that the remarkable stability of colloidal suspensions is often due to electrical charges, which prevent close particle–particle approach, or to adsorbed surface coatings such as a surfactant.

Principles of AC Electrocoagulation

The alternating current electrocoagulation process, invented by Moeglich et al. [17,18,19], utilizes colloidal chemistry principles, AC power and electrophoretic metal hydroxide coagulation. The process employs two main principles:

- electroflocculation, whereby small quantities of metal ion are dissolved from the electrodes and generate metal hydroxides, which assist in flocculation of the suspended particles
- electrostriction, whereby the suspended particles are stripped of their charges by subjection to alternating electric field conditions

Electroflocculation

The theory of electroflocculation, or metal ion flocculation, is well established. Iron and aluminum ions have been widely used to clarify water. Recently, Parekh et al. [20,21] developed a coagulation model involving the use of metal hydroxides and fine particles. They reported that the optimum coagulation of a metal ion–particle system takes place at the isoelectric point of the metal hydroxide precipitate. In general, optimum coagulation does not occur at the point of zero charge, since other mechanisms such as bridging are important [22].

Electrostriction

The theory of electrostriction is not very well understood. In 1962 Schwan [10] reported that colloidal suspensions exhibit variable, and often extremely high, dielectric constants (ϵ) in alternating electric fields. Not only did the dielectric constant of the suspension decrease as the frequency increased, but the dielectric loss factor (ϵ') went through a maximum, and the conductivity (k) of the suspension increased dramatically, as indicated in Figure 2. The peak of the dielectric loss factor curve appears at the characteristic frequency, f_o, for the diagram, which is proportional to $1/R^2$, where R is the radius of spherical colloidal particles. For rod-shaped particles, $f_o = 1/R^3$. At the characteristic frequency, adsorbed ions, which give the colloidal particle a surface charge, are least tightly held and can move freely over the particle surface in response to the electric field.

Process Description and Operation

Introduction

System designs for alternating current coagulation vary depending on the characteristics and quantity of the solution being treated, treatment objectives, and location. Suspension characteristics, such as particle size, conductivity, pH, and constituent concentrations, dictate operating parameters of the coagulator. The quantity and flow rate of the raw solution will affect total system sizing, coagulator retention time, and mode of operation (recycle, batch, or continuous). Treatment objectives will establish the type of gravity separation system to use, set recovery criteria, identify the utility of side stream treatment, define effluent standards to be met, and determine the advantages of recycle or multiple staging. Treatment objectives may be to simply precondition a solution prior to using an existing process or to use it as a polishing step after treatment. Location will impact design by imposing physical size constraints and pumping requirements. In-plant industrial applications, for example, may be configured differently than a mobile on-site system used for remediation or treatment of ponded water.

Basic Process

The basic process of alternating current coagulation is depicted in Figure 3. The key function of electrocoagulation is to provide liquid/liquid and solid/liquid phase separation without the use of expensive polyelectrolytes and chemical aids, while improving the performance of conventional mechanical separation systems.

Coagulation and flocculation occur simultaneously within the coagulator and in the product separation step. The onset of coagulation occurs within the AC

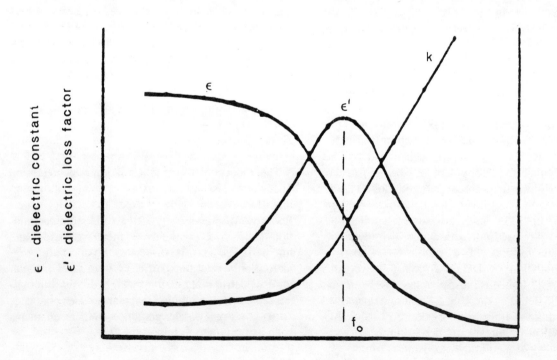

FIGURE 2. Effects of AC power efficiency.

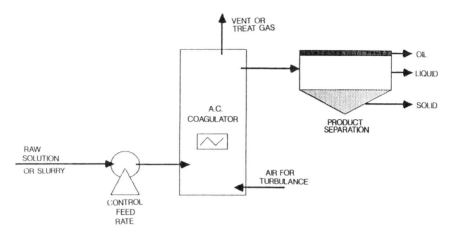

FIGURE 3. Alternating current coagulation basic process flow.

coagulator as a result of both exposure to the electric field and catalytic precipitation of metal hydroxide from the electrodes. This activity occurs rapidly (within 30 seconds) for most aqueous solutions. Once this occurs, the electric field treatment is complete, and the solution may be transferred by gravity flow to the product separation step.

Product separation can be accomplished in conventional gravity separation and decant vessels. Continued coagulation and flocculation occur in this step until complete phase separation is achieved. Generally, the rate of separation is faster than methods that employ chemical flocculants or polyelectrolytes, and for some applications, the solid phase is denser than the solids resulting from chemical treatment. Product removal is accomplished using standard practices of surface skimming, bottom scraping, and decanting.

In many applications, the electrocoagulator retention time may be reduced and performance improved by agitating the solution as it passes through the electric field. This turbulence can be induced by using a torturous path static aerator concept or by diffusing small bubbles of air or nitrogen through the solution in the space between the electrodes. Air has been used in full-scale applications treating pond waters and removing fines from coal washwaters. Bottled nitrogen and bottled air have been used in the laboratory to conduct treatability tests. Since the gas used to create turbulence may also contribute to stripping of volatile organics, it is necessary to analyze the vent gas stream, especially when treating hazardous wastes. When appropriate, the vent gases can be collected and treated using available conventional technologies and thus control air emissions within acceptable limits.

After the product separation step, each phase (oil, water, solid) is removed for reuse, recycle, further treatment, or disposal. A typical decontamination application, for example, would result in a water phase that could be discharged directly to a stream or to a local wastewater treatment plant for further treatment. The solid phase, after dewatering, may be shipped off-site for disposal, the dewatering filtrate being recycled. Any flotable material would be reclaimed, re-refined, or otherwise recycled or disposed.

Examples of typical integrated systems are depicted in Figures 4 and 5. Although simplistic, they illustrate waste minimization by reusing water, recovering by-products, and reducing the quantity of waste to be disposed of. Inherent is the improved performance of conventional or in-place mechanical dewatering devices.

Operating Requirements

The ACEC operates at low voltage, generally below 110 VAC. It is designed to work at atmospheric pressure and is vented to preclude gas accumulation. As previously mentioned, an air emission abatement apparatus may be added, if necessary.

The internal geometry allows for free passage of particulates up to one-fourth inch. While normal operation is relatively maintenance free, some problems can be encountered if process upsets allow heavy particulates to inadvertently enter the lines. In this case, material buildup could retard flow. No permanent damage has been experienced in these cases, and the problem has always been alleviated by reverse flushing or minor disassembly and cleanout.

Electrode deterioration is relatively modest. Minor etching occurs on the electrode surfaces. The alternating current operating mode minimizes the electrode erosion that has been experienced in DC systems. Reasonable electrode life has been proven

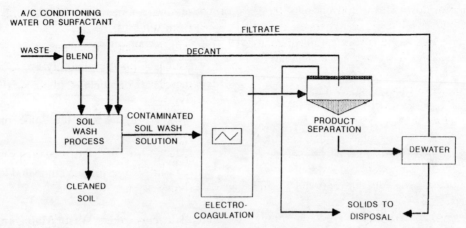

FIGURE 4. Integrated system—soil washing.

with ACEC. Electrodes were replaced after four months of continuous operation (twenty hours per day) in a 250-gpm commercial unit.

Electrical energy costs depend on the characteristics of the solution being treated and the specific application. Commercial units have treated coal washwaters for $0.40 per 1,000 gallons at power costs of $0.50 per kwh. This cost is more than offset by savings in the chemical costs associated with alternative methods, which require the use of polyelectrolytes and chemicals to adjust pH [2].

Residual Effectiveness

Bench-scale tests [23] and full-scale field applications [2] have demonstrated a phenomenon referred to as residual effectiveness. Once the solution has passed through the coagulator and settling is complete in the product separation stage, the separated products can be remixed, and subsequent phase separation will recur without further treatment through the electrocoagulator. This phenomenon is important in that mixing and pumping can be accommodated after coagulation, if so dictated by other system design conditions, without losing the phase separation effectiveness. This also indicates that in some applications only a portion of the total contaminated solution would need to be treated. For example, in removing constituents from a fixed quantity of solution in a vat or pond, a portion of the solution may be treated and returned to the vat until the desired phase separation results. Phase separations have been accomplished by passing as little as 25% of the total volume through the electrocoagulator.

ADVANTAGES AND BENEFITS

Benefits in Industrial Applications

Electrocoagulation may be used as a pretreatment strategy to provide source control while improving

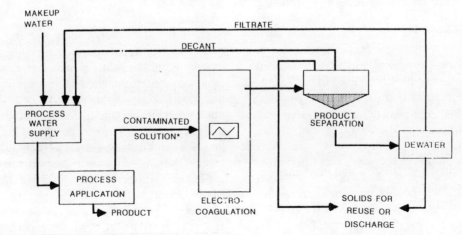

FIGURE 5. Integrated system—process water (process water or washwater containing unwanted particles or emulsions).

the performance of downstream conventional treatment systems. Industrial users can expect a cost benefit when compared to the cost of chemical coagulation with metal salts. This cost advantage can be seen in three areas. First, the operating costs of AC electrocoagulation are moderated by the ability to treat only a small sidestream of the flow. It has been observed [23] that as little as 25% of the flow can be treated with AC electrocoagulation and is then returned to the mainstream to cause particle destabilization in the mainstream. This sidestream treatment option is not available with conventional coagulation technologies.

Second, electrocoagulation will reduce sludge handling costs. Levin [24] estimated the cost of sludge handling to be 25–50% of waste treatment costs. Ultimate disposal of hazardous waste sludges is expected to be even more expensive, since these wastes must be interred in a secure landfill or be thermally destroyed.

Third, electrocoagulation will reduce chemical usage and transportation costs. Groterud and Smoczynski [25] found that they could achieve 98% phosphate reduction by adding 1.4 mg aluminum per mg phosphate by DC electrocoagulation. Precipitation with alum generally requires 2.4 to 3.0 mg aluminum per mg phosphate for 95% removal efficiencies [26]. Jageline and colleagues [12] found that DC electrocoagulation reduced the alkali dose required to treat electroplating wastes. Vik and coworkers [13] estimated that the amount of chemicals that have to be transported for the coagulation of lake water is ten times less for electrocoagulation than for alum coagulation.

Benefits in Remediation Projects

Table 3 provides a summary of on-site remedial strategies that can also benefit by integrating the AC electrocoagulation technology.

Benefits in Coal Industry

The following advantages were identified as a result of using the AC electrocoagulator in the coal industry [27]:

- improved fine coal recovery
- improved dewatering rate of fines
- reduced filtration time
- reduced recirculation of coal and clay fines in closed-loop water
- reduced buildup of fines and clays on dewatering screens
- neutralization of plant water pH
- removal of heavy metal and organic carbon from water
- reduced plant maintenance
- increased plant availability
- increased coal yields without sacrificing quality
- increased quality at the same or increased yield
- reduced freezing of treated coal
- reduced sludge volume compared to flocculant sludge

Benefits in Hazardous Waste Management

Electrocoagulation addresses hazardous waste management needs in three important ways. First, it accomplishes phase separation while minimizing sludge production. Chemical conditioners and coagulants used in conventional practice account for a significant portion of the solid hazardous waste load; electrocoagulation avoids the generation of this volume of solid residuals. Second, electrocoagulation promotes recycle and reuse by recovering product and allowing the reuse of process water. Third, this technology can be integrated with a wide variety of in-plant industrial processes. The small size of electrocoagulation equipment permits the phase separation of segregated waste streams and thus facili-

Table 3. On-Site Remedial Technologies Benefitted from Integrating AC Electrocoagulation

Dewatering of Pits/Ponds/Impoundments
- Enhance Liquid—Solid Phase Separation
- Enhance Liquid—Liquid Phase Separation
- Improve Dewatering Efficiency

Stabilization/Solidification of Sludges
- Separate Emulsified Oils Prior to Chemical Addition
- Eliminate Unnecessary Volumes of Water

Groundwater Treatment
- Phase Separate Suspended Solids Prior to Treatment
- Demulsify Aqueous Emulsions Containing Organically Contaminated Oils and Solids

Soil Flushing/Extraction
- Pretreat Aqueous Emulsions, Enhancing Reuse of Extracting Fluid While Purifying Soils

Mobile Incineration
- Pretreatment Emulsions and Aqueous Suspensions, Reducing Waste Quantities Requiring Thermal Treatment
- Pretreat Scrubber Solution

Run-Off
- Treat Truck Washwaters
- Treatment of Surface Run-Off

*TOC = total organic carbon.

tates waste minimization and waste management practices at the source.

Aqueous and organic liquid waste comprise a significant portion of hazardous wastes generated by industry in the United States. Effective separation of solids, dissolved inorganics, organics, and hydrocarbons from these aqueous solutions and emulsions is vital to effectively reduce and manage these hazardous wastes. This function is currently addressed using conventional mechanical phase separation equipment and a variety of polyelectrolytes and other treatment chemicals [28]. As mentioned previously, the chemicals themselves have the potential for transferring the hazardous waste characteristics to the resulting sludges, and thus add to the total volume of hazardous waste generated. A technology that can effect phase separation without employing chemicals and that can be integrated with upstream industrial production processes instead of being relegated to end-of-the-pipe treatment provides an important source reduction tool.

It is pertinent to note that current hazardous waste reduction strategies are focused on methods that impact aqueous process streams. The Chemical Manufacturers Association's 1986 Hazardous Waste Survey [7] observed that hazardous sludge resulting from wastewater treatment represented 12.8 percent of the total hazardous solid waste generated by the plants surveyed. Since a portion of this sludge was comprised of polymers and other treatment chemicals, a reduction in the generation of hazardous sludges can be achieved with technologies that do not rely on chemicals for treatment of aqueous wastes.

STATE OF DEVELOPMENT

This technology was invented by Moeglich and Hedgetts [17] and Moeglich [18,19] for removing colloidal or suspended particles from water. Subsequent development in the coal industry resulted in commercial coal cleaning applications in 1982. Based on this success and on an awareness of the need for cost-effective methods to manage hazardous wastes, efforts were focused in mid-1988 on new applications of this technology to address issues related to the treatment, management, and minimization of wastewater containing hazardous substances.

Hazardous wastes may differ from coal washwaters in many ways. Characteristics such as effective conductivity, pH, total dissolved solids, particle size, concentration of chemical constituents, and turbidity can be orders of magnitude different from those found in the coal industry. Also, hazardous waste streams differ dramatically from one to another.

Therefore, bench-scale studies are recommended to assess the suitability of ACEC for most hazardous waste applications. Treatability studies will be used to select optimum design parameters for many ACEC systems. Some applications may require on-site demonstration trials to confirm overall system design, especially when the ACEC equipment is integrated with other unit processes.

Currently (4th Quarter, 1988), ACEC is developed to the treatability phase for applications in hazardous waste management. For selected applications, a 13-gpm unit is available for on-site demonstration trials. For still other specific applications, the electrocoagulators that are used in the coal industry and are rated at 250 to 750 gpm may be appropriate as currently configured.

Bench-scale evaluations are under way using both "real world" and synthesized hazardous waste solutions to demonstrate the effectiveness of this technology. The body of knowledge needed to select optimally designed systems is growing. This growth is fed by interest in ACEC by various Fortune 50 companies and by consulting engineers who are looking for optimum solutions for the remediation of Superfund sites.

SUMMARY OF RESULTS

Electrocoagulation has been applied to hazardous waste management issues, including particulate removal (contaminated soil washings, coal fines), removal of organics (industrial wastes, waste oils, naturally occurring organics), metals removal (industrial streams, acid mine drainage), and enhanced dewatering of sludges. Examples of these applications are listed in Table 4. These examples demonstrate that electrocoagulation techniques are capable of greater than 90% removal of suspended solids, organics (total organic carbon), and metals. Table 5 lists the results of one study with ponded water. Dramatic reductions in suspended solids, organic carbon, and metals were observed.

AC electrocoagulation has been used to effect phase separation in water, slurries, and sludges, and for the removal of coal fines [29]. Berry and Justice [2] report that electrocoagulation increased the removal of fines, decreased chemical addition requirements, and increased sludge dewaterability.

Bench-scale tests [30] using synthesized and real world samples demonstrated suspended oil/particulate separation, soluble oil removal, and enhanced particulate settling rates. Creosote was successfully separated from a sample taken from a Superfund site. The dark brown, turbid, stable suspension was

Table 4. Applications of Electrocoagulation

Application	Result	Reference
1. Particulate removal		
a. Water from contaminated soil wash	>washwater clean enough to recycle to the ground	3
b. Clay colloids in ponded water	>99% removal of suspended solids	1
c. Removal of coal fines		1, 29
d. Other (paper industry, alcohol plant)		9
2. Removal of soluble organics		
a. Ponded water	>99% removal of TOC*	1
b. Creosote suspension	>98% removal of TOC	30
c. Emulsion of rolling coolant and waste oils	>90% removal of TOC	30
d. Lake water (DC electro-coagulation)	>95% removal of color	13
3. Metals		
a. Ponded water	>99% removal of Fe, Mn, Al	1
b. Acid mine waste (DC coagulation)	>98% removal of Fe, Cu, Al	34
4. Enhanced dewatering		
a. coal fines	>dewatering rate increased by 30–50%	1, 2

*TOF = total organic carbon.

separated into a clear liquid and a solid residue. The total organic carbon concentration was reduced by more than 98%. The treated sample was vigorously mixed (clear liquid and solid residue), and the separation recurred very rapidly without further treatment. This phenomenon, referred to as residual settling, indicates that "sidestream" treatment may be sufficient in some full-scale applications.

Recent tests [23] using a 1.5% solids colloidal suspension of smectite with one to two micron particles demonstrated that complete separation can be achieved by treating as little as 25% of the total volume. Operating costs will range from six to fifty cents per 1000 gallons treated depending on the portion of solution treated and assuming a power cost of $0.05/kwh.

A bench-scale test of oily soils (10% oil) was also conducted. After water washing, the colloidal washwater solution (5 to 10% solid) was subjected to the ACEC. A total organic carbon reduction of more than 90% was measured, and the oil and grease were reduced by more than 99%. These results indicate that this technology can be used in conjunction with solvent extraction techniques to effect the oil/soil

Table 5. Field Test Results—Electrocoagulation of Ponded Water

Parameter	Pond A		Pond B	
	Raw Water	After Treatment	Raw Water	After Treatment
pH	6.4	7.7	7.3	8.3
Suspended solids	197 ppm	1 ppm	195,000 ppm	15 ppm
Dissolved solids	—	—	7,212 ppm	3,344 ppm
Soluble Iron	88 ppm	0.13 ppm	—	—
Total Iron	285 ppm	263 ppm (in sludge)	3,500 ppm	0.18 ppm
Manganese	3 ppm	1.9 ppm	104 ppm	0.02 ppm
Aluminum	—	—	304 ppm	0.08 ppm
Alkalinity	—	—	48,500 ppm	400 ppm
TOC	—	—	11,000 ppm	30 ppm

Adapted from Plantes, Reference [1].

separation and to promote the recycle/reuse of process water.

On the basis of both these results and previous experience in the coal/mining industry, the electrocoagulation process is ideally suited for removing suspended/emulsified oil and particulate matter from aqueous solutions.

MORE INFORMATION

Electro-Pure Systems, Inc. (EPS) markets alternating current electrocoagulation as custom-designed systems or as standard units on either a lease or purchase basis.

Most applications will require a bench-scale treatability study, and some will require on-site demonstration trials to confirm design criteria for most effective treatment. EPS conducts the treatability studies and demonstration trials and uses the services of certified laboratories for all analytical testing. EPS also has access, through a consulting agreement, to a staff of analytical chemists and environmental scientists to assist in evaluations and to ensure optimum applications.

For more information, write or call: Electro-Pure Systems, Inc., Audubon Business Centre, 10 Hazelwood Drive, Suite 106, Amherst, New York 14150 (716)691-2600.

REFERENCES

1. Plantes, W. J. "Electrocoagulator—Removal of Colloidal and Suspended Solids," Report on Mine Pond Testing, Westinghouse Electric Corporation (July/August 1978).
2. Berry, W. F. and J. H. Justice. "Electro-Coagulation: A Process for the Future." Presented at the 4th International Coal Preparation Conference and Exhibition, Lexington, Kentucky (April 28-30, 1987).
3. EPS (Electro-Pure Systems, Inc.) "Alternating Current Coagulation Information Summary," unpublished report (1988).
4. ERM (ERM-Northeast, Inc.) *Hazardous Waste Facilities Needs Assessment Summary Report and Appendices.* Prepared for the New York State Department of Environmental Conservation, Division of Solid and Hazardous Waste (March 1985).
5. NYSHWTF (New York State Hazardous Waste Task Force). *Final Report* (September 1985).
6. DEC (New York State Department of Environmental Conservation, Division of Hazardous Substances Regulation). Draft New York State Hazardous Waste Siting Plan and Draft Environmental Impact Statement (June 1988).
7. CMA (Chemical Manufacturers Association). *1986 CMA Hazardous Waste Survey* (May 1988).
8. *Federal Register* (1988).
9. Westinghouse Corporation: Various classified internal reports and memoranda dated between 1975 and 1978.
10. Schwan, H. P. et al. "On the Low Frequency Dielectric Dispersion of Colloidal Particles in Electrolyte Solution," *J. Phys. Chem.*, 66:2626 (1962).
11. Schwarz, G. A. "Theory of the Low Frequency Dielectric Dispersion of Colloidal Particles in Electrolyte Solution," *J. Phys. Chem.*, 66:2636 (1962).
12. Jageline, I., M. M. Grigorovich and R. Daubaras. "Electromechanical Treatment of Electroplating Wastes. Effect of pH of the Solution on the Elimination of Copper ($+2$), Zinc ($+2$), and Chromium ($+6$) Ions during Electrocoagulation," *Liet. TSR Mokslu Adad. darb. Ser. B*, 65-72 (1979). As found in *Chem. Abstrac.*, 91, 26704m.
13. Vik, E. A., D A. Carlson, A. S. Eikum and E. T. Gjessing. "Electrocoagulation of Potable Water," *Water Res.*, 18(11):1355-1360 (1984).
14. Baker, A. F. and A. W. Deurbrouck. "Hot Surfactant Solution as Dewatering Aid during Filtration," 7th International Coal Preparation Congress, Australia (1976).
15. Sprute, R. H. and D. J. Kelsh. "Limited Field Tests in Electrokinetic Densification of Mill Tailings," USBM RI 8034 (1975).
16. Verwey, E. J. W. and J. Th. G. Overbeek. *Theory of Stability of Lyophobic Colloids.* New York:Elsevier (1948).
17. Moeglich, K. and H. L. Hodgetts. "Water Purification Method and Apparatus," U.S. Patent 4,053,378 (1977).
18. Moeglich, K. "Water Purification Method," U.S. Patent 4,094,755 (1978).
19. Moeglich, K. "Water Purification Method and Apparatus," U.S. Patent 7,176,038 (1979).
20. Parekh, B. K. "The Role of Hydrolyzed Metal Ion in Charge Reversal and Flocculation Phenomena," Ph.D. Thesis, The Pennsylvania State University (1979).
21. Parekh, B. K. and F. F. Aplan. "Flocculation of Fine Particles and Metal Ions," Annual AIME Meeting, Phoenix AZ (1988).
22. Jensen, J. Professor, Civil Engineering Department, State University of New York at Buffalo; discussion of 12/21/88.
23. Co-Ag Technology, Inc. Laboratory Report by J. H. Justice, 9/16/88. "Determine the Effect of Propagation by Mixing E/C-Treated Slurry with Various Percentages of Untreated Slurry."
24. Levin, P. "Disposal Systems and Characteristics of Solid Wastes Generated at Wastewater Treatment Plants," *Proc. 10th Sanit. Eng. Conf.*, University of Illinois Bulletin, 65:115, 21 (1968). As found in: Dick, 1972.
25. Groterud, O. and L. Smoczynski. "Phosphorous Removal from Water by Means of Electrolysis," *Water Res.*, 20(5):667-669 (1986).
26. Metcalf and Eddy, Inc. *Wastewater Engineering Treatment, Disposal, and Reuse.* New York:McGraw-Hill Book Company, p. 747 (1979).
27. Nickerson, F. H. "Electrical Coagulation: A New Process for Prep Plant Water Treatment," *Coal Mining and Processing* (September 1982).

28. Dick, R. I. "Sludge Treatment" in *Physiochemical Processes for Water Quality Control*. W. J. Weber, Jr., ed. New York:John Wiley and Sons (1972).
29. Goscinski, J. S. "Effects of A. C. Powered Electrocoagulation Related to Fine Coal Cleaning," W. F. Berry & Associates Report on Coagulator at Delta Industries Mine at Meyersdale, Pennsylvania (February 8, 1982).
30. Electro-Pure Systems, Inc., Laboratory Notebook, Data Sheets E788001 through E1288012, 1988.
31. Picton, H. and S. E. Linder. *J. Chem. Soc.*, 67:63 (1895).
32. Hardy, W. B. and H. Burton. "A Preliminary Investigation of the Conditions Which Determine the Stability of Irreversible Hydrosols," *Proc., Roy. Soc.*, London, 66:110–125 (1900).
33. Adamson, A. W. *Physical Chemistry of Surfaces*. New York:Interscience Publishers (1967).
34. Jenke, D. R. and F. E. Diebold. "Electro-Precipitation Treatment of Acid Mine Wastewater," *Water Res.*, 18(3):855–859 (1984).

EPP Process for Stabilization/Solidification of Contaminants

S. L. Unger,[1] H. R. Lubowitz[1]

INTRODUCTION

Stabilization/solidification is a process for treating contaminants to prevent their pollution of soil and aquifers. In this process, contaminants are immobilized by fixative materials. The monolithic compositions must withstand chemical, physical, and mechanical stresses characterizing toxic waste management. Contaminants may then be consigned to the earth for safe, cost-effective final storage.

Stabilization/Solidification Treatment of Contaminants

The task of developing stabilization/solidification processes is one of the most exacting tasks in toxic waste management. When toxic wastes are solidified, no ready recourse is available for correcting an unsatisfactory product. Incomplete toxic waste destruction by incineration can be corrected by reincinerating the residues. "Reworking" the product to obtain a more suitable one is also an employable operation in toxic waste treatment processes such as precipitation, photo-decomposition, pyrolysis, etc. In contrast, toxic waste solidification products are not reworkable without incurring, comparatively, appreciable expense. Thus, the task of stabilization/solidification is exacerbated by the requirement for exceptional reproducibility of high-performance products. In order to realize general utilization for stabilization/solidification, contaminants with variable compositions of matter must be employed.

At this point the rationale for the stabilization/solidification of toxic wastes is given specifically for the EPP Process. The task of employing fixative material to immobilize contaminants can be carried out in two ways:

1. The contaminants are chemically compacted into the chemical composition of the material.
2. The contaminants are embedded into the body of the material.

These options are limited in scope according to the nature of the solid state of matter.

Chemically incorporating contaminants into the chemical makeup of a material, notwithstanding the associated technical problems, gives rise to questions about the nature of the resultant solid state. Fixative material that demonstrates high performance under chemical, physical, and mechanical stresses reflects the merit of its inherent chemical composition for providing such performance. It is therefore questionable whether this performance will manifest itself when a material's composition is disturbed by extraneous compositions, such as high concentrations of contaminants.

Nevertheless, the chemical compaction of contaminants into compositions of matter that may yield tough solid states (compounds) is possible. However, their toughness is manifested in small volume elements of their solid states. Thus, their monolithic structures would likely be very weak. For example, one of the strongest materials, ceramics, is nevertheless a very brittle one in monolithic dimensions. Discontinuities in the strength of ceramics due to weak "tissue" interconnecting tough, small elements of material give rise to their brittleness.

Although it appears that embedding contaminants in the fixative material is the more technically viable option, the question remains whether the high performance of a material will be exhibited upon the in-

[1]Environmental Protection Polymers, Inc., Hawthorne, CA 90250

corporation of appreciable amounts of embedded contaminants, even though they are not chemically compacted into its composition. In the current treatment of toxic metals, insoluble metal hydroxides are precipitated from aqueous mixtures. Although the precipitates are tough materials, their toughness (as for ceramics) manifests over a short range. Thus, strong monoliths cannot be readily fabricated from metal hydroxides. The precipitated contaminants are therefore embedded within pozzolanics, fly ash, and cement, with aliquots of the aqueous mixture usually employed for solidification. In this way, manageable monoliths are fabricated, because viable strengths characterize large volume elements in the fixative material in contrast to only small ones in the precipitates. However, hydraulic inorganic materials, being compositionally brittle, may experience appreciable decreases in their inherent performance due to flaw formation aggravated by embedded contaminants. They are labile to chemical corrosion by leachates and groundwater, enhanced by flaw formation, a process that contributes to further weakening of the monoliths. The embedded contaminants delocalize from the degrading fixative material and rendered as dispersed particulate matter in the soil. Unshielded by the fixative, the contaminants are subject to the full force of earth stresses.

Plastics, metals, and vitreous inorganic materials such as glass are candidate fixative materials, replacing hydraulic inorganic materials in order to produce mechanically viable and chemically stable monoliths. Principally due to chemical corrosion and material processing considerations for metals and glass with regard to toxic waste management, we selected plastics as the most desirable materials for managing contaminants. Our technical approach of using plastics for toxic waste management follows the previous discussion for use of hydraulic inorganic materials in present technology. The contaminants are first rendered into insoluble materials versus leachates, and the acceptability of the materials' stabilities is determined according to regulatory guidelines and tests. But, rather than embedding the "insoluble" contaminant compositions into hydraulic inorganic materials, metals, and glass, they are embedded in plastics. Plastics, in contrast to other materials, manifest an unusually high degree of their inherent mechanical strength when they are highly loaded with embedded particulate matter. The resulting plastic monoliths make disparate toxic particulates manageable, and they remain functional in safeguarding contaminants when they are subjected to the stresses of handling, transportation, and final disposal in the earth. The vastly superior chemical and physical stability of plastics vis-à-vis earth leachates with respect to that of "insoluble" contaminant compositions generally advances the ability of the embedded contaminants to withstand delocalization by leachates and groundwater.

The setting forth of plastic monoliths to manage contaminants requires judicious design of monoliths and selection of organic resins which comply with criteria of cost conservation; high chemical, physical, and mechanical performance of monoliths; and ease in processing them. Monoliths of large plastic/contaminant ensembles are generally difficult to produce and may be costly, even though low-cost plastics are employed. Their facile processing depends upon the nature of the selected resins. They must be chemically stable and mechanically viable under severest transportation and final disposal stresses. They must also exhibit such performance when embedded with high concentrations of contaminants. Thus, effective stabilization/solidification of contaminants by plastics is an exacting operation, and, for that reason, the EPP Process was designed and developed.

Managing highly concentrated contaminants by EPP Process monoliths addresses not only the above-stated criteria, but also considerations regarding toxic waste storage in a hazardous waste landfill. The conservation of landfill space is an important consideration for addressing ecological and fiscal concerns in relegating contaminants to the earth. Another relates to the ready retrievability of monoliths by conventional material handling techniques should it be desirable to clear the landfill. This retrievability feature should markedly reduce premiums for liability insurance regarding toxic waste disposal in landfills, because it allows for safe, quick emergency response to correct untenable hazardous waste depositories.

In previous work we demonstrated the EPP Process effectively manages very high concentrations of contaminants, much greater than those realizable by current technology, and the Process yields monoliths with high performance and processing advantages. These features are the result of the unique composite configuration of the monolithic products. To fabricate the monoliths, the Process agglomerates particulate contaminant compositions with thermosetting resin and then encapsulates the agglomerates with thermoplastic resin. In exploratory work, polybutadiene resin was found to have exceptional properties for agglomerate formation, and tough agglomerates were found to form readily in the presence of high concentrations of many different types of contaminants. Polyethylene resin, a chemically stable and resilient plastic, was selected as the most effective resin for encapsulation. Encapsulation is effected by fusing polyethylene resin pellets onto the surfaces of the agglomerates. The pellets are commercial products and are employed in blow molding of polyethylene drums.

The monoliths are characterized by a core that holds contaminant compositions in chemically cross-linked polybutadiene, and a thick, seamless, polyethylene jacket encapsulates the core. Significant performance advantages stem from the composite configuration. Landfill overburden is carried by the core rather than the jacket; thus, monoliths provide greater load-bearing capability for overburden than containers. Monolith deformation and creep due to compressive loads are mitigated by the chemically cross-linked nature of the core resin. Polyethylene resin encapsulation of the core markedly improves the monolith's mechanical properties—particularly important is the impact strength. Chemical stability is achieved by virtue of the chemical compositions of polybutadiene and polyethylene. Hence, monolith corrosion is addressed, notwithstanding the multiplicity of leachate compositions. Water penetration of monoliths is mitigated by the hydrophobic nature of the resins, which is a result of their aliphatic compositions. Thus, monoliths effectively stabilize difficult-to-manage hydrophilic contaminants such as spent ion-exchange resins and ceramics. The advantages above are realized in the context of desirable product performance and cost parameters:

- high performance monoliths of toxic wastes
- minimal volume increase relative to bulk volume of contaminant compositions
- simple processing operations
- processability of many different toxic waste types
- excellent product reproducibility
- employment of low-cost, commercial resins

The goal of our present work is to demonstrate the general applicability of the EPP Process for the treatment of real-life toxic wastes without incurring extensive preliminary operations. By showing that the Process abbreviates such operations, its economic viability is enhanced. In earlier work we demonstrated the Process management of many diverse compositions of toxic wastes and their aqueous sludges. In the present work we will demonstrate the Process tolerance of such compositions when they are blended with oil, as well as with water.

LABORATORY-SCALE INVESTIGATIONS

Resin modules with unique composite structures are fabricated to stabilize toxic wastes. They are cylindrical monoliths that are 7.6 cm (3 inches) in diameter and 10.2 cm (4 inches) in height. Seamless resin jackets encapsulate cores that hold contaminants up to 90% by weight. Cores are formed by coating pollutant particles with polybutadiene resin and then agglomerating them by a thermosetting reaction. The encapsulation is carried out by fusing polyethylene pellets onto the surfaces of agglomerates, thereby jacketing them with 0.6 cm (1/4 inch) thick high-density polyethylene (HDPE). An important aspect of this operation concerns the heat distortion temperature of the core material. Due to the polybutadiene binder, the heat distortion temperature is greater than the polyethylene resin fusion temperature. Thus, proper polyethylene encapsulation can be carried out. The resin jackets mechanically reinforce the agglomerates and improve the performance of waste modules versus harsh testing conditions that simulate handling and environmental stresses of toxic waste management.

Sludge Management

Toxic waste sludge treatment involves two steps. In the first step the contaminants are chemically stabilized and the sludge solidified. Solidification converts the sludge into a solid state that is free of mobile liquid. In the second step the solidified sludges are particulated, and the particles are agglomerated by polybutadiene binder and encapsulated by high-density polyethylene. Figures 1 through 4 show the products of fabricating laboratory-size modules from heavy metal sludges. Table 1 lists the wastes managed on a laboratory scale.

FIGURE 1. Aqueous sludge in container holding mineral and small amounts of organic contaminants.

FIGURE 2. Particulated sludge solidified by lime.

FIGURE 3. Sludge particles agglomerated by polybutadiene.

FIGURE 4. Agglomerate reinforced by polyethylene encapsulation.

Table 1. Examples of Toxic Waste Sludges Treated by Surface Encapsulation

Source	Major Contaminants
Salt cake from wastewater evaporators	radionuclides
Ion-exchange resins for wastewater demineralization	radionuclides
Spent activated carbon	radionuclides, heavy metals
Ion-exchange glasses for wastewater demineralization	radionuclides
Pesticides	monosodium methanearsonate
Electroplating sludge	Cu, Cr, Zn
Nickel–cadmium battery sludge	Ni, Cd
Pigment production sludge	Cr, Fe, CN$^-$
Chlorine production brine sludge	Hg, Na, Cl$^-$
SO$_x$ scrubber sludge, double alkali process, western coal	Cu, Na, SO$_4^=$/SO$_3^=$
SO$_x$ scrubber sludge, limestone process, eastern coal	Cu, SO$_4^=$/SO$_3^=$, heavy metals
SO$_x$ scrubber sludge, lime process, eastern coal	Ca, SO$_4^=$/SO$_3^=$, heavy metals
SO$_x$ scrubber sludge, double alkali process, eastern coal	Na, Ca, SO$_4^=$/SO$_3^=$, heavy metals
SO$_x$ scrubber sludge, limestone process, western coal	Ca, SO$_4^=$/SO$_3^=$, heavy metals
Calcium fluoride sludge	Ca, F$^-$
Mixed waste sludge	heavy metals, radionuclides

The chemical stabilization of contaminants addresses the requirements for delisting toxic wastes. Although the modules exhibit excellent stabilization of contaminants under harsh leaching stresses, delisting requires grinding of the modules; therewith showing contaminant stability in the leachates. Therefore, the sludges are chemically treated to stabilize their contents of heavy metals and organic pollutants versus leachates specified in the regulations. The precipitation of heavy metals as sulfides and polysilicates has generally been successful for rendering pollutant compositions that address Extraction Procedure (EP) toxicity and Toxicity Characteristic Leaching Procedure (TCLP) requirements. Absorbents are employed to stabilize organic liquid pollutants.

Low-cost dehydrating agents, such as lime, kiln dust, or Portland cement, are employed for sludge solidification. This operation does not significantly increase the volume of the sludges, because the quantity of such reagents needed to dewater the sludges is not appreciable. The small amounts of reagent employed do not give rise to load-bearing solids; the cured mixtures are very friable and they particulate readily. The particles are agglomerated by polybutadiene binder, and the agglomerates are encapsulated by polyethylene.

Binder Properties

The polybutadiene binder resins are of the 1,2 stereoconfiguration and they were formed to thermoset in the presence of many different contaminants to yield a chemical structure analogous to chemically cross-linked polyethylene. The resins show excellent chemical stability and withstand in the cured state mechanical stresses without distortion due to creep. Their low surface energy in the liquid state facilitates good wet-out of particulated filler material; thus, easing the processing of high filler content products.

The ability of polybutadiene resin to react in the presence of contaminants and to withstand pH and compositional variations of input material is due to the resin's water incompatibility (hydrophobicity) and high functionality. Since the resin does not absorb polar and mineral impurities, these reaction-interfering materials do not develop at the reaction sites to impede the curing of resin/waste mixtures. Furthermore, the high functionality of the resin gives rise to three-dimensional resin networks at relatively low extents of reaction; thereby readily yielding dimensionally stable agglomerates.

The nonpolar nature of the resin's functional groups provides additional processing benefits. In contrast to polar functional groups of resins such as epoxy and urea-formaldehyde, whose reactivities are activated over a broad temperature range, including atmospheric temperatures, the nonpolar groups are stable up to a threshold temperature that is moderately above atmospheric temperatures; thereon, they are triggered into reactivity by free radical initiators. This feature permits the mechanical blending of resin and particulated contaminants to be carried out at atmospheric temperatures without being impeded by increasing viscosity and premature gelling of the mixture. By alleviating such operations from time-sensitive constraints, one gains the options of storing, adjusting, and inspecting the mixture prior to the formation of dimensionally stable agglomerates. In addition, the mixture can be made by pouring resin onto a bed of contaminants; thus eliminating mechanical blending.

Upon the initiation of the agglomerating reaction for mixtures of contaminants and polybutadiene, the temperature of the mixtures does not significantly increase in the thermosetting process. This is due to the curing of mixtures depending upon the heat transfer rate. The mixtures cure incrementally, and the surrounding material provides a heat sink for heat generated at the reaction site. In contrast, material fixatives, such as thermosetting polyesters, epoxides, and inorganic cements, begin to react upon the addition of the reactants; consequently, the resins cure *in toto* rather than incrementally. In fabricating large monoliths, the simultaneous release of chemical energy causes a significant increase in the temperature of the mixtures, thereby making solidification and product reproducibility difficult to control. Polybutadiene mixtures, on the other hand, exhibit reaction exotherm temperatures that do not significantly exceed the reaction initiation temperatures.

Module Properties

An examination of modules shows the 0.6-cm (1/4-inch) thick HDPE jackets to be free of flaws. The modules exhibit dimensional stability under compressive loads of about 12.4 MPa (1,800 psi) [1]. Appreciable distortion of the modules by severe compressive loads does not produce jacket flaws; the modules remain watertight and the contaminants secure. Figures 5 to 8 show module behavior under uniaxial compression, partially compressed, and then greatly compressed to 20% of the module's original height. The jacket still remains intact. The compressed modules typically recover to 90% of their original height upon release of the compressive load. Figure 9 exhibits a module greatly compressed laterally—a severe mode of compression. In this mode of compression as well, the jacket remains intact. The released modules appreciably regain the original dimensions of their unconstrained state [1].

The uniaxial compression strengths and toughness of resin modules holding considerable amounts of pollutants exceed those of pozzolanics with low pollutant content. The modules' cores sustain the applied mechanical load and resist creep due to the chemically cross-linked structure of the resin. Dropped onto a steel plate, the modules remain intact on impact. It requires extraordinary impact, according to our evaluations, to rupture a module. Even then, there is no spillage of waste contents because they are fixed in the module core by a high-strength resin matrix. When penetrating modules with a pointed steel probe, they resist great force because the resin jackets are tough and reinforced by impregnation onto the surfaces of the cores.

The modules are dimensionally stable in leachate immersion, notwithstanding the waste type or concentration. During the immersion of ion-exchange resin (IER) laden modules, no change was noticeable. The dimensions of immersed specimens were

FIGURE 5. Onset of vertical compression of module.

FIGURE 6. Partial compression of module.

FIGURE 7. Module compressed to 20% of original height with no jacket rupture.

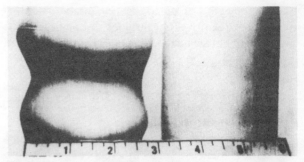

FIGURE 8. Comparison of module after release of compressive load (left) to unperturbed module (right).

monitored analytically. Figure 10 shows module stability during a forty-five-day water immersion period. These modules hold 70% by weight IER. In comparative tests, IER-laden modules, fabricated with asphalt and cement, underwent appreciable change: the former swelling and the latter cracking and spalling. These modules contained only 20% and 8% IER, respectively [2,3].

In Environmental Protection Agency (EPA) sponsored work, static leaching of modules holding concentrated contaminants was carried out in leachates that are harsher than those used in tests currently specified by the EPA or the Nuclear Regulatory Commission (NRC). The leaching solutions included strong mineral acids and bases, simulated seawater, an organic solvent, and organic acid. The studies showed excellent retention by the modules for highly mobile heavy metals [1].

The Army Corps of Engineers conducted column leaching tests for a two-year period on modules and comparative ones for other solidified wastes from various laboratories. The modules exhibited excellent performance [4]. Testing of the modules was also carried out to assess their physical and chemical performance. The modules were subjected to wet/dry and freeze/thaw cycling. They were frequently water-soaked and then rapidly dried. They were cycled from a $-10°C$ salt/ice bath to $+100°C$ boil-

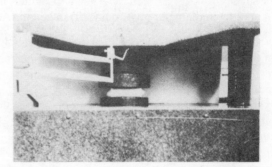

FIGURE 9. Module greatly compressed laterally with no jacket rupture.

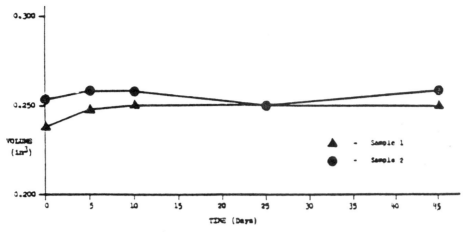

FIGURE 10. Dimensional stability of modules immersed in water. Sample volume versus duration of immersion in water (stabilized ion-exchange resin powders).

ing water; the cycles were repeated every 15 minutes with 16 cycles/day. The modules were stored in a freezer overnight. The modules showed no visible signs of damage, and their compressive strengths were within 95% of their original values.

SCALE-UP DEVELOPMENT

Under Department of Energy (DOE) Small Business Innovative Research Program (SBIR) sponsorship, we designed and constructed a unique apparatus for fabricating commercial-size modules [5]. Large modules give rise to cost-effective management of pollutants, but size is limited by product processing and handling considerations. The first pollutant managed in scale-up development is sodium sulfate salt, a low-level radioactive contaminant that accumulates in evaporators in nuclear plants.

Commercial-size modules are cylindrical in shape and 182 liters (48 gallons) in volume. The module shape approximates that of a 55-gallon drum; thus making modules manageable with conventional drum handling equipment. Figure 11 shows a module core consisting of agglomerated sodium sulfate. Figure 12 shows the complete module, which is 61 cm (2 ft) in diameter by 61 cm (2 ft) in height. Additional module specifications are:

- weight: 325 kg (715 lb)
- content: 87% by weight salt
- exterior: 0.95-cm (3/8-inch) thick HDPE, with 2% carbon black

Toxic Waste Processing

Figure 13 provides a schematic for managing sludges. Contaminants are stabilized and sludges dewatered chemically. The particles are loaded into the agglomerating mold by an auger, and polybutadiene binder is simultaneously introduced by a metering pump. With "kiss" pressure and moderate temperature, agglomerates of particulated wastes are fabricated in approximately two hours. The agglomerates are then encapsulated by molding polyolefin resin onto their surfaces. This is carried out in

TOXIC WASTE MODULE FABRICATION
(Module weight and size: 715 lbs, 24 inches × 24 inches)

FIGURE 11. Na_2SO_4 agglomerated by polybutadiene resin.

a matched die mold providing interstitial space for introducing powdered or beaded resin.

Apparatus

Our prototype apparatus is employable for managing limited quantities of wastes, such as the low-level radioactive waste output from commercial power plants. Figure 14 pictures the apparatus and operator stand. The apparatus capacity is 1,000 tons of wastes per year. The utility requirements are electricity (440V, three-phase) and water for cooling the molds. The apparatus features:

- molds mounted on a single frame
- an indexing table for product manipulation
- hydraulics for mold actuation

The molds for agglomeration and encapsulation have electrical band heaters and drilled-in channels for water cooling. The agglomerating mold is a cylindrical steel shell 60 cm (23.5 inches) in diameter. A platen fits inside the mold body to confine the waste/binder mixture. A linear positioning table transports the fabricated agglomerate to the encapsulating mold. The jacketing mold is split vertically to facilitate product demolding.

FIGURE 12. Agglomerate encapsulated by 1/4-inch thick black high-density polyethylene.

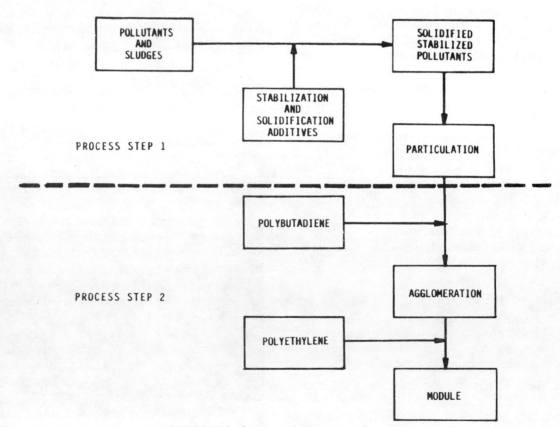

FIGURE 13. Surface encapsulation process schematic.

FIGURE 14. Pollutant stabilization process apparatus.

The apparatus can be operated by remotely with a microprocessor that controls the following functions: actuating and heating the molds, maintaining mold temperature, regulating cooling water flow, indexing the agglomerate to the encapsulating mold, and indexing the module to the unloading station.

Large-Scale Processes

The rate-determining step for product throughput by the apparatus is that of agglomerate formation. Consequently, for large-scale waste stabilization/solidification, such as the management of toxic bottom ash and fly ash stemming from toxic waste incinerators, agglomerates would be mass-produced independently. Agglomerate formation may be carried out in an oven capable of processing multiple agglomerate units. They would then be mounted on the apparatus for encapsulation. The output of the apparatus would thereby markedly increase, producing about one module per fifteen minutes.

Material Cost Estimate

The EPP employs low-cost, mass-produced materials. The resins, which stem from ethylene and butadiene monomers, are marketed at about $0.20 and $0.50 per pound, respectively. The monomers are well-known, are produced commercially in large amounts, and yield resins for fabricating polyethylene containers, polyethylene film, and polybutadiene vehicle tires. For module encapsulation, we employ polyethylene that is marketed to the rotomolding and blow molding industries for producing holding and transportation containers for fertilizers and harsh chemicals. The polybutadiene employed is the type used in the United States and Japan for making corrosion-proof steel frames for automobiles by resin-coating them in electroplating vats. The cost of resins for fabricating a 182-liter (48-gallon) module is estimated to be approximately $80.00. Since crude polybutadiene, rather than the refined polybutadiene for electrodeposition, is employable for module fabrication, the potential exists to significantly reduce the costs of materials.

Material costs for state-of-the-art low-level radioactive waste (LLRW) fixation and containerization are compared to those for the EPP Process. Table 2 presents material costs for fixation processes employing cement, asphalt, polyester, polybutadiene/polyethylene (EPP Process), and high-integrity containers (HIC). The costs of materials and HIC are vendor quotations. The values given for the ratios of materials to wastes represent those reported in the technical literature for managing spent ion-exchange resins.

The cost advantage for the EPP Process stems from the minimal amount of materials employed. The cost advantage is evident for managing many other toxic wastes, as well as spent ion-exchange resins. In addition to cost advantages for materials, the EPP Process yields products with lower transportation and disposal costs due to their significantly reduced volumes.

Table 2. Comparative Material Costs for LLRW Management

Materials	Costs $/lb	Materials/Ion-Exchange Resins Wt. Ratio	Management Cost for Ion-Exchange Resins $/lb
Cementitious Sodium Silicate	0.06	10:1	0.60
Asphalt	0.20	3:2	0.30
Polyester	0.75	2:3	0.50
EPP Process	1.25	2:10	0.24
High-Integrity Containers	225 (per 55-gallon drum)	420 lb/drum	0.60

CONCLUSIONS

The EPP Process provides a unique option for high-performance, cost-effective stabilization/solidification of toxic wastes. In its present state of development, laboratory investigations and scale-up have been successfully concluded. The work to date concludes Phases I and II of the SBIR program. The program is postured for Phase III: pilot plant studies. In these studies the processing parameters will be defined for positioning the process in the field.

ACKNOWLEDGEMENTS

The authors wish to thank the USEPA Hazardous Waste Engineering Research Laboratory (Cincinnati, Ohio), Carlton Wiles, project officer with EPA; and the Department of Energy Small Business Innovative Research Program, James Turi, project officer with DOE, for their support of the work described in this chapter.

REFERENCES

1. Telles, R. W. et al. "Contaminant Fixation Process," *Proceedings from the 40th Annual Purdue Industrial Waste Conference*, pp. 685–691 (May 1985).
2. Matsura, H. et al. "Improvement of Asphalt Waste Products in Leachability," JAERI-M8664, Japan Atomic Research Institute, Tokai (1980).
3. Columbo, P. and R. M. Neilson, Jr. Waste Form Development Program Annual Progress Report, BNL-51614 (Sept. 1982).
4. Malone, P. G., R. B. Mercer and D. W. Thompson. "The Effectiveness of Fixation Techniques in Preventing Loss of Contaminants from Electroplating Wastes," *First Annual Conference on Advanced Pollution Control for Metal Finishing Industry*, AES and USEPA (Jan. 1978).
5. Unger, S. L., R. W. Telles and H. R. Lubowitz, U.S. Patent 4,756,681 (July 12, 1988).

2.9

Electromembrane Process for Recovery of Lead from Contaminated Soils

William F. Kemner,[1] E. Radha Krishnan[1]

INTRODUCTION

Numerous Superfund sites throughout the United States are contaminated with toxic metals. Battery reclamation, lead smelting and lead-based paint manufacturing are examples of processes which could result in lead contaminated soils. Soils from defunct battery reclamation sites have been found to average about 5 weight percent of lead (Pb). Quantities of contaminated soils range from less than 5,000 cubic yards per site to almost 100,000 cubic yards. Many of the sites are located over key underground aquifers in populated areas, raising concerns for contamination of water supplies.

The concentration range of lead in soils found at 436 contaminated sites has been reported to be 0.16 to 466,000 ppm, compared with the natural background level of 2 to 200 ppm [1].

This chapter presents the results of a research program to optimize, through bench-scale tests, the process variables for the chelation and electroplating operations of an innovative electromembrane process employing a chelating agent for recovery of lead from contaminated soils. It also discusses scale-up design criteria and comparative economics for the process.

Ethylenediaminetetraacetic acid (EDTA) was used as the chelating agent for the process. Metals treatability tests conducted in the laboratory on soil samples from two different sites showed that the optimum EDTA/lead molar ratio for the chelation reaction is in the range of 1.5 to 2.0 for the two materials, and that the chelation reaction is essentially complete within one hour. A pH of 12 effectively prevented the chelation of other metals such as iron. Tests conducted on a bench-scale electromembrane reactor (EMR) unit showed lead removal efficiencies approaching 90 percent for chelate solutions with initial lead concentrations greater than 1 percent. The lead removal efficiency decreased with decreasing lead concentration. The pH of the chelate solution had no apparent effect on plating efficiency. Higher current densities resulted in faster plating rates and a spongy lead deposit on the cathode. The as-plated material was analyzed at over 75 weight percent lead in two experimental runs.

APPLICABILITY

The process appears to be applicable to a wide variety of lead-containing soils and wastes such as slags, dusts, and sludges from industrial processes. Economic analyses for the overall soil washing electromembrane process based on a preliminary process design show its cost to be competitive with off-site disposal options.

The applicability of the process is highly site-dependent. Factors such as soil fines content, clay content, and lead solubility can strongly influence the cost and performance of the process. Consequently, both the soil treatability (chelation) and electroplating tests were conducted on a variety of samples in order to make preliminary assessments of process applicability.

PROCESS DESCRIPTION

Although research has focused on the chelation and plating steps of the soil-washing process, the

---
[1]PEI Associates, Inc., 11499 Chester Road, Cincinnati, OH 45246

design factors necessary for scale-up must be considered for the overall process. The four major process operations are solids handling, EDTA reaction/washing, lead plating (EMR), and water treatment.

Solids Handling

Initial solids processing depends upon the specific site characteristics. The material may be processed via screening, magnetic separation, and/or crushing. Metal and other bulk material must be removed. If crushing is required, the material may be rescreened and stockpiled for later feed to the system.

EDTA Reaction/Washing

The purpose of the EDTA-reaction step is to thoroughly mix the soil and EDTA solution to chelate the lead. After chelation, the lead complex is washed from the solids in a series of dilution steps.

The major parameters governing the operation of this phase are the lead concentration leaving the system as a final product and the moisture content of the solids as they move through the system. Dewatering characteristics of the material are critical in this step. The amount of water to be used must be optimized through the use of multiple stages.

Lead Plating (EMR)

The primary control variables in the EMR process are current density, lead concentration in the chelate, and cathode solution pH. Higher current density generally produces a lower quality plated metal, but plated metal quality is not of paramount importance. Effective operation at both high and low lead concentrations is extremely important in order to accommodate various levels of contamination in soil or waste materials. The EMR should be able to function well at both low and high pH.

Water Treatment

Water from the plating step is sent to a waste treatment system. When the lead concentration decreases to a low level in the EMR, it will probably be cost-effective to reconcentrate the water to maximize lead recovery. Eventually, dissolved solids will build up and a blow-down stream will have to go to a waste treatment system. It is essential to the economics of the process to recover and reuse the chelating agent prior to final discharge of the water.

Economic Analysis

Comparative economics for cleanup of a given site are highly dependent upon site location, lead concentration, and nature of the material. Some sites contain lead only, and others are contaminated with multiple pollutants, both inorganic and organic. In addition, the dewatering characteristics of various materials vary widely, which in turn affects processing cost. The comparative economics of soil washing versus other alternatives must be determined specifically and individually for each site.

A computerized cost model was developed to evaluate the effect of site-specific process variables. Table 1 lists the variables included in the model. The current cost model is based on the use of mobile equipment, including cement mixers, trailer-mounted pressure filters, and trailer-mounted EMR units. Preliminary cost analyses indicate the cost of the soil washing process to be competitive with landfilling. The process has an economy of scale, being more competitive as the quantity of material to be treated increases above approximately 2000 yd^3. Assumptions used for the cost computations must be refined based on the specific considerations associated with each site.

ADVANTAGES OVER CONVENTIONAL TREATMENT OR DISPOSAL

The cleanup of sites contaminated with heavy metals has traditionally involved excavation of the wastes and contaminated soils with subsequent disposal at an off-site, Resource Conservation and Recovery Act (RCRA)-approved landfill. In addition to increasing costs and dangers to public safety from large-scale transportation of wastes, long-term environmental liability is also a concern associated with the landfilling approach. Many experts have characterized this approach as simply "moving the problem" instead of solving it. Thus, there is great incentive for the development of alternative methods for cleanup of contaminated sites.

Figure 1 summarizes the alternatives available for treating lead-contaminated soil. It should be noted that only the soil-washing option actually removes the lead from the contaminated soil. This chapter describes research conducted to investigate the process characteristics, design, and economics of a soil-washing process employing an electromembrane reactor (EMR) for treatment of contaminated soils for recovery of heavy metals such as lead. Figure 2 provides a highly-simplified overview of the soil-washing process. The process uses ethylenediamine-

Table 1. Variables in Cost Model

Total Material (cubic yards)	Lead Credit, $/lb
% Material Dry Process	Blow-down rate, %
Equipment Rental, $/mo	Water Treatment, POTW, $/1000 gal
Mixers	Trans. to POTW, $/loaded mile
Screens and conveyers	Distance to POTW, miles
Filters	Avg. Operating Labor Rate, $/h
EMRs	Avg. Maint. Labor Rate, $/h
Tankage	Labor Overhead, %
On-site Trailer	Per Diem, $/d
Others	Electricity, $/kWh
Operation, h/d	Supply water, $/1000 gal
Number of Rinses	Connected Load, hp
Water/Soil Ratio	Lighting Load, kW
Filtering Rate, gph/sq ft	Plating Load, kWh/lb Pb
Plating Rate, h/2000 gallons	Total Crew Size (operating)
Reaction Time, h/batch	Maintenance Crew Size
Batch Size, yd	Pounds lead recovered
Analytical, $/batch	Lead in Soil Processed (lbs)
Operating Supply %	Supply Water, gallons
Maintenance Supply %	Filtering Time, h
Lead in Soil, %	Reaction Time, h
Lead Recovery, %	Plating Time, h
EDTA/Lead Ratio	Number of Units
EDTA Recovery, %	Mixers
Capacity Utilization, %	Screens and Conveyers
Cost of EDTA, $/lb	Filters
Cost of Caustic, $/ton	EMRs
Cost of Sulfuric acid, $/ton	Tankage
Cost of Sodium Carbonate, $/ton	On-site Trailer
Cost of Sodium Sulfide, $/ton	Others
	Filter Size, sq ft

POTW—Publicly-owned treatment works.

tetraacetic acid (EDTA) as the chelating agent and recovers lead by electrodeposition. The primary objective of the research was to optimize, via bench-scale tests, the process variables for the chelation and electroplating (EMR) operations of the process. The classification and dewatering steps, though crucial to the overall process, represent existing technology and were not studied specifically during this research. This process results in a lead product containing about 90 weight percent lead at optimum process conditions.

STATE OF DEVELOPMENT

The bench-scale research has shown the feasibility of the two essential process steps of an innovative soil-washing process: chelation and electro-deposition.

A long-term pilot-scale demonstration at several actual sites is necessary to develop the data required for commercialization. The research will examine several critical issues that have been identified through the bench-scale experimental studies and the preliminary engineering design/economic evaluation. These include:

1. The concentration of low-lead-content wastes to achieve lead levels in the EMR of 1 percent or more
2. The dewatering of fine materials
3. The water balance for the washing and rinsing process, i.e., degree of recycle possible, blow-down requirements, and total supply of water required

The completion of the pilot-scale research will culminate in:

- a final process flowsheet
- a complete material balance
- a plot plan and equipment layout
- mechanical flowsheets
- equipment specifications
- design calculations
- detailed operating procedures manual
- a refined cost model

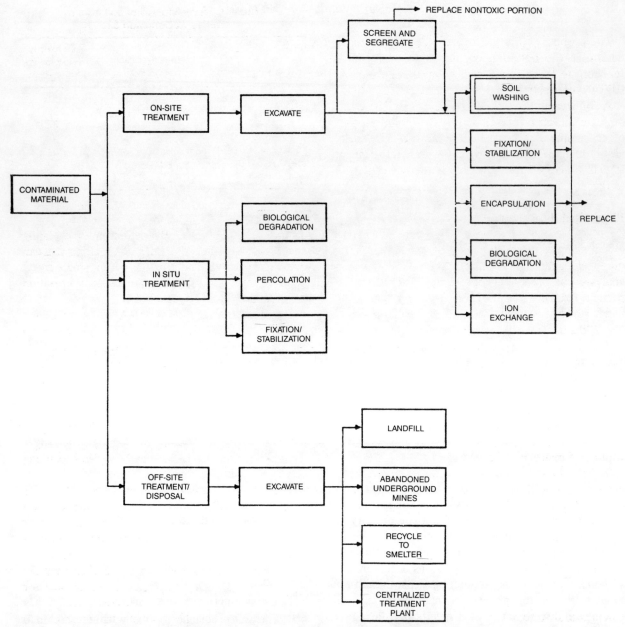

FIGURE 1. Treatment alternatives for lead-contaminated soils.

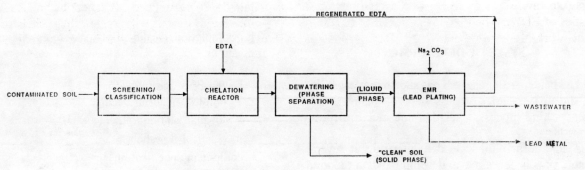

FIGURE 2. Simplified process flow diagram of overall soil washing process.

RESULTS

The purpose of the soil treatability testing was to determine the optimum conditions for soil-EDTA reactions to 1) maximize lead chelation, 2) minimize EDTA consumption, and 3) minimize reaction time.

A soil treatability test procedure was developed to evaluate the effect of pH, EDTA consumption, and reaction time at a constant temperature. The treatability testing involved physical and chemical characterization of the raw material followed by chelation testing for lead recovery/metals interference.

Physical and Chemical Characterization

Soil samples typically consist of varying amounts of gravel, sand, silt, clay, and organic matter. A sieve analysis was used to determine the distribution of particle sizes in the soil. The exact test is described under ASTM Designation D 422. Material passing a No. 200 sieve tends to be composed largely of clays and silts, and is generally difficult to dewater. Screening of the material prior to reaction separates the material into fractions which can be analyzed to determine the particle size distribution of the material. Screening has shown a tendency for higher lead content material to segregate in the fine fractions. Consequently, screening may be used to reduce the volume of material to be treated.

Samples of soil from two sites were screened and extraction procedure (EP) toxicity tests performed on each fraction to determine if a toxicity gradient existed based on physical sizing. The results shown in Tables 2 and 3 illustrate the tendency for lead to segregate in the fine fractions for these soils. A similar relationship, however, may not be expected for all soils.

Chelation Testing

Before describing the chelation tests in detail, it is helpful to review briefly the properties and characteristics of EDTA. There are many forms of EDTA. In this work, the tetrasodium salt of EDTA was used as the chelating agent. By definition, a chelating agent is a compound containing donor atoms that can combine by coordinate bonding with a single metal atom to form a cyclic structure called a chelation compound or, simply, a chelate.

A range of molar ratios of EDTA/lead were used at a selected pH condition to determine the minimum ratio necessary for essentially complete chelation. Liquid chelate was sampled from the soil-EDTA reactor at specified time intervals to determine chelation as a function of time.

These tests provide information on lead recovery, iron interference, reagent needs, and feasibility of treating a particular waste by chelation. The ranges for pH, time, and EDTA use can be varied depending on the particular soil.

The soil treatability procedures developed for this study were performed on lead-contaminated soil samples from two Superfund sites (Arcanum near Troy, Ohio, and Lee's Farm in Woodville, Wisconsin) [2]. Table 4 provides the analysis of the metals content of these two soils. Figure 3 illustrates the relationship of chelation efficiency versus time for two test runs on the Lee's Farm soil and one test run on the Arcanum soil. It is apparent that the chelation reaction is essentially complete within one hour for both the Lee's Farm soil and Arcanum soils at each of the EDTA/Pb molar ratios. It cannot be predicted that other wastes or soils will necessarily be chelated so rapidly. Figure 4 presents the final chelation effi-

Table 3. Sieve Analysis of Soil from a Battery Reclamation Site

Size fraction	Total Pb, (%)	EP toxicity value for Pb, mg/L
>10 mesh	1.5	7
>20 mesh	3.0	22
>35 mesh	4.4	37
>70 mesh	4.8	42
>100 mesh	4.5	51
>200 mesh	6.0	49
<200 mesh	6.2	55

Table 2. Sieve Analysis of Waste from an Industrial Site

Size fraction	Percent in size fraction		EP Toxicity value for Pb, mg/L
	Range	Mean	
+20 mesh	45–63	54	67
(−20) +35 mesh	9–12	11	186
(−35) +100 mesh	18–29	23	174
(−100) +200 mesh	5–8	7	248
−200 mesh	5	5	344

Table 4. Chemical Analysis of Test Soils
(μg/g on as-received basis)

Element	Soil source	
	Arcanum	Lee's Farm
Cadmium	4	1
Calcium	59630	47340
Chromium	19	14
Iron	20790	22010
Lead	78950	38670
Zinc	110	81

ciency as a function of EDTA/Pb molar ratio. The optimum EDTA/Pb molar ratio appears to be approximately 1.5 to 2.0 for both the soils tested. The optimum EDTA/Pb ratio may be different for other materials. Chelation efficiencies exceeding 90 percent were observed for the Lee's Farm soil at an EDTA/Pb ratio above 1.5. The apparent lower chelation efficiency for the Arcanum soil may be due to the presence of either metallic lead (as opposed to ionic lead) in the sample, or microencapsulation of lead.

Metallic lead is digested in the analysis procedure for total lead but is not chelatable. It should also be noted that metallic lead is not extracted in the EP toxicity procedure used to determine leachability characteristics. The EP test is conducted at a pH of 5 using acetic acid. Since the basic purpose of the chelation process is to render the soil nonhazardous, lead recoveries must be based on the ability of the chelation process to produce a residue that has an EP toxicity lead content of less than 5 mg/L (the federally allowable standard) rather than the total lead removal.

EMR Tests

Previous research on the electromembrane reactor (EMR) has been performed in the context of regen-

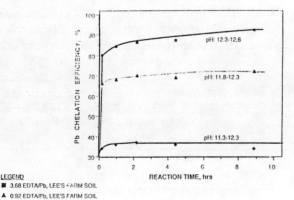

FIGURE 3. Chelation efficiency as a function of time.

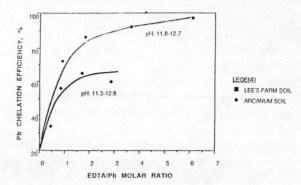

FIGURE 4. Chelation efficiency as a function of EDTA/Pb molar ratio.

erating ion-exchange resins [3]. The current research expanded upon this application. Several variables are of importance in the experimental design of the EMR tests.

Electrode Potential

The extent of chemical reaction occurring in an electrolytic cell is directly proportional to the quantity of electricity passed into the cell. For example, it requires two moles of electrons to produce a mole of copper from Cu^{2+} and three moles of electrons to produce a mole of aluminum from Al^{3+}:

$$Cu^{2+} + 2e^- \rightarrow Cu$$
$$Al^{3+} + 3e^- \rightarrow Al$$

The electrical charge on a mole of electrons is called a Faraday (F), equivalent to 96,500 coulombs. A coulomb is the quantity of electrical charge passing a point in a circuit in 1 sec when the current is 1 ampere. Therefore, the number of coulombs passing through a cell can be obtained by multiplying the amperage and the elapsed time in seconds.

Current Density

Current density is calculated in milliamps (ma)/cm² (amps/ft², etc.). Current density for the experiments was determined by computing the ratio of the current flow on the power supply unit to the cross-sectional area of the membrane.

pH

The pH in the electromembrane reactor is a very important process condition which influences both the removal of metal from the solution and the recovery of the chelating agent by regeneration. The pH at the anode and the cathode varied during the EMR experiments due to the production of hydrogen ions at the anode and hydroxide ions at the cathode;

the pH, however, was not adjusted during each experiment.

Current Efficiency

The energy requirement for ionic transport in the electromembrane process is a function of the electrical resistance of the solutions and the membranes and the back electromotive forces caused by concentration gradients. The current efficiency can be calculated according to the following equation:

$$\text{Current efficiency} = \frac{\text{Metal ion removed (meq)} \times 96.5 \text{ (C/meq)}}{\text{Time(s)} \times \text{applied current (C/s)}} \times 100\%$$

where

meq = milliequivalent
C = coulomb
s = second

The current efficiency was determined as a function of time for the tests.

Chelate Concentration

The concentration of the lead chelate in the cathode chamber of the EMR affects current efficiency. As the concentration decreases, the power requirements to plate a given mass of lead increase.

Experimental Procedure

Figure 5 depicts the reactor system used for these experiments. The rectangular unit was constructed from a commercial glass aquarium with 1/4-inch-thick plexiglass. It was divided into two chambers by two 1/8-inch-thick plexiglass pieces. The frames served as supports for the cation exchange membrane. The membrane was glued into place and the joints sealed with silicone rubber sealant to prevent leakage between chambers.

The membrane used was manufactured by Ionics, specifically 61CZL386 modacrylic fiber-backed cation transfer membrane. The membrane has low electrical resistance and excellent resistance to physical and chemical stress. Most importantly, it has the ability to allow sodium ions to pass from the anode to the cathode chamber while preventing ionic transport in the opposite direction.

Lead electrodes were used in the EMR system. Both electrodes had dimensions of approximately 7 × 10 × 1/16 inch. They were mounted on wooden

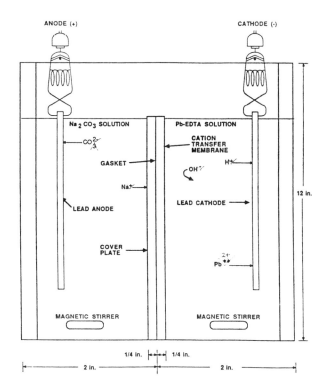

FIGURE 5. Schematic illustration of EMR test unit.

dowel rods suspended across the top of the aquarium. The power source supplied a potential of up to 40 volts and a direct current of up to 30 amps.

Once the reactor was operational, each run was started by addition of a 5 percent by weight of sodium carbonate solution to the anode chamber. In addition, an appropriate amount of metal chelate complex solution was placed in the cathode chamber. Each electrode was then placed in the EMR by suspending it approximately one inch from the membrane surfaces. The test began when voltage was applied and the current set at the proper amperage. The voltage across the circuit was allowed to vary in such a fashion that the current was maintained at a desired setting.

Considering the overall reactions involved in the reactor system, the major reaction of concern was the one resulting in the removal of metal from solution; thus, the metal concentration and reaction time were monitored regularly. This was done by taking samples from the cathode chamber at regular time intervals. To enhance mass transfer, a magnetic stirrer was placed in both chambers to cause mild turbulence throughout the operational period.

Experimental Design

The three primary control variables of interest in the EMR bench-scale experiments were current density, lead concentration in the chelate, and cathode solution pH. Higher current density generally produces a lower quality plated metal, but plated metal quality is not of paramount importance in the soil-washing process as long as its quality is not so inferior that it would inhibit sale of the product. The maximum current density for the experiments was kept below 30 ma/cm². Effective operation at both high- and low-lead concentrations is extremely important in order to accommodate various levels of contamination in soil or waste materials. Solution pH is of interest because an elevated pH is needed to inhibit iron chelation in high-iron wastes. The EMR should thus be able to function well at both low and high pH.

The source of lead chelate solution for the experiments was actual chelate produced at the Lee's Farm site. This material contained about 3 percent Pb and portions were diluted with water to create nominal 1 percent and 0.2 percent solutions. The solutions were adjusted to the desired pH using sulfuric acid or sodium hydroxide. Table 5 summarizes the actual lead content and pH of the feedstock solutions.

Five experiments were performed on the 0.2 percent Pb solution, and two experiments each were performed on the 1 percent and 3 percent Pb solutions. A partial factorial experimental design was adopted to evaluate the effects of lead concentration, current density, and pH.

Theoretical plating time was calculated based on Faraday's law.

$$Pb^{2+} + 2e^- \rightarrow Pb$$

Two moles of electrons (2 faradays) are required to plate one mole (2 equivalents) of lead. The grams of lead plated in one hour at one ampere (A) at 100 percent current efficiency can be calculated as follows:

$$\text{grams of Pb} = (1 \text{ h}) (1 \text{ A}) \frac{(3600 \text{ s})}{\text{h}}$$

$$\times \frac{(1 \text{ coulomb})}{\text{A-s}} \frac{(1 \text{ faraday})}{96,500 \text{ coulomb}} \frac{(1 \text{ mol Pb})}{2 \text{ faradays}}$$

$$\times \frac{207 \text{ g Pb}}{\text{mol}}$$

$$= 3.86 \text{ g Pb/A-h}$$

Given the total amount of Pb in solution and the desired current density, theoretical plating time (at 100 percent current efficiency) was determined. Current densities were calculated based on the 400 cm² area of the membrane.

Table 5. Lead Content and pH of Feedstock Solutions Used in EMR Experiments

Feedstock No.	Nominal lead content, mg/L	pH
1	30,000	11
2	30,000	4
3	10,000	11
4	10,000	8
5	10,000	4
6	2,000	11
7	2,000	8
8	2,000	4

EMR Test Results

Figures 6 through 9 illustrate plating efficiency (i.e., lead plated as a percent of total lead in solution) as a function of time. As expected, increased lead is plated with increasing time in all cases. Extremely high lead recoveries and current efficiencies are observed for the 3 percent and 1 percent lead solutions during the experimental time period. It appears, however, that current efficiency (and subsequent lead removal) at the starting lead concentration of 0.2 percent is low regardless of pH or current density. Figures 6 and 7 show that lead recoveries are below 40 percent at the 0.2 percent lead level for the experimental time period. Greater time periods should result in higher lead removal efficiencies for the low-lead solutions. Figures 8 and 9, however, show lead removal efficiencies approaching 90 percent for the 1 percent and 3 percent lead solutions. Figure 8 shows

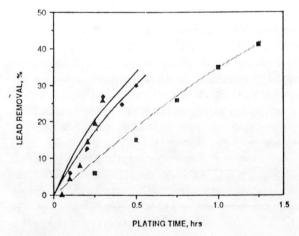

LEGEND
▲ pH 11 CURRENT DENSITY 25 ma/cm²
◆ pH 11 CURRENT DENSITY 15 ma/cm²
■ pH 11 CURRENT DENSITY 5 ma/cm²

FIGURE 6. Lead plating efficiency as a function of time (0.2 percent Pb in initial solution, pH = 11).

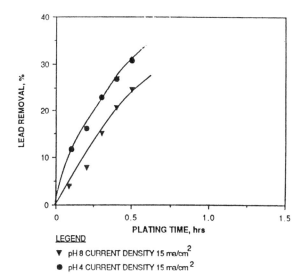

FIGURE 7. Lead plating efficiency as a function of time (0.2 percent Pb in initial solution, pH = 4, 8).

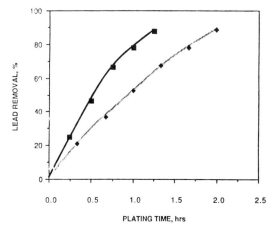

FIGURE 8. Lead plating efficiency as a function of time (1 percent Pb in initial solution).

the effect of current density at constant pH for a 1 percent lead solution. As expected, the higher current density produces a faster plating rate. It should also be noted that higher current density produces a spongy lead deposit on the electrode. Figure 9 illustrates the high-plating efficiency achievable at higher initial lead concentrations. The effect of current density on plating rate is again confirmed by the results shown in Figure 9. There is no apparent effect of initial cathode solution pH on plating efficiency.

There was no noticeable difference in the visual appearance of the lead product from the various experiments of a given initial lead concentration. In the 0.2 percent lead experiments, the plated lead was not visibly discernible on the electrode, but was confirmed by analytical results and the increase in the weight of the cathode.

Based on the experiments on the 0.2 percent lead liquor, the current efficiencies are higher at lower current densities, decreasing from 40 percent at a current density of 5 ma/cm² to approximately 20 percent at 25 ma/cm². There is no apparent effect of pH on this relationship. In the full-scale process, the current efficiency should not be a controlling factor in the economics because power costs are insignificant compared to other cost elements. Time, however, is an important factor because it relates to labor cost. Consequently it is desirable to run as high a current density as possible.

Table 6 provides an analysis of the plated lead product for those experiments where sufficient deposit could be scraped off the cathode. The plated metal analyzed over 75 weight percent lead in Runs 1 and 2. As shown, the amount of other metals plated is insignificant compared to the lead. Although not shown, the moisture content of the product is the other main constituent. After drying, therefore, the lead product is expected to have a purity in excess of 90 percent.

Hydrogen is generated at the cathode as a product of the electrolysis of water. The hydrogen generation rate was not measured, but the pH increase detected during the experiments in the cathode chamber indicated a decrease in hydrogen ion concentration.

Table 6. Analysis of Plated Metal from EMR Experiments (all μg/g as-received basis)

Experiment No.	1	2	8	9	9 (duplicate)
Cd	6.7	3.1	4.0	2.1	2.4
Ca	1128	1751	499	2709	3015
Cr	<1.2	<1.2	<1.2	<1.2	<1.2
Cu	264	226	175	265	259
Fe	35.1	25.7	23.9	48.7	51.2
Pb	787700	755700	497500	669500	672000
Mg	74.1	292	70.2	180	182
Zn	54.0	43.1	56.8	75.8	84.7

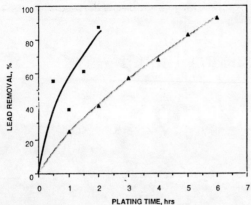

LEGEND
■ pH 11, CURRENT DENSITY = 25 ma/cm^2
▲ pH 4, CURRENT DENSITY = 15 ma/cm^2

FIGURE 9. Lead plating efficiency as a function of time (3 percent Pb in initial solution).

CONTACTS FOR MORE INFORMATION

William Kemner, PEI Associates, Inc., Cincinnati, Ohio 513/782-4700; Radha Krishnan, PEI Associates, Inc., Cincinnati, Ohio 513/782-4700; Ric Traver, U.S. EPA, Edison, New Jersey; Robert Fox, IT Corporation, Knoxville, Tennessee 615/690-3211.

REFERENCES

1. Sims, R. and K. Wagner. "In-Situ Treatment Techniques Applicable to Large Quantities of Hazardous Waste Contaminated Soils," *Proceedings of Management of Uncontrolled Hazardous Waste Sites, Hazardous Materials Control Research Institute (HMCRI), Silver Spring, MD.* Library of Congress, Catalog No. 83-82673, 1983.

2. Castle, C. et al. "Research and Development of Soil Washing System for Use at Superfund Sites," *Proceedings of Management of Uncontrolled Hazardous Waste Sites, Hazardous Materials Control Research Institute (HMCRI), Silver Spring, MD.* Library of Congress, Catalog No. 81655 (1985).

3. Tseng, Dyi-Hwa. "Regeneration of Heavy Metal Exhausted Cation Exchange Resin with a Recoverable Chelating Agent," a thesis submitted to the Faculty of Purdue University, Dr. James E. Etzel, School of Civil Engineering, Aug 1983.

Electrolytic Treatment of Waste Pickling Liquor

Dr. David W. Scarooson,[1] **Ed Flemming**[1]

This chapter describes a new process for treating waste pickling liquor to recover the acid used in the pickling process for return to the pickling bath and produces a harmless and possibly useful solid by-product.

THE PROBLEM

Iron and steel products, after fabrication, often acquire an undesirable coating of rust. This is removed in an operation called pickling by immersion of the products in an acid bath. Eventually the bath becomes too dilute in acid and too concentrated in iron salts to be effective. The used content of the bath is discarded as waste pickling liquor. It is typically a liquid containing about 10 percent unreacted acid and 5 percent dissolved iron and is a very troublesome hazardous waste.

CURRENT TECHNOLOGY

Four methods are currently in use to deal with waste pickling liquor:

1. It is sent to a hazardous waste disposal site. Sometimes it receives further treatment by a waste handling firm or at the disposal site. The option of land disposal of this hazardous waste is an obvious problem. Such disposal will be increasingly restricted as the laws now enacted are put into force.
2. It is neutralized with caustic. This sometimes occurs at the waste handling firm's facility or at the disposal site. This alternate works but is a waste of resources. The resulting iron hydroxides are harmless but messy. The by-products sodium sulfate or sodium chloride are harmless but may present disposal problems in some watercourses.
3. The solution is treated by evaporation or freezing, which permits recycling of the unreacted acid that has not been consumed in the pickling process and recovers ferrous sulfate in a partially hydrated state. There is some market for the resulting hydrated ferrous sulfate. The drawbacks to this alternate are a relatively high energy cost, the corrosiveness of the hydrated ferrous sulfate, and the limited market for hydrated ferrous sulfate.
4. A small proportion of waste pickling liquor is sold to water treatment plants for use as a flocculating agent. However, only a small portion of the waste pickling liquor generated can be used for this purpose. Further, the ferrous compounds produced are not nearly as desirable as flocculating agents as the corresponding ferric compounds. One alternate of the process proposed below would yield acid-free ferric compounds instead of the present mixture of ferrous compounds and free acid.

THE PICKLING PROCESS

Although the vast majority of pickling is done with sulfuric acid, a significant amount is carried out with hydrochloric acid, which acts faster and at a lower temperature. More hydrochloric acid might be used were it not for its higher cost. The successful development of a process for recycling the acid, such as the one outlined below, might well encourage the use of more hydrochloric acid.

---
[1]Flemming & Wickett, Inc., P.O. Box 1300, Issaquah, WA 98027

The following reactions take place in the pickling bath:

For sulfuric acid:

$$FeO + H_2SO_4 \rightarrow FeSO_4 + H_2O$$
$$Fe + H_2SO_4 \rightarrow FeSO_4 + H_2$$
$$Fe_2O_3 + 3H_2SO_4 \rightarrow Fe_2(SO_4)_3 + 3H_2O$$

For hydrochloric acid:

$$FeO + 2HCl \rightarrow FeCl_2 + H_2O$$
$$Fe + 2HCl \rightarrow FeCl_2 + H_2$$
$$Fe_2O_3 + 6HCl \rightarrow 2FeCl_3 + 3H_2O$$

A small amount of pickling is done with nitric and/or hydrofluoric acids. These are considered beyond the scope of this paper.

The ferric compounds produced as shown are usually reduced to ferrous compounds in the pickling bath:

$$Fe_2(SO_4)_3 + Fe + 3e^- \rightarrow 3FeSO_4$$

PROPOSED PROCESS

The proposed process is electrolytic. An electrolytic cell (or multiplicity of cells) is divided into anode and cathode chambers by a membrane permeable only to cations. The anodes and cathodes are graphite.

The treatment process for hydrochloric acid is slightly different from that used to treat sulfuric acid. As it is somewhat simpler, it will be discussed first.

Hydrochloric Acid

The waste hydrochloric acid pickling liquor is fed to the anode chamber, and a solution of sodium chloride is fed to the cathode chamber to provide conductivity.

The reactions in the anode chamber are:

$$Fe^{2+} \rightarrow Fe^{3+} + e^-$$
$$2Cl^- \rightarrow Cl_2 + 2e^-$$

Hydrogen ions, ferric ions, and a small amount of ferrous ions migrate from the anode chamber to the cathode chamber because they are drawn to the cathode and can pass through the cation permeable membrane.

Chloride ions that are not discharged as chlorine gas and the few hydroxyl ions present in the acid solution are retained in the anode chamber because they are not drawn to the cathode and cannot pass the cation permeable membrane.

In the cathode the reactions are:

$$3H_2O + 3e^- \rightarrow 3/2H_2 + 3OH^-$$

Fe^{3+} (from the anode chamber) $+ 3OH^- \rightarrow Fe(OH)_3$

Fe^{2+} (from the anode chamber) $+ 2OH^- \rightarrow Fe(OH)_2$

The sodium chloride (which has been successfully replaced experimentally by sodium sulfate, sodium hydroxide, and sodium carbonate) is present only to provide conductivity in the cathode chamber.

The overall cell reaction is closely approximated by:

$$2FeCl_2 + 6H_2O \rightarrow 2Fe(OH)_3$$
$$+ 2HCl + 2H_2 + Cl_2 + 2e^-$$

The overall cell process is shown in Figure 1.

The liquid effluent from the anode chamber will contain some hydrochloric acid and be depleted in iron content. The liquid effluent from the cathode chamber will be filtered to remove the iron hydroxide precipitate and then recycled to the cathode chamber.

The gaseous effluent from the cell consists of hydrogen and chlorine gases in the ratio of 2 moles of hydrogen to one mole of chlorine. The hydrogen and chlorine can be reacted to form HCl by the reaction:

$$H_2 + Cl_2 \rightarrow 2HCl$$

This leaves the net reaction:

$$2FeCl_2 + 6H_2O \rightarrow 2Fe(OH)_3 + 4HCl + H_2 + 2e^-$$

The reaction of hydrogen and chlorine to produce HCl is a standard reaction carried out on a large scale in industry.

Sulfuric Acid

In theory, the treatment of sulfuric acid pickling liquors is similar to that of hydrogen chloride pickling liquors. The only, but critical, difference is that oxygen is produced at the anode. The generation of oxygen at this point attacks the anode at a significant rate.

The overall cell reaction is:

$$2FeSO_4 + 9H_2O \rightarrow 2Fe(OH)_3 + 2H_2SO_4 +$$
$$4H_2 + 3/2O_2 + 2e^-$$

Various methods to mitigate the problem of anode corrosion were considered: platinum or titanium electrodes, linseed oil treated graphite, and the use of graphite chips as sacrificial material. Before examining these plausible, but not attractive, alternatives, it was decided to try a process similar to, but somewhat different from, the process described above for hydrochloric acid pickling liquors.

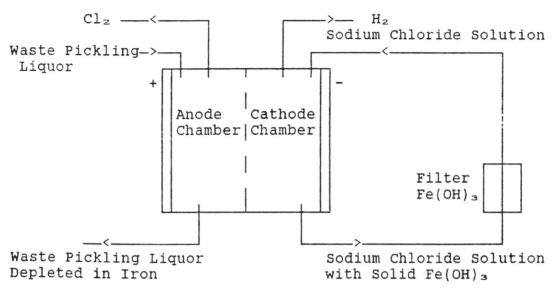

FIGURE 1. Cell Schematic.

A three-chamber cell, comprised of a graphite anode, an anode chamber bounded by the graphite anode and a cation permeable membrane, a middle chamber bounded by two cation permeable membranes, and a cathode chamber bounded by the second cation permeable membrane and a graphite cathode, was constructed.

The anode chamber is fed with a solution of hydrochloric acid (approximately 4 percent). The middle chamber is fed the waste sulfuric acid pickling liquor. The cathode chamber is fed a solution of sodium chloride as in the case of the hydrochoric acid pickling liquor cell.

Since there are no electrodes in the middle chamber, there are no electrolytic reactions. The middle chamber receives hydrogen ions from the anode chamber and transfers both hydrogen ions and iron ions to the cathode chamber. The sulfate ions remain in the middle chamber. The liquid in the middle chamber becomes enriched in acid and depleted in iron. This liquid is then recycled back to the pickling bath. The system is sized to remove iron as fast as it is generated in the pickling bath.

In an ideal situation only iron would be transferred from the middle chamber to the cathode chamber. In practice, however, hydrogen ions also transfer in this direction. The amount of hydrogen which is transferred in this manner depends on the ratio of acid to iron in the middle chamber. Depending on this ratio, from 30 to 60 percent of the current used moves iron from the middle chamber and the remainder moves hydrogen.

Thus there are two cell reactions competing for the available current:

$$2HCl \rightarrow H_2 + Cl_2$$

and

$$FeSO_4 + 2HCl + 2H_2O \rightarrow Fe(OH)_2 + H_2 + Cl_2 + H_2SO_4$$

Although the reaction that produces only hydrogen and chlorine is not desirable, it is inevitable. The gaseous hydrogen and chlorine can be recombined, as described above, giving a net reaction of:

$$FeSO_4 + 2H_2O \rightarrow Fe(OH)_2 + H_2SO_4$$

This overall reaction gives a process that recovers both the acid consumed in the pickling operation as well as the unreacted acid and discharges the iron as a harmless, and possibly useful, by-product of ferrous hydroxide. The cost to run the process is water, energy, and a relatively small amount of make-up hydrochloric acid.

PROCESS ADVANTAGES

From an environmental and economic standpoint, the advantages of this process are obvious. Landfill disposal of waste pickling liquor costs approximately $2.00 per gallon in the Pacific Northwest and is continually rising. This process is also more advantageous than neutralization because it eliminates the cost of the caustic soda as well as the cost to handle and dispose of a large volume of material.

Compared to freezing, for which a commercial process is available, the electrolytic process works well with hydrochloric acid pickling solutions, which are too soluble to be separated by freezing.

Note that hydrochloric acid pickling solutions are more efficient than sulfuric acid solutions. Further, the electrolytic process allows sulfuric acid solutions to operate at a higher acid and lower iron concentration than the freezing process, as ferrous sulfate is still relatively soluble at the operating temperatures of the freezing process. The higher acid/lower iron concentrations provide for more efficient pickling.

Evaporation is an energy consuming process that would require a very expensive installation to avoid the corrosion problems associated with handling a highly acid solution. We are aware of no commercial applications of evaporation for treating waste pickling liquors.

STATE OF DEVELOPMENT

After limited preliminary testing, we have been awarded an EPA grant to develop the process. Work to date has confirmed the basic electrochemistry described above.

Yet to be completed are:

- establishment of accurate material and energy balances for various operating conditions

- definition of specific process conditions to recombine hydrogen and chlorine and form a weak hydrochloric acid solution for recycle to the anode chamber

- finalization of scale-up criteria and design of a model facility

CONTACTS FOR INFORMATION

For additional information contact: Flemming and Wickett, Inc., P.O. Box 1300, Issaquah, WA 98027.

2.11

Demonstration Results for Three Innovative Technologies

Howard M. Feintuch, Ph.D.[1]

INTRODUCTION

Due to the increasing need for reliable methods to treat hazardous waste, the U.S. Environmental Protection Agency (EPA) established a program in 1986 to promote the development of innovative treatment technologies to be used at Superfund and other sites [1,2]. The name of this program is the Superfund Innovative Technology Evaluation (SITE) Program. It was authorized by the Superfund Amendments and Reauthorization Act (SARA) of 1986 to provide for testing of ten innovative technologies per year through 1991, at a cost of up to $20 million per year. The purpose of the SITE Program is to help provide information on the treatment technologies necessary to implement permanent remedies at Superfund sites. The major focus of the program is to carry out a demonstration for each technology being tested at an actual site. It is through this demonstration that the SITE program is able to provide sound engineering and cost data on selected technologies.

In March of 1987 Foster Wheeler Enviresponse, Inc. was selected by the EPA to evaluate the first group of ten technologies to be demonstrated by the newly established SITE Program. These technologies may be classified into four categories of treatment, namely:

- thermal
- stabilization/solidification
- physical/chemical
- biological

This chapter will describe a portion of the results obtained from the first three completed SITE program demonstrations. These involved a thermal, a stabilization/solidification, and a physical treatment technology.

DEFINITIONS

EPA's mandate is to maximize use of alternatives to land disposal in cleaning up hazardous waste sites. A technology that can be used as an alternative to land disposal, regardless of the technology's state of development [3], is known as an "alternative technology."

An "alternative technology" may belong to one of three categories depending upon its state of development [3]:

1. Available Alternative Technology—A technology that is fully proven and in routine commercial or private use (e.g., several forms of incineration)
2. Innovative Alternative Technology—Any fully developed technology for which cost or performance information is incomplete, thus hindering routine use at Superfund hazardous waste sites. An innovative alternative technology requires full-scale field testing before it is considered proven and can be made available for routine use.
3. Emerging Alternative Technology—An alternative technology in an earlier stage of development; the research has not yet successfully passed laboratory- or pilot-scale testing.

Figure 1 [3] summarizes these three definitions in a convenient graphical form.

Based on the definition given above, this chapter will provide information on the SITE Program's full-

[1] Director, Technology and Design Operations, Foster Wheeler Enviresponse, Inc., 8 Peach Tree Hill Road, Livingston, NJ 07039

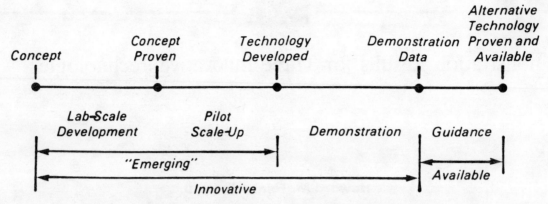

FIGURE 1. Development process for alternative technologies.

scale field testing of three "Innovative Alternative Technologies."

FIRST THREE COMPLETED DEMONSTRATIONS

The following is a brief description of the technologies and sites involved in the first three completed SITE Program demonstrations:

Shirco Infrared Systems, Inc.
(A Thermal Technology)

The first innovative technology site demonstration was completed at the Peak Oil Superfund site in Brandon, Florida [1,4]. This demonstration took place during a removal operation by EPA Region IV. The site contained a waste oil sludge contaminated with PCBs and lead. The Shirco infrared incineration process uses rows of electrically powered silicon carbide rods to bring the waste to combustion temperatures and then to incinerate any remaining combustibles in an afterburner. The four-component system, shown in Figure 2, can process from 100 to 250 tons of waste a day, depending on the waste characteristics. The first component, the primary furnace, generates temperatures up to 1850°F. Waste moves through the furnace on a woven wire mesh belt that transports it through modules lined with layers of lightweight ceramic fiber blanket insulation. While the ash from the combustion process is deposited into a receptacle, the exhaust gases pass through a secondary processing chamber where temperatures can reach 2300°F. Remaining gases are cooled and cleaned in a scrubber. A processing center controls and monitors the operation. Because electricity is used in the primary chamber rather than fuel, less stack gas is produced by this system, as compared to conventional incineration.

HAZCON, Inc.
(A Stabilization/Solidification Technology)

The HAZCON solidification process blends contaminated soil or sludge with cement, pozzolans, and a proprietary ingredient called Chloranan, which aids in the solidification of organics [5]. The Chloranan neutralizes the inhibiting effect that organic contaminants normally have on the crystallization of pozzolanic materials. For this treatment, the wastes are immobilized and bound by encapsulation into a hardened, leach-resistant, concrete-like mass.

For the demonstration test, HAZCON provided their Mobile Field Blending Unit, shown in Figure 3, along with cement and water supply trucks. The mobile unit consists of soil and cement holding bins, a Chloranan feed tank, and a blending auger to mix all the components. Water is added as necessary, and the resultant slurry is extruded into molds.

The location for the demonstration test was the Douglassville Superfund site along the Schuylkill River near Reading, PA. This site is a 50-acre rural area that is contaminated with high levels of oil and grease and low levels of polychlorinated biphenyls and volatile organics from the operations of an oil reprocessing plant. Soil samples from six different plant areas were treated to test the process's capability to handle diverse feedstocks containing both organic and inorganic contaminants.

Terra Vac, Inc. (A Physical Technology)

The Terra Vac in situ vacuum extraction technology, illustrated in Figure 4, uses a vacuum to extract volatile contaminants from soils [6]. The process consists of installing subsurface wells and introducing a negative pressure gradient through the use of a vacuum pump. The resulting contaminant-laden airstream that comes from the wells is passed through a

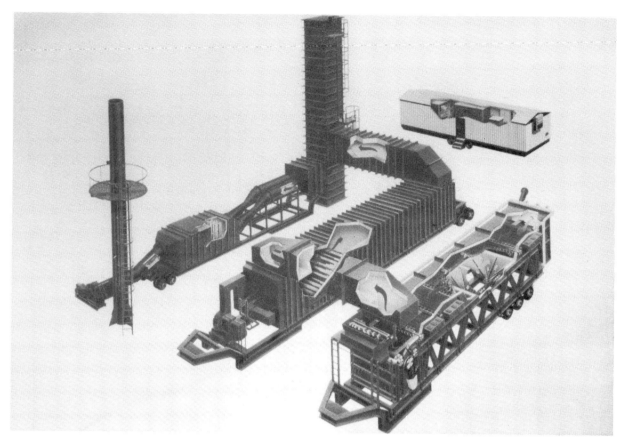

FIGURE 2. The Shirco infrared incineration system.

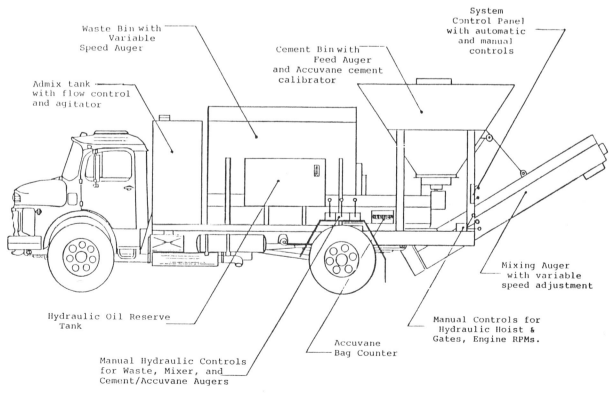

FIGURE 3. The HAZCON mobile field blending unit.

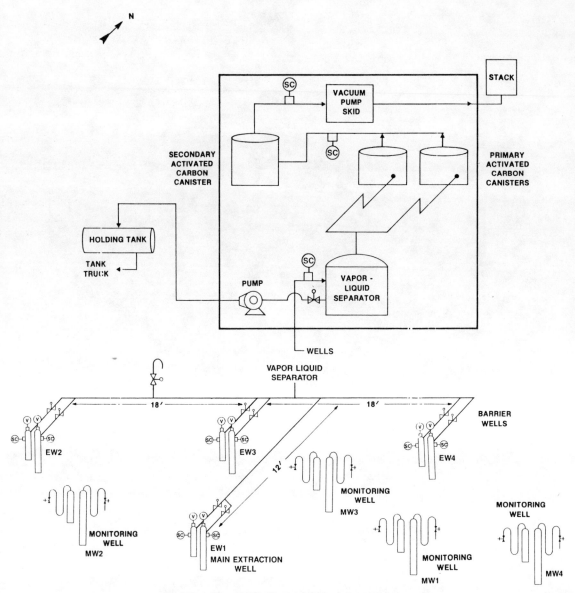

FIGURE 4. Terra Vac in situ vacuum extraction technology.

vapor/liquid separator to remove water and through carbon canisters to remove the volatile contaminant, before the air is vented to the atmosphere. The process has been applied to a wide range of volatile compounds, as well as organic and chlorinated solvents. The process is capable of removing volatile contaminants from the vadose zone, the layer of soil below the surface and above the water table.

The site for this demonstration was the Groveland Wells Superfund site in Groveland, Massachusetts. An operating machine shop on the site is the source of contamination caused by the disposal of degreasing solvents. The soil is contaminated with volatile organic compounds, principally trichloroethylene, with lesser concentrates of 1,2-*trans*-dichloroethylene and tetrachloroethylene. Most of the contamination occurs beneath the machine shop. Because this is an active manufacturing site, excavation of the soil beneath the facility would be very disruptive and costly. The use of vacuum extraction allows for removal of the volatile contaminants, without the need to excavate the soil.

SHIRCO DEMONSTRATION [4]

Overview

The Shirco evaluation at the Peak Oil site included a determination of toxins in the waste material being incinerated, as well as in the effluent streams comprising ash, air emissions, and scrubber water.

Table 1. Destruction Removal Efficiency (DRE) of PCBs from Shirco Peak Oil Demonstration Test

Date of run	DRE for PCBs (%)
8/1/87	99.99967
8/2/87	99.9988
8/3/87	99.99972
8/4/87	99.99905
Average	99.99931

These streams were sampled and analyzed for heavy metals, organics, dioxins, furans, NO_x, and inorganic acids. Leaching tests were performed on the ash.

During the four-day test period, the Shirco unit was fully operational at a 100 ton-per-day capacity. The feed stream and all effluent streams were evaluated during this period of operation.

Summary of Results

Based on the test data for this four-day period, the following results were obtained:

1. The Shirco Infrared unit successfully reduced the PCB content of the feed to less than 1 ppm in the resulting ash, which was the object of the removal action. The feed to the unit contained a low level of only 3 to 6 ppm PCBs, because the lagoon contents, with up to 100 ppm PCBs, was mixed with soil surrounding the lagoon.
2. The destruction and removal efficiency (DRE) for PCBs in the gaseous exhaust stream averaged 99.99931 percent (see Table 1).
3. Essentially all of the PCBs were destroyed in the burning process, demonstrated by the absence of PCBs in the scrubber water or the ash after burning.
4. Particulate emissions for two of the four test periods during the characterization were lower than 0.08 grains/sdcf.
5. The EP Toxicity leaching test results indicated that lead had a leachability of 31 mg/L compared to a regulatory value of 5 mg/L. These results are shown in Table 2.

   All the other metals surveyed in the leachability tests did not exceed regulatory levels for EP Toxicity (EP Tox) and Toxicity Characterization Leaching Procedure (TCLP).
6. An economic analysis indicated the cost of using the Shirco Infrared System could be as low as $200 per ton of contaminated material, and as high as $800 per ton depending on the amount of material preparation required and the equipment usage rate. A projected realistic average cost per ton of contaminated feed material is $425 per ton.

HAZCON DEMONSTRATION [5]

Overview

The major objectives of the HAZCON Demonstration were to determine the following:

1. The ability of the process to immobilize a wide range of organic and heavy metal contaminants in soils with oil and grease concentrations ranging from 1–25 percent by weight

Table 2. Leaching Test Results from Shirco Peak Oil Demonstration Test

Parameter	EP Toxicity		TCLP Analysis	
	Average level, mg/L	Regulatory level, mg/L	Average level, mg/L	Regulatory level, mg/L
Arsenic	0.02	5.0	0.007	5.0
Barium	1.35	100.0	0.25	100.0
Cadmium	0.099	1.0	0.008	1.0
Chromium	0.037	5.0	0.037	5.0
Lead	31.0	5.0	0.011	5.0
Mercury	0.0015	0.20	ND	0.2
Selenium	ND	1.0	0.031	1.0
Silver	0.031	5.0	0.059	5.0
Detected TCLP organic compounds				
Acrylonitrile			0.0130	5.0
Methyl chloride			0.0200	8.8
Toluene			0.0020	14.4
1,1,1-Trichloroethane			0.0006	30.0
Trichloroethane			0.0006	0.07

ND—not detected.

2. The potential long-term integrity of the solidified soils
3. The field reliability of the equipment used to process the wastes at the site
4. The costs for applying the technology to commercial-size operations or Superfund sites

Six different contaminated soils at the Douglassville, PA site were processed and are designated as lagoon north (LAN), lagoon south (LAS), filter cake storage area (FSA), drum storage area (DSA), plant facility area (PFA), and landfarm area (LFA). The intent was to process sufficient soil to yield a treated volume of 5 yd³ from each of five areas and to carry out an extended duration run for a sixth area. This extended run, using contaminated soil from LAS, was to determine the reliability of the operating equipment; approximately 25 yd³ of treated soil was prepared.

Soil samples were taken before treatment, as a slurry exiting the Mobile Field Blending Unit for analysis after seven days, and as cores from the buried blocks after twenty-eight days of curing. For the 5-yd³ blocks, two untreated soil composite samples were taken along with three sets of slurry and solidified core samples. For the extended run at LAS, additional samples were taken for analysis. Figure 5 shows a photograph of one of the sampling activities.

The specific analyses performed on soil, slurry, and core samples included measuring these physical properties:

- grain size of untreated soil
- bulk density of treated and untreated soil
- moisture content of treated and untreated soil
- permeability of treated and untreated soil
- unconfined compressive strength of the solidified cores
- weathering test properties for the 28-day cores (wet/dry, freeze/thaw)

Chemical analyses were performed to identify the volatile organic compounds (VOCs), base/neutral acid (BNA) extractables (semivolatiles), polychlorinated biphenyls (PCBs), and heavy metals contaminants in the processed soil. In addition, three leaching tests were performed:

- Toxicity Characteristic Leaching Procedure (TCLP)—EPA-developed leaching procedure used for measuring leachability of organic and metal contaminants

FIGURE 5. HAZCON solidification technology sampling.

Table 3. Concentration of Metals in TCLP Leachates from HAZCON Demonstration Test

Soil Location	Metal Concentration—mg/L					
	Pb	Cr	Ni	Cd	Cu	Zn
Soil						
DSA	1.5	<0.008	0.02	<0.004	<0.03	0.07
LAN	31.8	<0.008	0.07	0.02	<0.03	1.1
FSA	17.9	0.27	0.11	0.13	<0.3	23.0
LFA	27.7	<0.008	0.06	0.03	<0.08	6.7
PFA	22.4	<0.008	0.05	0.01	<0.03	1.4
LAS	52.6	<0.008	0.07	0.04	0.13	4.8
7-Day Cores						
DSA	0.015	<0.07	<0.15	<0.04	<0.06	<0.02
LAN	<0.002	<0.07	<0.15	<0.04	<0.06	<0.02
FSA	0.07	0.02	<0.008	<0.003	<0.03	0.02
LFA	0.04	<0.07	<0.15	<0.04	<0.06	0.04
PFA	0.01	<0.07	<0.15	<0.04	<0.06	0.02
LAS	0.14	<0.008	<0.008	<0.003	<0.05	0.04
28-Day Cores						
DSA	0.007	<0.007	0.020	<0.004	0.023	0.037
LAN	0.005	0.007	<0.015	<0.004	0.010	0.017
FSA	0.400	<0.070	<0.15	<0.040	<0.060	0.037
LFA	0.050	0.009	0.015	<0.004	0.080	0.013
PFA	0.011	<0.007	<0.015	<0.004	0.027	0.030
LAS	0.051	0.015	0.025	<0.004	0.055	0.258
Regulatory Levels	5.0	5.0	—	1.0	—	—

Notes: Where the symbol < is used, indicates values below detection limits of quantity shown. The detection limits vary between metals and from analysis to analysis.
Where two of three values are above detection limits, three values were averaged, assuming the one below detection limits is zero. If only one of three values are above detection limits, the results are reported as below detection limits.
Regulatory levels given have not been passed into law; they are proposed levels, listed in the Federal Register, 6/13/86.

- ANS 16.1 (American Nuclear Society)— simulates leaching from the intact solidified core with rapidly flowing groundwater
- MCC-1P (Materials Characterization Center, Richland, WA)—simulates leaching from the intact solidified core in relatively stagnant groundwater regimes.

Summary of Results

Based on the test data, the following results were obtained:

1. The concentration of lead in the leachates for the treated soils was reduced by a factor of about 1000 compared to untreated soil leachates, from 18–53 mg/L to less than 0.1 mg/L. Treated soil cured for seven days produced results similar to samples cured for twenty-eight days.
   The contaminated soils compositions and the TCLP leachate analyses are shown in Tables 3 and 4 for metals and volatiles, respectively.
2. The concentrations of VOCs in the TCLP leachates from the treated soils were approximately the same as for leachates of untreated soils, with both typically in the range of 0.1–1.0 mg/L (see Table 4).
3. PCBs were not detected in any leachates.
4. The results of the special leach tests, MCC-1P and ANS 16.1, showed leachate concentrations greater for MCC-1P than for ANS 16.1. The TCLP leachates had concentrations equivalent to MCC-1P for VOC, BNA, oil and grease, and lower concentrations for the priority pollutant metals. Thus with respect to contaminant migration, it appears that the longer MCC-1P leach time, up to 28 days and at 40°C, more than compensated for the increased surface area involved in the TCLP test, which is performed for an 18-hour leach period at ambient temperatures.
5. The freeze/thaw and wet/dry weathering tests showed that the treated soil maintained its structural integrity.
6. System mixing was not highly efficient, since significant amounts of unhydrated cement existed, brownish aggregates did not disaggregate, and many pores were seen. However, more efficient mixing equipment is available, which would eliminate or reduce the noted deficiencies.

Table 4. Volatiles in TCLP Leachates from HAZCON Demonstration Test

Volatile Organic	VOC Concentration—µg/L					
	DSA	LAN	FSA	LFA	PFA	LAS
Untreated Soil						
Toluene	915	10	245	5100	1100*	10
Xylenes	<50	7	525	<230	<180	35
Trichloroethene	<20	2.4	165	<95	<76	8
Tetrachloroethene	<40	<4	19	<210	<160	5
Ethyl benzene	<70	<7	80	<360	<290	<7
7-Day Cores						
Toluene	380	<6	220	210	350	<15
Xylenes	3.5	6	340	5	20	15
Trichloroethene	<10	<2	105	<2	<5	<5
Tetrachloroethene	<20	<4	11	<4	<10	<10
Ethyl benzene	<40	<7	60	<7	<20	<20
28-Day Cores						
Toluene	370	40	230	370	670	50
Xylenes	6	8	330	<6	170	40
Trichloroethene	<9	2	100	<9	<9	8
Tetrachloroethene	<6	3	20	<6	<6	10
Ethyl benzene	<3	2	60	2	<3	4

Notes: * Two values <60 and 2200 µg/L.
Where the symbol < is used, indicates value below detection limits. Within a sampling area the detection limits may change between samples. For these, the highest detection limit is shown.
Proposed regulatory levels (Federal Register 6/13/86): toluene 14.4 ppm, trichloroethene 0.07 ppm, and tetrachloroethane 0.1 ppm by wt.

7. The economic analysis assumed a remediation of part of the Douglassville, PA, Superfund site. The HAZCON processing unit at the operating conditions used during the test was the starting basis. To determine a range of operating costs, two system improvements that are likely in a future unit were assumed: a larger unit and lower chemical consumptions. The analysis considered three variables: on-stream factors of 70 percent and 90 percent; two chemical additive rates, one at demonstration test levels and the other at two-thirds of that; and two operating capacities, 300 lb/min (size of demonstration test unit) and 2300 lb/min (planned future unit). The cost to process the feedstock ranged from $90–200/ton. The process costs were found to be influenced mostly by the costs of labor and chemical additives. At a large-scale commercial remediation, values close to the lower value would be anticipated.

TERRA VAC DEMONSTRATION [6]

Well drilling and equipment setup were begun on December 1, 1987. A mobile drill rig was brought in, and equipped with hollow-stem augers, split spoons, and Shelby tubes. The locations of the extraction wells and monitoring wells had been staked out previously based on contaminant concentration profiles from a previously conducted remedial investigation and from bar punch probe soil gas monitoring.

Each well drilled was sampled at 2-foot intervals with a split spoon pounded into the subsurface by the drill rig in advance of the hollow stem auger. The hollow stem auger would then clear out the soil down to the depth of the split spoon, and the cycle would continue in that manner to a depth of 24 feet.

The sampling and analytical program was conducted in five separate periods: the pretreatment period, the commissioning period, the active treatment period, the midtreatment period, and the posttreatment period.

Soil borings were taken during the pretreatment, midtreatment, and posttreatment periods. Twenty-eight days into the active treatment period, treatment was halted, and following a brief midtreatment period, active treatment was resumed for an additional twenty-eight days. The posttreatment period started after the end of the active treatment period. Figure 6 shows a photograph of one of the soil sampling activities.

Soil borings were analyzed for volatile organic compounds using headspace screening techniques, purge and trap GC/MS techniques, and the TCLP procedure. During the active treatment period, gas sampling was conducted on a periodic basis.

Other measurements were taken routinely during the course of the active treatment period, including the following:

FIGURE 6. Terra Vac in situ vacuum extraction technology sampling.

- flowrates from each well section, with a portable rotameter
- vacuum at each well section
- moisture content of well-head gas, using Modified Method 4 sampling train
- temperature
- wattmeter readings on vacuum pump skid

Activated carbon samples were taken from all spent carbon canisters to check for VOC loading. This was done both as a check on the calculations for the quantity of VOC extracted from the soil and to gather data on adsorption at various inlet concentrations of VOCs.

The major objectives of the demonstration were to determine the ability of the technology to reach an acceptably low level of contaminant concentration in the soil, to assess the effectiveness in various soil types, to gather capital and operating costs, and to gain performance and reliability information. Specific results derived from the Terra Vac SITE Demonstration are being evaluated by the EPA and will be available in the near future.

CONCLUDING REMARKS AND FURTHER RESULTS

This paper has provided a few highlights and a portion of the results from the first three completed SITE Program demonstrations. Table 5 provides information on individuals who may be contacted to

Table 5. Contacts for Further Information on Technologies

Developer	Technology	Developer Contact	EPA Contact
HAZTECH/EPA Region 4 Atlanta, GA	Shirco Infrared Thermal Destruction	Fred Stroud (Reg. 4) 404-347-3931	Howard Wall 513-569-7691
HAZCON, Inc. Katy, TX	Solidification/ Stabilization	Ray Funderburk 713-391-1085	Paul de Percin 513-569-7797
Terra Vac, Inc. Dorado, PR	In Situ Vacuum Extraction	James Malot 809-723-9171	Mary Stinson 201-321-6683

receive further information on these technologies. Final results will be available in the form of Technology Evaluation Reports and Applications Analysis Reports obtainable from the EPA.

REFERENCES

1. "Superfund Innovative Technology Evaluation (SITE) Program Brochure," HMCRI (Hazardous Materials Control Research Institute) Conference, Washington, DC, November 16–18, 1987.
2. "Superfund Innovative Technology Evaluation (SITE) Program Brochure," HWERL* Symposium, Cincinnati, OH, May 9–11, 1988.
3. Superfund Innovative Technology Evaluation (SITE) Strategy and Program Plan, EPA/540/G-86/001, OSWER 9380.2-3, Washington, DC, U.S. Environmental Protection Agency, December, 1986.
4. Wall, H. and S. Rosenthal. "SITE Demonstration of the Shirco Infrared Incinerator," Superfund '88 Conference, Washington, DC, November 28–30, 1988, Silver Spring, Maryland, HMCRI.
5. De Percin, P. and S. Sawyer. "SITE Program Demonstration of HAZCON Technology," Superfund '88 Conference, Washington, DC, November 28–30, 1988, Silver Spring, Maryland, HMCRI.
6. Stinson, M. and P. Michaels. "Terra Vac In Situ Vacuum Extraction Process SITE Demonstration," HWERL* Symposium, Cincinnati, OH, May 9–11, 1988, Silver Spring, Maryland, HMCRI.

---

*HWERL is now the Risk Reduction Engineering Laboratory, RREL.

2.12

Vitrokele®, High-Efficiency, Selective, Metal-Chelating Adsorbents: Integrated Use with Electrolytic Metal Recovery for Closed-Cycle Elimination of Metal Wastes

Bruce E. Holbein, Ph.D.[1]

INTRODUCTION

Hazardous wastes containing various toxic heavy metals represent a substantial portion of all wastes of environmental concern from industries involved in metal surface finishing, electronics parts manufacturing, and nuclear materials processing.

While conventional approaches to waste metal treatment, such as precipitation and disposal of metal-containing secondary wastes, have proved adequate in the past, more efficient and environmentally-sound treatment technologies are now required. New, more stringent regulations for metal discharge exceed the treatment capabilities of conventional technologies in the case of some heavy metals. Increasing costs for by-product waste disposal and secondary environmental problems associated with conventional primary to secondary waste transformation have made technologies providing waste minimization or elimination (recycling of recovered metals) much more attractive and necessary.

Newer manufacturing processes, in some instances, generate wastes that are more recalcitrant to conventional treatment means and demand higher efficiencies or specificities for treatment. Some wastes of the nuclear industry contain small quantities of highly toxic metal within a complex waste mixture of interfering, less toxic materials, thus demanding high degrees of selectivity for metal removal.

This chapter describes the use of new chelating metal adsorbents which provide higher efficiencies and selectivities for metal recovery, and how they can be successfully utilized in concert with electrochemical metal recovery to provide closed-cycle metal recovery thereby eliminating the secondary by-product waste metal.

Process Development

The metal adsorbents comprise a series of chemically distinct materials that adsorb metals with varying degrees of affinity and selectivity by their incorporated metal-coordinating ligands [1-7]. These materials, marketed as Vitrokele®, are employed in process equipment similar to that developed for ion-exchange resins. The processing equipment allows continuous treatment of wastewater containing relatively low target metal contents (<1,000 ppm) usually in conjunction with electrochemical metal recovery, as shown in Figure 1. This closed-loop treatment approach, employing bulk metal removal from concentrated waste (plating baths, etc.), polishing of electrowinning tails through Vitrokele®, and bulk metal recovery of Vitrokele®-polished metal, provides complete secondary waste elimination in most cases. We have successfully employed various types of electrolytic metal recovery equipment, including Chemelec® (Bewt Water Engineers Ltd., U.K.) and Retec® (Eltech Research Corporation, U.S.A.). The inexpensive reticulate-type cathodes of the Retec® units provide both excellent metal recovery and service as well as reduced operating costs.

The integrated treatment, as shown in Figure 1, is a prime example of synergistically acting companion technologies. Primary treatment by electrolytic metal recovery from concentrated wastes allows the use of more compact Vitrokele® metal polishing systems which are necessary due to the relative inefficiency of electrolytic recovery means for meeting discharge

---

[1] President, Tallon Metal Technologies Inc., 110 Leacock Road, Pointe Claire, Quebec, CANADA H9R 1H1

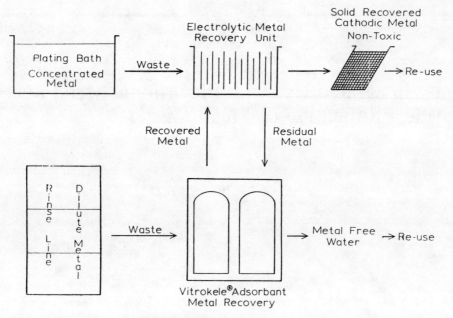

FIGURE 1. Closed-loop treatment process.

limits for metals. Vitrokele, due to its selectivity and affinity for a given target metal, can readily polish residual target metal from electrolyzed feed waste and more dilute rinse waters which often contain substantial amounts of nontarget inorganic components. Vitrokele®-recovered metal then becomes an ideal feed for final electrolytic metal recovery. A relatively small system can therefore treat both concentrated spent bath wastes and dilute rinse waters achieving low residual metal levels in the treated wastes. We generally employ Vitrokele® in three column systems that provide continuous duty with two columns in service at any time (leading-trailing configuration). The third column is either being treated for metal recovery after service or is on standby for the next service cycle. The leading-trailing process allows maximal loading of the lead column while ensuring compliance due to the trailing-polishing column. The new column put into service at each service cycle mode change becomes the trailing column.

Waste Types Amenable to Vitrokele®—Electrolytic Metal Recovery

Table 1 summarizes various waste types from the metal finishing industry for which the technology has either been successfully used commercially or has

Table 1. Examples of Metal Finishing Wastes Amenable to Vitrokele—Electrolytic Recovery

Waste Type	Vitrokele Metal Recovery	Electrolytic Recovery
Electroless nickel (Ni)	rinse waters electrowinning residues*	spent bath Ni, Vitrokele-recovered Ni
Electroless copper (Cu)	rinse waters electrowinning residues	spent bath Cu, Vitrokele-recovered Cu
Chrome Plating/Anodizing	rinse waters	not applicable
Cadmium (Cd), Copper (Cu), Zinc (Zn)-Cyanides	rinse waters electrowinning residues	spent bath metals Vitrokele-recovered metals Cn destruction
Electrolytic Cu	rinse water electrowinning residues	spent bath Vitrokele-recovered Cu
Electrolytic Zn	rinse water electrowinning residues	spent bath Vitrokele-recovered Zn
Electrolytic Ni	rinse water electrowinning residues	spent bath Vitrokele-recovered Ni

*Residual metal after electrolytic metal recovery.

proven useful in laboratory or pilot-scale testing. Note that recovery of chrome is not possible by electrolytic means either as a pretreatment to Vitrokele® polishing or as a final recovery means for Vitrokele®-recovered chrome. Chrome containing rinse waters can, however, be efficiently polished using Vitrokele® to produce a recovered chromate by-product of sufficient purity for reuse in some instances. Efficiencies for electrolytic metal recovery are sufficiently high during concentrate pretreatment or final recovery from Vitrokele®-polished recovered product to provide very effective overall metal recovery. In the case of metal recovery from cyanide plating, Vitrokele® is used directly to recover metal-cyanides without prior cyanide destruction. Electrolytic recovery of the metals recovered from the Vitrokele units also destroys residual cyanide electrochemically. Residual cyanide in the treated effluents from the Vitrokele® system which are metal-free can be readily destroyed by in-line chlorination.

Table 2 summarizes other waste types for which Vitrokele® has been proven useful. Plutonium recovery from wastes with very high inorganic contents and trace levels of plutonium has been independently tested and is under development.

The treatment of mining effluents has seen extensive laboratory testing and is currently being prepared for commercial-scale piloting.

Advantages of Vitrokele for Waste Treatment

Vitrokele adsorbents are a family of metal-chelating materials elaborated on both inorganic and organic substrate carrier materials. Their physical properties allow them to be used in water treatment equipment similar to that developed for ion-exchange resins. Due to their various selectivities for metals and their high-affinity metal-binding mechanisms, these materials offer a number of advantages when used alone or in conjunction with electrolytic metal recovery.

Table 2. Nonplating Waste Applications of Vitrokele Adsorbents

Waste Type	Vitrokele Application
Incinerator scrub waters	high-efficiency mercury removal
Nuclear fuels reprocessing wastes	plutonium recovery from high salts
Mining effluents	residual precious metals recovery, toxic metal and cyanide recovery, electrolytic metal recovery used for Vitrokele recovered metals

Vitrokele® adsorbents allow high-efficiency removal of target metals when target metals are complexed with organic agents such as those found in electroless plating wastes or when the target metal is present in materials with high backgrounds of other inorganic competing materials, such as, nuclear fuels reprocessing waste. Thus, efficient metal removal can be obtained in wastes containing chelated metals or wastes carrying high inorganics with relatively small installations. The specificity obtainable with Vitrokele® allows recovered metal to be reutilized directly because it is substantially free of inorganic contaminants or to be recovered as relatively pure cathodic metal using electrolytic recovery. The latter allows reuse of metal directly or in its off-site recycling. Vitrokele® adsorbents are used repeatedly for hundreds of loading-recovery cycles with lifetimes similar to ion-exchange resins.

The combined use of pre-electrolytic recovery, Vitrokele® polishing, and final electrolytic metal recovery allows closed-loop metal recovery capabilities for both concentrated and dilute waste streams with the same equipment that largely reduces or completely eliminates secondary waste products requiring expensive off-site treatment. Off-site waste treatment, apart from its escalating cost, carries varying degrees of contingent liabilities. Closed-cycle metal recovery greatly reduces or eliminates such liabilities and provides a returned value for recovered metals.

State of Development of Vitrokele Technology

Vitrokele® technology has been developed over the last eight years and is subject to a number of patents covering various adsorbents and processes [1–7]. Some applications of the technology have been used commercially for four years, particularly with respect to metal finishing wastes including the combined use of Vitrokele® and electrolytic metal recovery. The examples shown in Table 1 have either been adopted commercially, are in the process of being installed as commercial systems, or have gone through pilot-scale testing either in the laboratory or in the field. Of the examples shown in Table 2, nuclear applications are being developed, following successful preliminary testing, and mining applications are entering full-scale field pilot testing following two years of bench testing in various laboratories in the hands of consulting engineers and metallurgists [8–11].

The engineering associated with the implementation of the technology is also well developed. For waste water treatment, we generally employ automatic, computer-controlled (on board) equipment to provide continuous treatment over the range of 5 to 100 gallons per minute. Mining applications, such as cyanide recovery, employ fluidized bed contactors for counterflow treatment. The latter are custom designed and similar to the hardware now employed in the mining industry for pulp and slurries. Mining decant solutions can be treated as for waste waters.

CASE STUDIES AND PILOTING RESULTS

Closed-Cycle Zinc Recovery from Electroplating Wastewater

A large multinational company has in one of its operating divisions in Canada a captive zinc chloride plating operation for which an ageing conventional precipitation/sludge disposal system was replaced with a Vitrokele® zinc (Zn) removal and electrolytic recovery unit some four years ago. A diagram of the treatment system is shown in Figure 2. The Vitrokele® polishing equipment automatically treats some 300,000 litres of rinse water per month containing an average of 170 ppm Zn and delivers treated water with less than 0.5 ppm residual Zn for sewer discharge. Zinc recovered from the polishing unit is batch recovered by electrolytic treatment to yield some 50 kg of zinc per month for reuse as anode material in the plating baths. Residual zinc, not recovered electrolytically, is repolished through the Vitrokele system prior to discharge. This installation has totally eliminated by-product waste for off-site disposal. Capital costs for the system are approximately Can$80,000 with average total operating costs of Can$800 per month (not including credits for recovered zinc).

Closed-Cycle Nickel (Ni) Recovery from Electroless Ni Wastes

A large multinational company with an operating division in Canada is completing a waste minimization program with the aim of reducing waste disposal costs and avoiding, where possible, off-site treatment of toxic waste. An intermediate pilot-scale test program has now been completed involving the treatment of various electroless Ni associated wastes. Because this firm has stringent manufacturing specifications, relatively large quantities of spent plating baths and other wastes representing 74 cubic meters per year have been disposed of off-site. A treatment approach as shown in Figure 3 has now been shown to effectively treat all wastes resulting in >99.9 percent recovery of all Ni (108 Kg per year), while producing effluents for biological treatment for removal of the residual COD and BOD, with <1 ppm (typically 0.5 ppm) residual Ni.

Capital cost estimates for the treatment system shown in Figure 3 are Can$140,000 with yearly operating costs of approximately Can$2,500. Based on these estimates, payback versus current disposal costs will be realized in under five years. The added benefits to this treatment approach are built in expansion capabilities to approximately a 100 percent

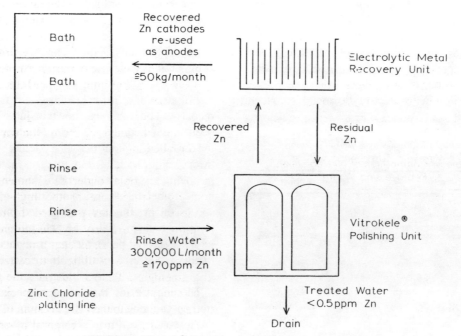

FIGURE 2. Vitrokele® zinc removal and electrolytic recovery unit.

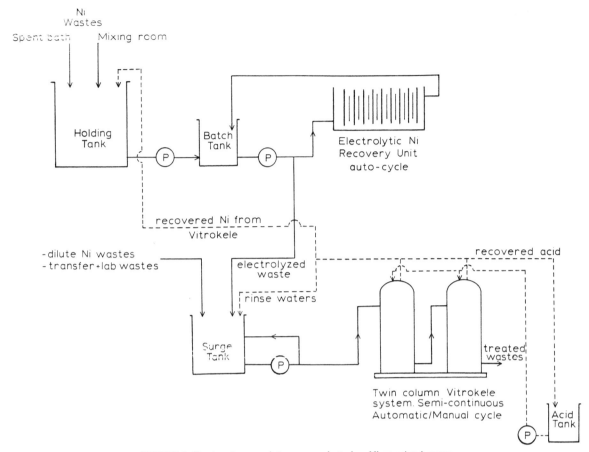

FIGURE 3. Treatment approach to process electroless Ni associated wastes.

increase in waste volume for treatment without alterations and the elimination of liabilities associated with off-site disposal of the wastes.

Chrome Recovery from Chrome Plating/Anodizing

A large Canadian aerospace manufacturing firm generates some 55 gallons per min rinse water from their chrome plating and anodizing lines and until recently did not have waste treatment provisions. A three column automatic Vitrokele® polishing unit is being installed for continuous duty to recover up to 5 kg of chrome per day and provide a chrome concentrate for reuse after evaporative reconcentration. Capital costs for the treatment system were Can-$132,000 with yearly operating costs of approximately Can$5,000. The Vitrokele®-recovered chrome has sufficient purity for reuse although it needs upgrading from 35 g per litre, as recovered, to 150 g per litre prior to reuse.

Plutonium Recovery from Nuclear Fuels Reprocessing Waste

This example, under development, demonstrates the specialized use of the high-affinity selective Vitrokele®. Wastes produced from nuclear fuels reprocessing are very high in inorganic constituents and can contain significant amounts of residual Pu which has proven difficult to extract with existing technologies. Table 3 illustrates the general makeup of such wastes that require special handling and treatment due to their high cesium (Cs) radioactivity. Despite successful Cs removal by zeolite polishing, the wastes would still be hazardous due to the pres-

Table 3. Composition of Nuclear Fuels Reprocessing Waste

Specific gravity (20°C)	1.32
Total suspended solids	84 ppm
Total dissolved solids (%)	39.5%
Water (%)	60.6%
Nitrates, Nitrites, Sulfates, Sulfites } % Salts	97%
Cations (Na:K)	36:1
Cs^{137}	2.8×10^{-3} Ci/g
Pu	62 nCi/g

Table 4. Removal of Pu from Fuels Waste by Vitrokele®

Vitrokele® Type	Test Sample	Pu Extraction Efficiency, Rd*
V-553	actual waste (3× diluted)	3,300
V-553	simulated waste (3× diluted)	12,820
V-638	simulated waste (3× diluted)	>690,000

*Rd = Vitrokele® -bound Pu: Free Pu in solution (at equilibrium).

ence of transuranic elements such as plutonium (Pu). Pu removal would reduce the ultimate stabilization and storage costs for these wastes. Vitrokele® has been tested for Pu removal characteristics in Battelle's laboratories through a Department of Energy funded program for the West Valley Project. These preliminary results as shown in Table 4 indicate excellent potential for the use of Vitrokele® to remove the small amounts of Pu from such wastes. Pu is present in these wastes at approximately 10^{-6} M prior to treatment, and ion-exchange resins cannot remove it due to the high inorganic (cations/anions) content. This application is currently undergoing further test and development work by West Valley Nuclear Services Inc.

Contacts for More Information

Various aspects of Vitrokele® technology are well developed for applications in industrial waste treatment and mining applications. Tallon is also engaged in the development of other areas of the technology including biotechnology applications, precious metal applications, and other hydrometallurgical applications. Further information can be obtained directly from Tallon for these and waste treatment applications. Other contacts include:

1. Jasmetech Ltd. Pty (Australian licensed company for mining applications)
   P.O. Box R356, Royal Exchange
   Sydney, N.S.W. 2000
   Australia
   Telephone: (02) 233-1557
   Fax: (02) 221-8269
   Jack McCarthy, Managing Director

REFERENCES

1. Holbein, B. E., D. Brener, C. Greer and E. N. C. Browne, U.S. Patent 4,654,322 (March, 1987).
2. DeVoe, I. W. and B. E. Holbein, U.S. Patent 4,530,963 (July, 1985).
3. DeVoe, I. W. and B. E. Holbein, U.S. Patent 4,585,559 (April 1986).
4. DeVoe, I. W. and B. E. Holbein, U.S. Patent 4,626,416 (December 1986).
5. DeVoe, I. W. and B. E. Holbein. "Newer Methods for the Removal of Trace Metals from Aqueous Solution," *Proceedings Annual. Chem. Congress, Royal Society of Chemistry, 1986, University of Warwick.*
6. Huber, A. L., B. E. Holbein and D. K. Kidby. "Metal Uptake by Synthetic and Biosynthetic Chemicals," in *Biosorption of Heavy Metals.* B. Volesky, eds. Chapter 2.6, CRC Press (in press).
7. Holbein, B. E. "Immobilization of Metal Binding Compounds," in *Biosorption of Heavy Metals.* B. Volesky, ed. Chapter 3.2.3, CRC Press (in press).
8. Holbein, B. E., A. L. Huber and D. K. Kidby. "Integrated Gold and Cyanide Recovery with Vitrokele," *Proceedings Randol Gold Forum, Phoenix, 88, Randol International Ltd., Colorado, 1988.*
9. Holbein, B. E. and A. L. Huber and D. K. Kidby. "Vitrokele Performance for Selected Ores: Gold, Silver and Cyanide Recoveries," *Proceedings Randol Gold Conference Perth 88, Randol International Ltd., Colorado, 1988.*
10. Elvish, R. D. and A. L. Huber. "The Use of Cyanosave Detoxification and Cyanide Recovery Process for Cyanide Tailings," *Proceedings Australian Institute of Mining Engineers, Sydney, 1988.*
11. Holbein, B. E. and M. J. Noakes. "The Use of Aurosave Adsorption Process for Gold and Precious Metals," *Proceedings Australian Institute of Mining Engineers, Sydney, 1988.*

2.13

Recovery and Disposal of Nitrate Wastes

John M. Napier[1]

INTRODUCTION

The Oak Ridge Y-12 Plant is located in Oak Ridge, Tennessee, and is owned by the U.S. Department of Energy (USDOE) and is operated by Martin Marietta Energy Systems, Inc. The plant's major responsibilities are: (a) producing components for weapons systems and supporting DOE's design laboratories, (b) processing special materials, (c) supporting other Energy Systems installations, and (d) supporting other DOE plants and other agencies of the U.S. government. One of the plant's programs involves the purification and recycle of nonirradiated enriched uranium. One of the unit operations in the uranium recycle process is solvent extraction for the removal of impurities. For this operation, some of the recycle uranium or uranium recycled material is dissolved in nitric acid which is then solvent extracted. Two types of acidic nitrate wastes are generated from the extraction process, one of which is a dilute nitric acid waste, called condensate that is produced from several evaporation operations. This waste may contain up to 10 wt percent nitric acid with trace amounts of impurities such as organic carbon, chlorides, and fluorides. The acid must be concentrated to 30 or more wt percent, and the impurities require removal before the nitric acid can be reused in the plant. A second waste, called raffinate, is also produced from the extraction processes. The raffinate waste contains high levels (more than 10 wt percent) of aluminum nitrate which is used in the extraction process. It also contains most of the impurities extracted from the uranium. Most of the impurities must leave the extraction process in either the condensate or raffinate stream.

In the early 1950s, four unlined waste ponds were constructed to receive the acidic wastes from the extraction process. Other acidic wastes from nonenriched uranium processes, such as pickling baths, were also discharged into the ponds. The four ponds are located near the headwaters of two small streams: Bear Creek and the east fork of Poplar Creek. The ponds had no surface drainage but relied on underground seepage to prevent direct overflow to the creeks. The ponds were located close to the beginning points of both creeks where the surface flow rate in either creek is usually low to nonexistent. Even at a distance of several hundred yards from the ponds, the water flow in Bear Creek is zero for much of the year. About two miles from the ponds at the exit point of the plant, the east fork of Poplar Creek has a typical daily flow rate of 7 million gallons of water per day (18,400 L/min) but most of the water is due to discharge from cooling systems within the plant.

In the early 1970s, a decision was made to develop and install systems to recover as much acid and aluminum nitrate as practical from the wastes generated by the solvent extraction processes. The remaining wastes that could not be recycled would be treated to remove the environmentally objectionable compounds. A later decision in the early 1980s was made to develop and install in situ treatment processes for the liquid wastes in the four ponds and for similar wastes that might be generated in the future.

DESCRIPTION OF WASTE STREAMS

One of the waste streams was a dilute nitric acid stream (condensate) that contained trace amounts of

---

[1]Chemistry and Chemical Engineering Department, Development Division, Oak Ridge Y-12 Plant, P.O. Box 2009, Oak Ridge, TN 37831

Table 1. Typical Analysis of Condensates, Raffinates, and Mother Liquor

Condensates	
Nitric Acid	0.5 to 9 wt %
Water	99.5 to 91 wt %
Chloride	14 mg/L
Fluoride	9 mg/L
Uranium	0.8 mg/L
Raffinates	
Nitric Acid	6.0 wt percent
Total Nitrates	31.1 wt percent
Aluminum	3.0 wt percent
Chloride	67 mg/L
Fluoride	3000 mg/L
Uranium	1 mg/L
Water	60 wt percent
Mother Liquor	
Nitric Acid	6.0 wt percent
Total Nitrates	50.0 wt percent
Aluminum	4.4 wt percent
Chloride	81 mg/L
Fluoride	6500 mg/L
Uranium	2 mg/L
Water	39.6 wt percent

Table 2. Typical Analysis of a Waste Pond in 1978* (pH 1.0)

	Concentration (mg/L)
Nitrate	46,000
Fluoride	12
Uranium	227
Aluminum	2,260
Boron	23
Barium	0.6
Calcium	438
Cadmium	79
Cobalt	0.8
Chromium	48
Copper	28
Iron	467
Potassium	195
Lithium	22
Magnesium	411
Manganese	16
Sodium	2,887
Nickel	91
Strontium	2
Thorium	53
Zinc	5

*Data on the Northeast Pond (September 1978). The nitrate concentrations in the other three ponds ranged from 9,000 to 45,000 mg/L.

impurities (Table 1). In order to recycle the acid, it had to be concentrated to at least 30 wt percent, and most of the impurities had to be removed. The increase of the impurities (i.e., chlorides and fluorides) during the evaporation and concentration of the dilute nitric acid caused major materials problems in constructing the distillation equipment. Also, the trace quantities of certain organics created potential safety problems due to their reactivity with the hot, concentrated nitric acid product.

The raffinate waste had a typical analysis as also shown in Table 1. A process to recycle part of the aluminum nitrate was desired. Nonrecyclable quantities of aluminum nitrate waste had to be treated to produce an environmentally acceptable product. This waste from the aluminum recycle system, called mother liquor, contained high levels of aluminum nitrate, free nitric acid, and other impurities as noted in Table 1.

As previously noted, the four waste ponds contained materials from the extraction process as well as wastes from other processes within the plant. The wastes were placed in one of the two ponds located on the northeast or northwest side and these two ponds had interconnecting pipes to the southeast and southwest ponds. A typical analysis of a pond waste is shown in Table 2.

PROCESS DESCRIPTION

As shown in Figure 1, the unit operations for the recycle of nitric acid from the condensate stream are evaporation and distillation. The process for the recycle of aluminum nitrate is evaporation and recrystallization. The remaining unit operations for the nitrates, including the ponds, are biological degradation to reduce the nitrates to nitrogen gas. Each process will be briefly discussed in the following paragraphs.

Recycle of Nitric Acid

The condensate waste contains up to 9 wt percent nitric acid plus minor levels of fluoride and chloride. The selection of the materials for constructing the process equipment required a development program. The dilute nitric acid waste had to be concentrated to at least 30 wt percent in order for acid to be useful to the plant. In the distillation process, the minor levels of chloride and fluoride impurities are also concentrated to high levels at some location in the distillation column. A metal column, such as stainless steel, corrodes rapidly where these impurities are concentrated. It is therefore required that the chloride and

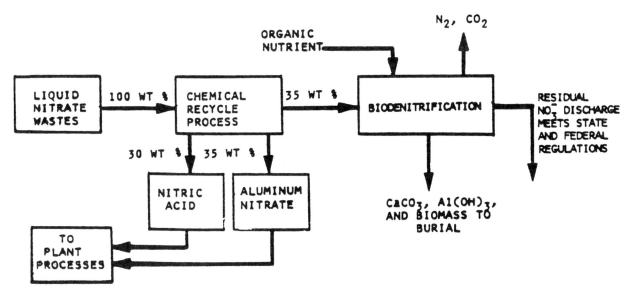

FIGURE 1. Treatment processes for nitrate wastes.

fluoride ions be removed before the distillation process.

A molten salt trap was developed to remove most of the fluoride ions from the incoming condensate wastes and also to serve as a heater for vaporization of the incoming feed to the column [1,2]. The salt trap contained 23.8 wt percent aluminum nitrate nanohydrate, 52.9 wt percent calcium nitrate tetrahydrate and 23.3 wt percent water. The incoming dilute condensate is passed through the molten salt bath that is heated to a nominal temperature of 110°C and the fluoride ions react with aluminum ions until a 5 to 1 aluminum to fluoride weight ratio is obtained (i.e., 4800 mg/L fluoride ion concentration is usually a maximum in the trap). The hot fluoride-free vapors from this trap are then fed to a glass distillation column. The dilute acid is concentrated in the column to a nominal 30 wt percent nitric acid. The overhead vapors are cooled to produce water containing less than 100 mg/L of nitrate ions, and the water is subsequently discarded to the environment. The acid product from the column contains most of the chloride and organic ions. The 30 percent nitric acid product is ozonated in a batch operation until all of the organics are converted to carbon dioxide, and the chloride ions are removed in the overhead vapor stream which is discarded to the environment. The purity of the 30 percent nitric acid product is better than commercially obtained nitric acid.

The fluoride salt bath is constructed of 304L stainless steel, and equipment corrosion is not a problem because the fluoride ions react with the aluminum ions in the salt bath. The glass distillation column has a tantalum heater to reflux the acid during distillation, and no corrosion of the heater has occurred. Other construction materials for the heater were investigated [1,2] and found to be unsuitable. The column and other parts of the distillation equipment are constructed of glass. No corrosion has been observed after several years of operation.

Recycle of Aluminum Nitrate Nanohydrate

Aluminum nitrate in the raffinate waste stream is concentrated using a forced-circulation vacuum evaporator with an operating temperature of 60°C. It is constructed of 304L stainless steel. The density of the raffinate in the evaporator is controlled at 1.46 g/mL when measured at 60°C. A stream from the vacuum evaporator is fed to a cooler (30°C) and then into a crystallizer where the aluminum nitrate nanohydrate crystals increase in size. The solution is then put into a batch type centrifuge to remove most of the crystals. The crystal-free solution is returned to the evaporator. Part of the solution is taken off as a waste stream to a holding tank where it is diluted with water to about 30 wt percent nitrate ion level and becomes the feed solution for the biological reactor. This waste solution, called mother liquor, contains the impurities from the uranium extraction process. The overhead condensate from the forced circulation vacuum evaporator contains up to 6 wt percent nitric acid that is cooled and returned to the condensate recycle system for recovery of the acid. Approximately 40 to 60 wt percent of the aluminum nitrate, and all of the free nitric acid, is recycled. The purity of the product is equal to or better than aluminum nitrate nanohydride prepared by the standard procedure (i.e., dissolving aluminum oxide trihydrate in 50 wt percent nitric acid).

Biological Denitrification of Wastes

The mother liquor waste stream is not suitable for recycle, and a treatment process was developed to render it environmentally harmless. A typical analysis, after addition of water, would be 4.4 wt percent aluminum, 1 to 6 wt percent nitric acid, nitrates plus minor levels of other metallic impurities. The mother liquor waste from the crystallization process is collected in a 304L stainless steel holding tank. A second holding tank serves as a feed tank for calcium acetate (17 wt percent solution). Calcium acetate is the organic substrate that is used by denitrifying bacteria. These liquids are injected at a controlled rate into a mild steel 100,000 liter (25,000 gallon) stirred-tank reactor. Any free acid is neutralized in the reactor by excess calcium carbonate which is always present in the reactor. Trace quantities of phosphate ions are also added to the reactor, and cooling coils are located inside the reactor to prevent the liquor temperature from rising above 45°C since the neutralization and biological reactions are exothermic. The reactor design is based on a design used for softening water except that it has a closed top to maintain neutralizing conditions. The off-gases (nitrogen and carbon dioxide) are vented through a monitoring instrument that measures the amount released to the environment. All liquids and solids that are injected or formed in the reactor are taken out once per day. Prior to 1983, these solids were then placed into four acid waste ponds to begin neutralization of the waste materials. Four unlined waste ponds were used since the mid-1950s to collect nitrate and other liquid wastes from the plant. After closure of the ponds, the liquids and solids were placed into large reactors (500,000 gallon) where other nitrate wastes were biologically reduced. In 1983, an in situ treatment process for the ponds was developed and the wastes were biologically decomposed into nitrogen and carbon dioxide gases. The equipment used consisted of large recirculating pumps which served as a mixer for the addition of calcium carbonate and hydroxide plus acetic acid. When denitrification was complete, the water was pumped through a precipitation and flocculation process to remove trace amounts of solids and uranium. This process involved reducing the pH to 2 to remove carbonate ions, and then adding ferric sulfate and raising the pH to 10 to induce co-precipitation of ferric hydroxide and other hydroxides or oxides. Solids were returned to the ponds, and the purified water was released to the environment. The ponds were subsequently filled with rocks to stabilize the bottom and covered with an approved multilayer cap. A process to treat future nitrate wastes was installed in place of the ponds which included 500,000-gallon bioreactors. After denitrification, the liquids and solids from the bioreactors are flocculated to remove trace amounts of solids and uranium. The solids are presently stored, but plans to delist these solids are being made. A process change will also be made to separate the solids before denitrification. When this change is made, the remaining solids can then be recycled (calcium carbonate) to neutralize incoming wastes.

Conventional Treatment or Disposal Methods

Nitrate wastes such as those produced from uranium recycle facilities contain trace amounts of uranium that may restrict disposal options. Conversion to calcium nitrate to be used as fertilizers was examined and found to be impractical because of the uranium contamination and large acreage required. Heating to destroy the nitrates into nitrogen oxide or nitrogen gas was believed to be impractical because of its high equipment and operating costs. Recycle and recovery of part of the waste was believed to be cost-effective and was therefore chosen as the primary option. The remaining wastes, because of the impurities, could not be recycled, and the biological destruction process was chosen as the most practical method to destroy the nonrecyclable nitrate waste materials.

State of Development

The recycle processes have been in operation since 1976. The stirred-tank denitrification process has successfully operated since 1976 and the larger batch denitrifiers have operated since 1985. The in situ treatment of the ponds began in 1983 and was completed in 1985. Final closure of the waste ponds is to be completed in 1989.

PROGRAM RESULTS

Recycle of Nitric Acid

As previously stated, the condensate stream contained chloride and fluoride impurities, and tests [1,2] showed that the fluoride ions, as well as most of the metallic ions, could be removed by passing this stream through a hot bath of aluminum-calcium nitrate. Fluoride-free acid vapors from this bath could then be fed to a glass distillation column to concentrate the condensate into a useful product (see Figure 2). The molten salt bath also trapped the trace quantities (0.1–0.5 mg/L) of enriched uranium pres-

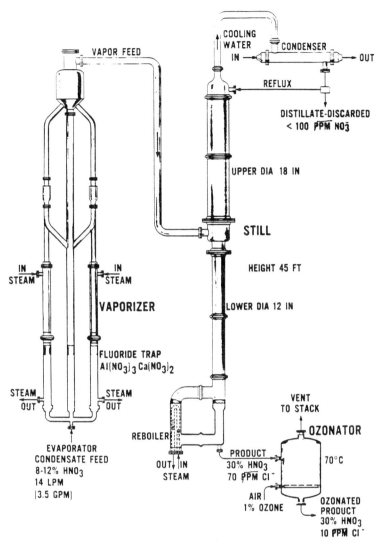

FIGURE 2. Nitric acid recycle.

ent in this waste that is also recycled. Condensate containing 200 to 300 mg/L fluoride ions are reduced to less than 1 mg/L until a 5 to 1 aluminum to fluoride weight ratio is obtained. In the plant, the bath size is large enough to permit thirty or more days of operation before it is changed.

The vapors from the salt bath are fed to a 50-ft-long glass distillation column which has a tantalum heater in the reboiler. Originally the acid was to be concentrated to 50 wt percent, but trace amounts of organics from the solvent extraction proess were found to be present. To prevent an organic-hot nitric acid reaction in the column, a decision was made to produce a 30 wt percent nitric product. No organic reaction is known to occur under these conditions. Most of the chloride ions remain in the 30 wt percent nitric acid product, and an ozonation step was added to remove this impurity from the product. Ozone gas at 60 to 70°C is passed through a 250-gallon batch of 30 percent acid until all of the chloride ions are removed in the overhead vapors that are cooled and released to the environment. The ozonation step also removes all of the organic impurities that may be present in the product. Since 1976, most of the nitric acid condensate from the main process facility has been recovered. The amounts of nitric acid purchased by the plant decreased approximately 30 percent because of this recycle stream. The overhead vapors from the distillation column is cooled and the water is discharged through a permitted discharge point.

Recycle of Aluminum Nitrate

As noted in Figure 3, the raffinate waste is evaporated under vacuum at 60°C to concentrate the aluminum nitrate and to remove the free nitric acid in the overhead condensate. The acid condensate is

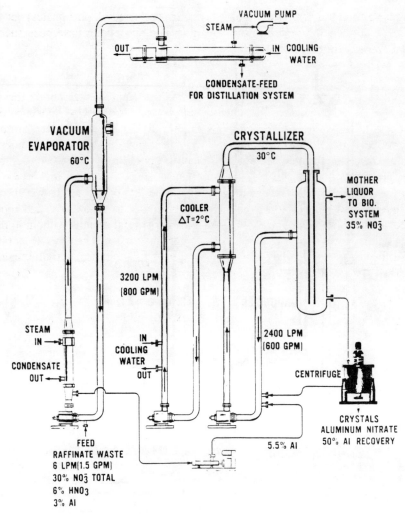

FIGURE 3. Aluminum nitrate recycle.

returned to the distillation process and the nitric acid is recycled to the plant. The concentrate from the evaporator is cooled to 30°C in a contact cooling-type crystallizer which produces aluminum nitrate nanohydrate crystals. These crystals are removed by centrifuging the stream, and are reused in the plant. The waste liquor, containing the bulk of the impurities, is sent to the biological process to destroy the remaining nitrates. Recycle of the aluminum nitrate has reduced the purchases of nitric acid and aluminum oxide by about 30 percent. Plans are being made to increase the amount of recycled aluminum nitrate and 20 percent more aluminum nitrate crystals are expected to be recycled without a significant increase in the impurities occurring in the product.

Stirred-Tank Bioreactor Process

Researchers conducting laboratory and pilot plant tests [2,4] developed a process using a stirred tank to biologically decompose the mother liquor wastes. Analysis of mother liquor is shown in Table 1. The liquor is diluted with water to a nominal 30 wt percent nitrate in order to prevent solidification when cooled to room temperature. Biological denitrification is nitrate reduction where the nitrate ions serve as the final electron (hydrogen) acceptor in the oxidation of an organic substrate by facultative bacteria. Many organic compounds can be used in the process, but laboratory studies showed that calcium acetate was an effective organic, and calcium carbonate was one of the waste products from the reaction. The calcium carbonate could be recycled to neutralize incoming acid wastes. The equation believed to be representative of this process is as follows: $0.55\ Ca(C_2H_3O_2)_2 + 0.8\ Ca(NO_3)_2 \rightarrow 0.04\ C_5H_2O_2N + 0.65\ CO_2 + 0.796\ N_2 + 1.35\ CaCO_3 + 1.45\ H_2O + 0.12\ (OH)$. The process equipment developed for the continuous-flow stirred-tank reactor is shown in Figure 4.

Laboratory studies investigated columnar systems as well as stirred tanks. Because of the high amount of insoluble compounds (aluminum hydroxide and calcium carbonate), frequent plugs developed in the column and the stirred tanks design was chosen for production operations.

Control of pH

In order for the process to function at high denitrification rates, several parameters must be controlled. The pH of the reactor solution must be above 6.8 with an optimum pH range of 7.3 to 7.5. The mother liquor stream was not neutralized before injecting it into the biodenitrification reactor. Neutralization of the free acid in the liquor was accomplished by its reaction with excess calcium carbonate ions that are formed in the denitrification process. Part of the reactor is used as a neutralization zone, and part is used as a biological reactor. Careful control of the reactor pH is required to maintain the biological reaction because it is sensitive to low pH conditions.

Agitation

Agitation of the solution must be present to prevent settling of the bacteria, but the rate of agitation does not appear to be important.

Phosphates

Trace amounts of phosphates are required for successful operation of the reactor. Because of the high calcium levels in the reactor, the soluble phosphates are usually 50 mg/L or less. Organic phosphates (trimethyl or triethyl phosphates) as well as several inorganic phosphates have been successfully used.

Temperature

Cooling coils are present in the continuous-flow stirred-tank reactor to maintain the temperature below 45°C. The upper temperature limit of the reactor is not known, but 45°C appears to be an effective upper limit. At 10°C or below, biological activity is reduced to very low levels, and practically no nitrates are destroyed. However, reactor solutions can be cooled to below 10°C and kept at this temperature for several months without impairing their biological reactivity. When the temperature is raised to above 10°C, biological activity begins within a very short period of time.

Organic Carbon

In laboratory and pilot plant tests, several types of organic compounds were investigated to obtain high denitrification rates. Ethanol, methanol, and calcium acetate solutions were found to produce high denitrification rates. Storage of concentrated ethanol and methanol feeds presented a potential fire hazard problem to the plant. Also, calcium ions are required for neutralization of the acid feed. Based on the results of the pilot plant tests, a decision was made to use a 17 wt percent calcium acetate solution as the organic feed solution. A carbon to nitrogen weight ratio of 1.03 to 1.08 must be maintained when calcium acetate is used. In one series of laboratory tests, a reactor was operated for several weeks using

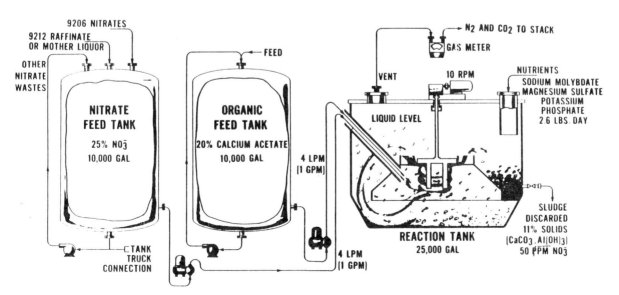

FIGURE 4. Stirred tank biodenitrification.

calcium acetate as the organic source. The reactor feed was changed to methanol and biological activity stopped for several weeks. Calcium acetate was again added to the reactor and the biological activity was immediately restored. These tests confirmed that bacteria, once acclimated to an organic compound, must become reacclimated to a second organic food source. A time delay of several weeks or longer for this reacclimation period may be considered to be normal under the operational conditions utilized.

Type of Bacteria

The type of bacteria used in the first laboratory tests was *Pseudomonas stutzeri* which were obtained from a commercial source. A large batch of these bacteria was cultivated and stored near $-20°C$ for use in the pilot plant tests and for the start-up of the two production-sized reactors. After eighteen months of continuous operations, one of the production reactors was sampled and the bacteria were identified [5]. In a sample, 1.2×10^9 colony-forming units per millimeter of reactor solution were found. There were sixteen different isolates that appeared to predominate. Fourteen of the sixteen isolates were capable of reducing nitrates to nitrogen gas. Fifteen of the sixteen isolates were nonfermenting, gram-negative rods characteristic of the family *Pseudomonadacea*. Five of the fifteen seemed to best belong to the fluorescent members of the Group 1 pseudomonads; eight appeared to belong to the stutzeri subgroup of nonfluorescent Group 1 pseudomonads, and one isolate could be identified only as a *Pseudomonas* species. The remaining bacterium could not be identified to the genus level. Each year, one or more samples are taken from the continuous-flow reactors and similar analyses are obtained. In all cases, a mixed population of bacteria has been found, and more than 100 species have been identified at various times during these studies.

Dissolved Oxygen

The dissolved oxygen (D.O.) level in the reactor must be less than 2 mg/L for destruction of the nitrate ions. In laboratory and pilot plant tests, a closed reactor design was used to provide anaerobic conditions. During start-up of the reactor, a dissolved oxygen level of 8 mg/L is common. However, the dissolved oxygen is quickly depleted from the reactor solution by the bacteria, and then the oxygen is stripped from the nitrates. Later, tests completed in 1983 showed that denitrification in open ponds could be accomplished, and the solution in the ponds could be maintained at dissolved oxygen levels of less than 2 mg/L. It is apparent that the presence of trace amounts of oxygen does not materially affect the biodenitrification process.

Stirred-Tank Development Data

In laboratory tests, four continuous-flow stirred-tank reactors were operated for a total of 453 days. A maximum denitrification rate of 11.2 g nitrate/day/liter of reactor solution was obtained. The denitrification rate is expressed in units related to the volume of the reactor and not to the number of bacteria because of the difficulties experienced in determining the bacteria count. Because of the high solids contents of the reactor solution (up to 9 wt percent solids), the number of bacteria could not be accurately determined using dry cell weight measurements.

The laboratory reactors operated 94.4 percent of the time even though many experiments were conducted to define operational parameters. The volume of gas evolved and its chemical composition was measured. The evolved gases were typically 58 wt percent carbon dioxide and 48 wt percent nitrogen. The total volume of gases was 0.45–0.48 liters of gas/gram of nitrate destroyed. The carbon dioxide concentration and the total amount of evolved gases are indicators of reactor performance and are used in the production reactors to determine operational problems. No nitrogen oxides were found in the evolved gases except during periods when the reactor was in an upset condition. For information purposes, when neutralized raffinate (calcium nitrate) was fed to the reactor, the carbon dioxide concentration was a nominal 38 wt percent because this feed did not require neutralization that releases carbon dioxide. When sodium nitrate was used as a feed to the laboratory column studies, a carbon dioxide level of 4 percent was typical in the evolved gases because most of the carbon dioxide gas remained in the effluent liquid stream as sodium bicarbonate or carbonate.

Production Reactors

Two 25,000-gallon reactors were installed in 1976 and have been successfully operated since that time. Each of these reactors normally decompose 4 grams of nitrate/day/liter of reactor solution (840 lbs NO_3/day/reactor) but rates as high as 6 grams/day/liter of reactor solution have been obtained. Occasional operational problems have been experienced with the feed pumps and pH control. However, the reactors operate more than 95 percent of the time, and all of the down time is normally related to the lack of waste liquor.

Solids Produced in the Reactor

During laboratory tests, the amounts of solids removed from the stirred-tank reactors were measured. When a total daily feed rate of 2242 grams/day of raffinate and calcium acetate were fed to an operating reactor, 275 grams of dried solids were produced. The chemical composition of the dried solids were typically 13.68 percent calcium, 10.14 percent aluminum, 0.51 percent magnesium, and 0.59 percent iron. The solids contained 6.23 percent organic carbon and calcium carbonate; the carbon dioxide equivalent in the carbonates was 21.14 percent. Trace quantities (less than 0.01 percent) of several other metals were identified in the solids.

Treatment of Waste Pond Liquid

As previously stated, four unlined waste ponds having a combined volume of 10 million gallons were used to receive nitrate wastes from several processes. The primary source of the nitrates were those emitted from the uranium purification processes which have previously been discussed. In 1983, a decision was made to eliminate the four ponds, and an in situ treatment process was developed [6,7]. Laboratory tests were conducted to develop this process, and one of the ponds was used as a pilot plant test. The other three ponds were treated using the same process. The developed process consisted of (a) neutralization of the ponds, (b) settling of the solids, (c) denitrification, (d) bio-oxidation to remove any excess organics, (e) flocculation to remove suspended solids, and (f) filtration and release through a permitted point. Any solids generated during these operations were returned to the empty ponds.

A typical analysis of the liquid wastes in the ponds is shown in Table 3 (1983 data). The first step in the treatment process was to conduct laboratory tests using open top 55-gallon drums to determine if the bacteria would denitrify in open containers. The laboratory tests were successful, and the southwest pond was used as a pilot test for the operations.

Pilot Test on Southwest Pond

The pond had an initial pH of 2.1 and a nominal 8000 mg/L of nitrates. A mixing system was designed and installed as shown in Figure 5. The system was composed of five valved intakes sub-

Table 3. Typical Analysis of the Ponds (1983 Data) (pH 1.0 to 2.7)

	Concentration (mg/L)
Nitrates	7,900 to 40,000
Southwest Pond Data	
Aluminum	583
Beryllium	.06
Cadmium	0.23
Chromium	4.8
Iron	26.2
Nickel	31.3
Lead	2
Zinc	4.3
Uranium	35
Total Organic Carbon	260

The data on the other three ponds were higher by about a factor of 4.

merged about 5 feet and distributed along the north wall of the pond. The flow from this manifold was pumped by two 4000 L/min centrifugal pumps to a discharge manifold composed of five valved outlets located on the west wall. Data indicate that the pond was mixed within 24 hours using this system. On the west side in an access road, a 0.5-m drain, originally used for liquid addition, was used for the introduction of chemical reagents into the pond. For the pilot test, a slurry tank, composed of a steel funnel fitted inside a 1.7-m cement culvert, was positioned over the drain and was used to add the calcium carbonate ($CaCO_3$). Flow pumped from the discharge of the inlet manifold pumps was continuously introduced into

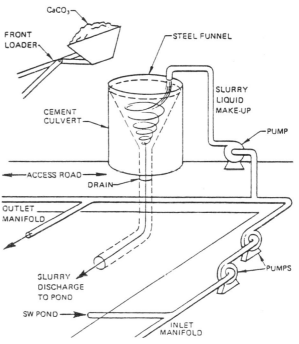

FIGURE 5. Mixing system for neutralizing and stirring the ponds.

the drain while intermittent loads of 150 to 200 kg of $CaCO_3$ were dumped from a front loader into the slurry tank. Acetic acid, sodium hydroxide, and the denitrification bacteria were added directly into the drain by tank truck. The phosphoric acid was dumped into the pond before neutralization.

Finely-ground and high-purity but damp calcium carbonate was used for the primary neutralization. The limestone was dampened to reduce dust problems during handling. The water content was estimated at about 10–20 wt percent. Food-grade acetic acid and phosphoric acid were also added. The final pH adjustments used 50 percent sodium hydroxide solution and calcium hydroxide. Biodenitrification sludge from the operating production reactor was added as bacteria seed. During neutralization, the first part of the calcium carbonate addition was very efficient, and a pH of 5.0 was quickly achieved. Several days later, acetic acid was added to the pond. The pH dropped to about 4.0, and a considerable amount of precipitated aluminum hydroxide redissolved. Sodium hydroxide was added as a 50 percent solution to finish the neutralization. The bacteria began to denitrify and, over the next twelve days, the pond pH rose from 6.8 to 7.1. The total required amount of acetic acid was not added initially because of shipment delays, and when it was added, the pH dropped to 5.9. Biological activity stopped, but in one week the pH rose to a stable value of 6.4. Calcium hydroxide was added over the next several days, until the pH rose to 7.0.

When the pH had been adjusted to 7.0, the bacteria count increased to 10^4 bacteria/mL in two days, and gas could be seen evolving throughout the pond. Unlike beaker experiments performed in the laboratory, the bacteria in the pond were acclimated almost immediately, showing significant denitrification (rates of 500 mg of nitrate destroyed per liter per day) within days after neutralization. As can be seen from Figure 6, the denitrification rate proceeded rapidly. During this time, the bacteria reached a concentration of 10^7 bacteria/mL. After five days, the final acetic acid needed for complete denitrification was added. The pH dropped to less than 6.0 and denitrification essentially stopped. During the first week, after adding the final amount of acetic acid, the pond was allowed to sit unperturbed to see if bacterial action could adjust the pH to 7.0. This change was slowly occurring to some extent as the pH increased to 6.4 before appearing to taper off. During this time, the dissolved oxygen level increased to about 2 mg/L, and the bacteria count started to decrease. During the second week, sodium hydroxide solutions were added over a period of four to five days to increase the pH. In a few days, the dissolved oxygen

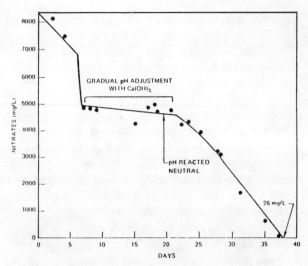

FIGURE 6. Decrease of nitrates in southwest pond.

began to decrease, and the temperature began to increase as denitrification restarted. The denitrification rate over the final 18 days was a nominal 300 mg of nitrate decomposed per liter per day until the nitrate level decreased to a value of 26 mg/L. At a nitrate level of 2500 mg/L, the pH increased from 7.2 to about 8.7 in one day. The dissolved oxygen remained steady at 0.5 mg/L, and the temperature, which had been on a gradual increase, appeared to peak and gradually began falling. During the rapid pH rise, a layer of white foam, about 1.5 cm thick, appeared on the top of the pond. The foam, which was probably composed of heavy metal carbonates and hydroxides saturated with bacteria, lasted for six days until the steady state pH of 9.3 was reached. At this point, the nitrate concentrations were less than 500 mg/L, and a faint sulfur-like odor could be detected. The solution color appeared to change from the usual green–brown to gray–brown. Within about two days, the nitrates decreased to 26 mg/L, and 510 mg/L of total organic carbon remained in the water.

The circulation pumps continued to operate in an attempt to bio-oxidize the remaining organic carbon. As shown in Figure 7, the first seventeen days showed little carbon decrease. The ponds did not obtain sufficient oxygen until an aerator was installed. The organic carbon was then consumed at a nominal rate of 37 mg/L of organic carbon per day for the next seven days until a final concentration of 250 mg/L total organic carbon (TOC) was obtained. At this point, the acetate carbon portion of the total organic carbon was less than 50 mg/L. During the aeration period, the dissolved oxygen remained less than 1 mg/L and began to increase only when the acetate ion concentration fell below 50 mg/L. During the final days of aeration, the adjacent northwest pond was neutralized. Within a week, the total organic carbon and nitrate concentrations in the treated southwest pond began to increase due to leakage from the northwest pond. Shortly thereafter, bacterial action began again in the southwest pond, and these nitrates were also decomposed.

Quality of Water After Denitrification

Samples were taken of the southwest pond and analyzed before and after the in situ denitrification treatment. The results are shown in Table 4. As noted, all heavy metals were reduced to very low levels.

Northwest Pond

In September 1983, treatment of the northwest pond was begun. The mixing system was moved to this pond and the procedure for introducing chemicals was the same as that of the southwest pond. The pH was less than 2, and the nitrate concentrations were over 40,000 mg/L (Figure 8). The treatment was begun when seasonal temperatures were decreasing. Denitrification decreased the nitrates from a nominal 40,000 mg/L to a nominal 28,000 mg/L in the first six weeks and then stopped. At this time, the ambient temperature had dropped to below 10°C. There was no denitrification during the winter months (November 1983 to February 1984) until ambient temperatures began to increase. In May, the pH of the pond decreased below 6.7 and the reaction stopped until the pH was increased. After the pH adjustment, all of the nitrates were destroyed at a rate of 300 mg NO_3 per day.

Northeast and Southeast Ponds

In September 1983, piping and pumps were installed in these ponds so that the waters in these two ponds could mix together. Neutralization of the two ponds was completed using the same procedure used for the northwest pond. The amount of chemicals added (calcium hydroxide and acetic acid) was similar to the amount added to the northwest pond. As noted in Figure 9, the results are very similar to those obtained for the northwest pond.

Solids Deposited in the Ponds

The neutralization step caused most of the metallic ions to form hydroxides or oxides that settled in the ponds. Excess calcium carbonate or hydroxide used during neutralization also precipitated. The biodenitrification process produced carbon dioxide gas and some of it reacted with the calcium ions in solution

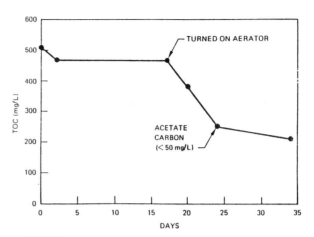

FIGURE 7. Total organic carbon decrease during bio-oxidation.

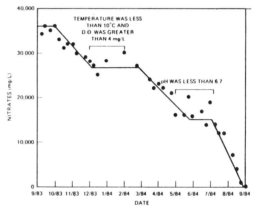

FIGURE 8. Northwest pond data.

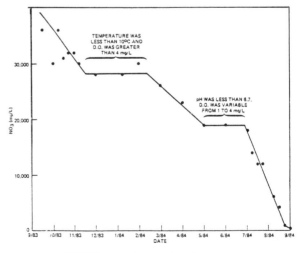

FIGURE 9. Northeast and southeast ponds.

Table 4. Southwest Pond Data

	Concentration	
	Before neutralization (mg/L)	After denitrification (mg/L)
NO_3	7900	47
TOC	260	100
U	35	2.6
Ag	<.006	<.009
Al	583	45
As	—	<.06
B	8	3
Ba	.27	<.2
Be	.06	<.005
Ca	847	—
Cd	0.23	<.007
Co	0.21	.009
Cr	4.8	.003
Cu	5.6	.078
Fe	26.2	.096
Mg	119	47
Mn	20.9	.013
Mo	.36	.16
Ni	31.3	.32
P	6.1	1.41
Pb	2	<.015
Th	.7	<.04
Ti	.5	.017
Zn	4.25	.009

Table 5. Solids Deposited in the Ponds (Dried Basis)

Chemical Element	Range (mg/Kg)
Ag	<.7–3
Al	11,400–21,100
B	13–43
Ba	39–130
Be	<1–2.9
Ca	29,200–47,000
Cd	2–7
Cr	95–140
Cu	78–100
Fe	1300–3000
K	100–1300
Li	36–77
Mg	1500–1700
Mn	62–95
Mo	9–19
Na	990–2900
Ni	230–570
P	470–990
Pb	16–36
Sn	18–35
Th	61–150
Ti	64–350
Zn	32–58
U	769–1040
U^{235} (%)	0.29–0.39
Organics	
Volatile (mg/L)	
Tetrachloroethene	.0016–.0036
Freon 113	.0036
All others total	<0.1
Acid base/neutral fraction	
Tetrachloroethene	1.2–6.6
Phenol	4.8–76
1,1,2,2-Tetrachloroethane	1.2–5.5
% moisture	78–82
Sludge layer (ft)	2–4.6

to form calcium carbonate, which also settled to the bottom of the ponds. After the pond treatment was completed, the pond bottoms were sampled. As noted in Table 5, the collected solids contained about 80 percent liquid and 20 percent solids. The liquid phase contained nitrate ions which was expected since these could have been trapped during the neutralization step. The solids in the ponds were stirred, and the liquid in the ponds dissolved most of the nitrate ions from the solids. The nitrate ions in the liquid phase were then redenitrified when an organic carbon (acetic acid) was added.

The solids contained 3 to 5 percent calcium that is believed to be present as calcium carbonate and 1 to 2 percent aluminum that is expected to be present as aluminum hydroxide.

Final Closure of the Ponds

The biologically treated water was then passed through a waste water treatment facility that consisted of a reactor clarifier, filters, and carbon absorption apparatus. The primary purpose of this treatment was to remove most of the suspended solids which may occur while pumping the water from the pond. Absorption of residual amounts of organics was also accomplished by the addition of activated carbon. The water was discharged via a National Pollution Discharge Elimination System (NPDES) permit and met all chemical parameters required by the permit plus passed two different biological tests. The biological tests involved the growth and survival of fathead minnows, *Primephales promelas* and the survival and reproduction of *Ceriodaphia dubia/affinis* over a seven-day testing period. The biological quality showed no problems in the 3 to 5 wt percent range; therefore, the water was discharged at a controlled rate, and the receiving stream passed both tests at 100 wt percent which means no degradation of the receiving water occurred using these two species as the reference species.

After the ponds were empty, 6-inch diameter or larger limestone rocks were added to stabilize the bottom of the ponds. A 24-inch clay cap was packed to have a permeability of less than 1×10^{-7} cm/sec. A flexible plastic liner and a geotextile fabric was placed on top of the compacted clay. Eighteen inches of compacted clay was placed on top of the fabric followed by three inches of crushed stone and three inches of asphaltic concrete. Several monitoring wells are located adjacent to the cap and if data from these wells show degradation of the groundwater quality, the water will be removed and subjected to the appropriate treatment.

Present Waste Treatment

The Y-12 plant continues to generate nitrate wastes. Several large tanks (500,000 gallon) have been installed and serve as biological reactors using the same process used on the ponds. When biologically treated, the water is passed through a settling tank to remove most of the solids and then through a waste water treatment process to remove the rest of the solids. These solids are currently being stored in large steel tanks until sufficient data is collected to delist them from the hazardous wastes category.

CONCLUSION

In conclusion, recycle processes for nitric acid and aluminum nitrate have reduced the purchases of new chemicals by about 50 percent. Biological treatment of the remaining nitrate wastes is being effectively performed in stirred-tank reactors. All waste nitrates are either being recycled or destroyed. Operation of the recycle and biological process has been conducted since 1976, and in situ biological treatment of four open ponds was started in 1983 and completed in 1985.

ACKNOWLEDGEMENTS

Appreciation is extended to the following co-workers for their valuable contributions to these studies.

1. F. E. Clark, retired, Oak Ridge Y-12 Plant
2. J. W. Strohecker, retired, Oak Ridge Y-12 Plant
3. H. C. Francke, retired, Oak Ridge Y-12 Plant
4. E. G. Laggis, retired, Oak Ridge Y-12 Plant
5. W. C. Dietrich, retired, Oak Ridge Y-12 Plant
6. C. W. Francis, Oak Ridge National Laboratory
7. R. B. Bustamante, Tennessee Technology University
8. I. W. Jeter, Oak Ridge Y-12 Plant
9. T. S. Higgins, Oak Ridge Y-12 Plant
10. H. R. Butler, Jr., Oak Ridge Y-12 Plant

REFERENCES

1. Dietrich, W. C. *Recovery of Nitric Acid from Raffinate and Condensate Solutions by Stream Distillation/Pyrohydrolysis and Fractional Distillation, Y-1886*. Oak Ridge, TN Y-12 Plant: Union Carbide Corporation, Nuclear Division (December 28, 1973).

2. Laggis, E. G. *Pilot Plant Nitric Acid Recovery by Distillation of Condensate Waste, Y-1950*. Oak Ridge, TN Y-12 Plant: Union Carbide Corporation, Nuclear Division (July 30, 1974).

3. Clark, F. E., C. W. Francis, H. C. Francke and J. W. Strohecker. *Denitrification of Acid Wastes from Uranium Purification Processes, Y-1990*. Oak Ridge, TN Y-12 Plant: Union Carbide Corporation, Nuclear Division (November 21, 1975).

4. McCarthy, P. L., Louis Beck and Percy St. Amant. "Biological Denitrification of Waste Waters by Addition of Organic Materials," *Proceedings of the 24th Industrial Wastes Conference, Purdue University, West Lafayette, Indiana*, pp. 1271-1285 (1969).

5. Taylor, P. A., B. R. Blevins, C. A. Baker and G. W. Claus. *Operation and Microbiology of Biological Denitrification System for Treating Highly Concentrated Nitrate Wastes, Y/DU-64 Preprint*. Oak Ridge, TN Y-12 Plant: Union Carbide Corporation, Nuclear Division. Preprint prepared for submission to *Biotechnology and Bioengineering Journal* (May 16, 1980).

6. Napier, J. M., M. A. Makarewicz, P. A. Taylor, P. F. Meredith and C. W. Hancher. *In situ Treatment of the S-3 Ponds, Y/DZ-112, Revision 1*. Oak Ridge, TN Y-12 Plant: Martin Marietta Energy Systems, Inc. (September 16, 1985).

7. Napier, J. M. Biodegradation of Nitric Acid Waste, Y/DZ-425. Preprint prepared for submission to *Engineering Foundation Conference on Hazardous Waste Management Technologies, August 7-12, 1988*. Oak Ridge, TN Y-12 Plant: Martin Marietta Energy Systems, Inc. (August 17, 1988).

Minimization of Arsenic Wastes in the Semiconductor Industry

Darryl W. Hertz[1]

INTRODUCTION

The California Hazardous Waste Reduction, Recycling, and Treatment Research and Demonstration Act of 1985 established a grant program to be administered by the California Department of Health Services. The program provides funding for research and development of hazardous waste reduction technologies. The Waste Minimization project was funded under this grant program and considered the components listed in Figure 1 as guidelines for process evaluations [1,2]. This project demonstrates that hazardous waste stream from semiconductor component manufacturing can be substantially reduced.

Hewlett-Packard (HP), like many other semiconductor companies, processes or manufactures gallium arsenide (GaAs) components, which create a substantial volume of waste containing hazardous levels of arsenic. Generally, the concentration of arsenic in wastewaters or sludges exceeds sewer and municipal landfill limits for disposal. These wastes usually require pretreatment and disposal at a Class I hazardous landfill. In 1987, HP generated about 30,000 kg (33 tons) of arsenic-laden calcium fluoride solids requiring disposal. This practice is not only expensive, but undesirable environmentally and economically as a long-term mechanism for waste disposal.

Based on the need to reduce arsenic waste volumes and disposal costs at HP's San Jose facility as part of their ongoing corporate waste reduction program, this arsenic waste minimization project was initiated.

In the first phase of this project, various processes were reviewed to reduce, recycle and/or treat arsenic-laden wastes on-site, if possible. The methods evaluated at various manufacturing stages included solid-liquid separation techniques, precipitation/fixation, ion exchange, and arsine generation. Filtration as a means of separating particulate GaAs solids from wastewater was selected as the most promising process. The findings of this project indicate that a filter system in combination with a filter press will recover the vast majority of arsenic from the wastewater stream. In addition to the benefit of recovering a valuable product (GaAs), the volume of arsenic waste will be reduced by more than 95 percent. This solid waste, calcium fluoride (CaF_2), is currently being transported as an extremely hazardous waste to a Class I landfill. This waste, once arsenic-free, could become a recyclable product.

In the project's second phase, the short-term evaluation of pilot-scale filtration and filter press equipment was carried out. Also, a more detailed evaluation of the arsenic waste streams at HP was completed that allowed a more complete design for full-scale installation, which is currently underway. Equipment evaluation during pilot testing was based on performance specifications from each manufacturer and the process needs at HP. Each vendor's equipment was given the same set of process conditions for testing so as to allow an objective engineering design to be prepared.

Filtration was selected as the most viable process for effective solids removal because arsenic waste was produced at HP primarily in the particulate form. Project findings indicate that filtration of gallium-arsenide particulates from wastewater streams prior to entering HP's hydrofluoric (HF) acid wastewater treatment facility will result in the following benefits:

---

[1]Envirosphere Company, 3000 W. MacArthur Blvd., Santa Ana, CA 92704

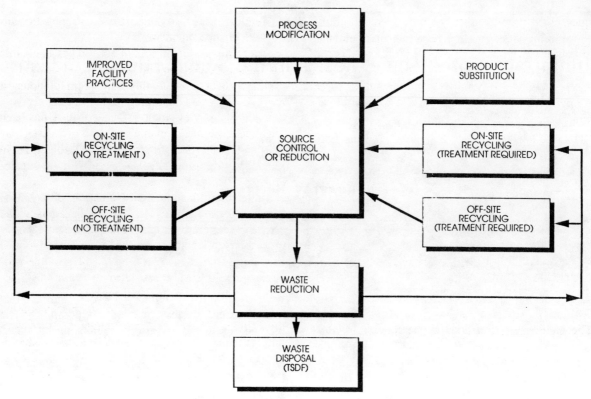

FIGURE 1. Waste minimization components.

- potential income of approximately $100,000 from the sale of treatment facility solids (calcium fluoride—CaF_2)
- $35,000 per year in disposal fees should be saved
- $5,000 per year in increased GaAs recycle revenues
- HP's maintenance and health and safety management costs should be reduced by $15,000 per year.

The information from this arsenic waste reduction project should serve as a model for installation of filtration systems in other facilities that produce particulate metal wastes in their wastewaters [1,2].

SEMICONDUCTOR COMPONENT MANUFACTURING PROCESS

Identification of all possible sources of arsenic by third party investigators required comprehensive process information in the form of written literature, lectures by HP's staff, demonstrations, and explanations by HP's process operating supervisors. This information exchange made possible the extensive chemical characterization with respect to arsenic at HP's San Jose facility.

Figure 2 illustrates how arsenic wastewater flows through HP's facility with respect to particulate gallium-arsenide. Because wafer cleaning and etching processes produce only minor amounts of soluble arsenic and these waste streams flow directly to the HF treatment system, they were not involved in the arsenic waste reduction project. Sample data indicated that these sources contributed less than five percent of the total arsenic produced at HP and did not significantly impact the arsenic waste production.

Laboratory data indicated that the GaAs-laden liquid overflowing to the HF treatment system was responsible for almost 100 percent of the arsenic in the sludge cake solids and approximately 95 percent of total arsenic produced.

Other wastewater streams were analyzed with data confirming that the arsenic level was below detection limits and did not present a potential arsenic contamination source. In fact, the wastewater treatment systems in place at HP's San Jose facility produce a wastewater stream to the sewer system at least twenty times lower than the discharge permit requires.

REQUIRED INITIAL STEPS FOR WASTE MINIMIZATION

Waste reduction processes may involve numerous chemical and physical treatment and separation techniques. Each process must be considered in light of the economic value of such recycling or treatment, and how it will impact the overall production of the facility's products.

The following are considered by the author to be the minimum necessary steps for the waste minimization projects undertaken at HP:

1. Detailed description of the manufacturing process
2. Chemical characterization of waste streams
3. Determination of minimum acceptance standards for recycled waste streams
4. Identification and evaluation of potential waste reduction processes
5. Determination of the environmental concerns related to the process

In performing a detailed review of the manufacturing process under study, a review must be made of how each waste stream enters and transits the manufacturing process. The major sources of wastes must be identified, and process steps examined to prioritize the economic and technical feasibility of waste reduction at each source. It is very likely that the manufacturing steps will exhibit widely different chemical conditions (i.e., nonsteady state), hence the capability to reduce wastes may also vary widely.

Chemical characterization of waste streams requires a detailed analysis of each step in the manufacturing process even though most waste streams have been characterized previously. Following this detailed characterization, the major waste sources, in descending order of importance, should be addressed in subsequent work.

Determination of minimum acceptance standards for wastes recycled on-site is vital in selecting a recycling process. While this may seem obvious, it must be addressed according to the process needs at the facility under evaluation. Once this task has been completed, the list of viable recycling processes may be a short one and can save considerable time in identifying the process(es) of choice for on-site recycling.

In the identification and evaluation of potential waste reduction processes, various data bases and scientific literature should be consulted to identify procedures and processes that have already been used for waste reduction. The list of possible processes should not be restricted to "commonly accepted" processes but should include consideration of new and innovative processes. This evaluation must, in many cases, give consideration to processes and equipment that are contrary to traditional chemical engineering. For example, some waste minimization processes may require numerous small self-contained units distributed throughout a facility rather than a large central process that would cost less, but would not be practical or acceptable.

The evaluation of potential waste reduction processes should be subjected to a selection process that includes at least the following criteria:

- Waste Reduction Efficiency. Each process should be judged upon its potential to reduce the greatest quantity of waste.

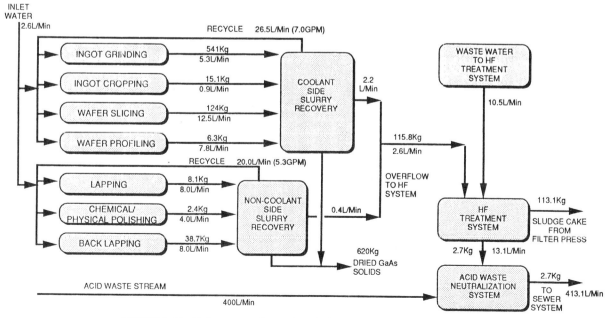

FIGURE 2. Liquid flow rates and annual particulate gallium-arsenide production rates.

- Operational Characteristics. The ease of operation, control, and downtime of the process should be evaluated [7].
- Health and Safety. Because the process deals with hazardous chemicals which can be highly toxic, effective protection for workers must be afforded substantial consideration [4,5].
- Cost. As in any engineering project, capital and operating and maintenance (O&M) costs must be considered. O&M costs may involve chemical additives, electricity, maintenance, and personnel to name a few [3].

The determination of the environmental concerns related to the process that is generating the waste is very important. Usually, these concerns are the driving forces behind waste minimization and must be addressed for each process. The following factors should be evaluated at a minimum:

- waste disposal costs
- mitigation of liability
- ease of process implementation
- personnel and community safety

RESULTS OF INITIAL WASTE MINIMIZATION STUDY AT HEWLETT PACKARD

The arsenic waste reduction project reviewed various processes to reduce, recycle, and/or treat arsenic-laden wastes in a manner such that costs and disposal to the land would be significantly reduced. The methods evaluated at various manufacturing stages included separation, precipitation/fixation, ion exchange, and arsine generation.

In HP's gallium arsenide (GaAs) semiconductor manufacturing process, a wastewater stream containing hydrofluoric acid (HF) is produced that also contains significant levels of arsenic as well as nonhazardous constituents. This wastewater is currently treated with lime that yields a calcium fluoride ($CaF2$) sludge containing high levels of arsenic.

Chemical analysis of the HF treatment system sludge cake indicated that this material is considered to be extremely hazardous solely because of its arsenic content. Tests indicate that this arsenic is in an extremely inert form (GaAs) and produces a nonhazardous leachate. It was concluded that a 50 percent reduction in arsenic content in the sludge cake would result in significant arsenic waste reduction at HP's San Jose facility.

Arsenic solids (GaAs) represent at least 95 percent of the arsenic contamination found in the sludge cake produced from the HF treatment process (see Figure 2). These solids are produced from physical sawing, grinding and polishing operations and are present in the GaAs wastewater stream. Removal of these solids from this water flowing to the HF treatment process by a filtration technique appeared to be the most direct and simple process from a chemical engineering point of view. Although filtration was deemed applicable, a review of other arsenic removal processes was important [1].

Identification of Potential Treatment Processes

The primary goal in hazardous waste reduction in a chemical process facility is to accomplish the task without impacting the process itself and to, if possible, generate a saleable product that has been generated from the waste reduction "treatment."

Table 1 lists each potential arsenic removal process evaluated in this study along with major evaluation criteria with respect to HP's GaAs processing facility. Because effective liquid–solid separation was required, filtration was selected as the process of choice.

Filtration can be an extremely efficient means to separate solid particles from liquid streams. This process has been in use for many years under almost every type of chemical and physical environment [3]. Figure 3 lists just some of the many types of filters available and their main selection criteria. For most liquid flow streams, filtration efficiency can be designed to whatever level is desired. For HP, a design efficiency of only 60 percent or better is required to eliminate the solid hazardous waste problem with respect to arsenic. Because of the ambient temperature and pressure conditions in the wastewater streams, design became dependent on the particle distribution of the solids present. In this case, a large portion of submicron particles were confirmed. Although successful installation of effective filtration equipment should be a matter of routine engineering practice, concern for significant wastestream changes cannot be ignored.

Operational characteristics of the required filtration equipment, like other liquid–solid separation processes, should be relatively simple in operation and offer excellent reliability with full automation [3]. This hands-off operation would reduce health and safety concerns to an absolute minimum.

Potential costs to HP for filtration equipment should be among the lowest of treatment processes because of: (1) high efficiency, (2) minimal operating and maintenance costs (i.e., no chemical addition or residence time constraints), (3) total automation capability, and (4) the favorable ambient conditions (low temperature and pressure) for filtration [1,2].

Table 1. Comparison of Potential Arsenic Treatment Processes [1]

Arsenic Removal Process	Form of Arsenic Most Effectively Treated	Possible Applicability to Hewlett Packard	Potential Waste Reduction Efficiency for Hewlett Packard
1. Liquid–Solid Separation			
• Filtration	Solid	Yes	Excellent
• Centrifugation	Solid	Yes	Excellent
• Settling	Solid	Yes	Good
2. Precipitation			
• Lime	Soluble	No	Low
• Sodium Sulfide	Soluble	No	Low
• Ferric Hydroxide	Soluble	No	Low
3. Coagulation			
• Ferric Sulfate	Soluble	No	Low
• Ferric Chloride	Soluble	No	Low
4. Solidification	Solid	No	Low
5. Adsorption			
Ion-Exchange Resins	Soluble	No	Low
Activated Carbon	Soluble	No	Low
Activated Alumina	Soluble	No	Low
Bauxite	Soluble	No	Low
6. Arsine Generation	Soluble	No	Low

Arsenic Removal Process	Operational Characteristics for Hewlett Packard	Potential Health and Safety Concerns	Potential Costs for HP Relative to Other Listed Procedures	Overall Potential Effectiveness in Reducing Hazardous Waste
1. Liquid–Solid Separation				
• Filtration	Excellent	Low	Low	Excellent
• Centrifugation	Excellent	Low	Low	Excellent
• Settling	Excellent	Low	Low	Good
2. Precipitation				
• Lime	Not Applicable	Low	Moderate	Minimal
• Sodium Sulfide	Not Applicable	Moderate	Moderate	Minimal
• Ferric Hydroxide	Not Applicable	Low	Moderate	Minimal
3. Coagulation				
• Ferric Sulfate	Not Applicable	Low	Moderate	Minimal
• Ferric Chloride	Not Applicable	Low	Moderate	Minimal
4. Solidification	Costly	Low	High	Minimal
5. Adsorption				
• Ion-Exchange Resins	Not Applicable	Low	High	Minimal
• Activated Carbon	Not Applicable	Low	High	Minimal
• Activated Alumina	Not Applicable	Low	High	Minimal
• Bauxite	Not Applicable	Low	High	Minimal
6. Arsine Generation	Not Applicable	Extreme	High	Minimal

PILOT TESTING STEPS FOR SELECTED WASTE REDUCTION PROCESSES

The following are the recommended steps for successful pilot testing of potential waste minimization processes. These steps were selected based on proven engineering principles and judgment that should provide the maximum waste reduction effect. Once the equipment type has been selected, a general specification must be prepared for an on-line demonstration of that equipment's reliable and economic operation [3]. In order to adequately demonstrate this equipment, a select few reputable vendors' equipment or processes should be tested by pilot unit operation to test the feasibility of their processes. Performance evaluations should be followed by specification of full-scale equipment. The support for installation and operation of the resulting system

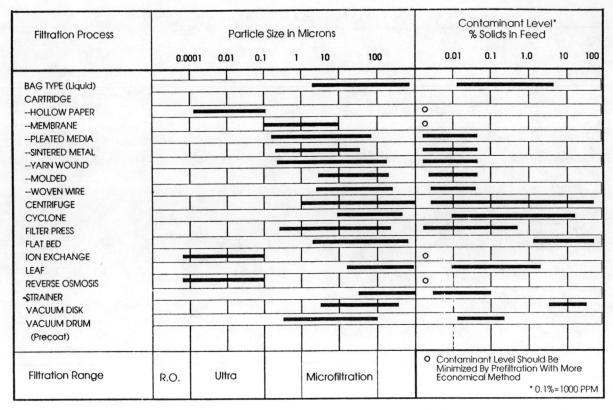

FIGURE 3. Selection of separation process by particle size & contamination level.

should be identified in terms of waste reduction primarily, and economic justification secondarily [2].

Recommend and Specify Waste Reduction Equipment

A recommendation of the best available and reliable waste reduction equipment should be made after evaluation of product literature of applicable equipment and associated waste separation processes. The criterion used to select the best overall equipment should include the following:

1. Waste separation or removal efficiency
2. Personnel health and safety
3. Capital and maintenance cost
4. Ease of operation and expansion

The best equipment or process may involve more than one unit in more than one location in the process plant building. Equipment placement within the manufacturing plant is a vital part of the equipment recommendation.

Once the exact type of equipment has been recommended, a general specification for the equipment installation must be completed. These specifications must include average and peak throughput rates, electrical requirements, overall physical dimensions, metallurgy, automation and control devices, and seal and elastomer specification.

Identify Vendors and Demonstration Criterion for Their Equipment

Vendor selection now becomes the most important criterion in engineering a successful waste minimization project. Choices should be based on past equipment reliability and service excellence. Operational guidelines must be established for each vendor selected prior to on-site demonstration of their equipment. These guidelines must comply with the minimum operating manpower required by the vendor and operating manpower available over the timespan of the equipment demonstration. Additionally, the health and safety requirements established should be superimposed on those already required by the equipment vendor if applicable.

Test and Document Process Operating Parameters

During the testing time period, not only should the vendor's equipment be scrutinized for verification of their operational efficiency claims, but also for the operation of the equipment under the ever-changing flow rate conditions in the process plant. Laboratory analysis is usually necessary to not only verify the vendor's equipment operation efficiencies, but to also verify that data compiled in the initial process evaluation is in fact representative of long-term operating conditions.

Vendor demonstrations should be scheduled such that minimum impact will be made on the plant process while allowing operation of the equipment during average production cycles. Control system's reactions to process changes must be evaluated so as to predict unit operations should plant conditions change in the future on a short-term and a long-term basis. This evaluation not only allows more realistic scaleup engineering to be done, but allows a relatively thorough, full-scale construction project recommendation to be made.

Delineate Final Process Equipment Specifications

Selection of the overall most effective waste reduction system must be made based on actual operation data and operation personnel input. Specifications for this equipment should include the following:

1. Equipment operation diagrams
2. Required minimum waste removal efficiency
3. Average and peak throughput capability
4. System turndown capability
5. Control equipment requirements
6. Installation diagrams
7. Metallurgy and elastomer requirements
8. Electrical requirements
9. Dimensional constraints
10. General installation procedures

These specifications should be complete enough to send to prospective equipment vendors for competitive biddings.

Calculate Expected Full-Scale Economics and Benefits

The benefits of waste reduction equipment must be thoroughly identified and discussed as for any engineered project. These benefits should focus on the proven reduction of waste by installation of the recommended equipment system. The waste reduction system operation must result in lower waste disposal costs, a mitigation of liability, and improvement in personnel and community safety.

RESULTS OF PILOT TESTING OF WASTE REDUCTION PROCESSES AT HEWLETT PACKARD

Pilot testing procedures for the available filtration equipment focused primarily on the parameters most important to the application to HP's facility GaAs wastewater cleanup (i.e., filtration efficiency, long-term backwash capability, and cycle time). Four filtration equipment vendors were selected to provide equipment for pilot testing. These vendors had long-standing reputations for providing not only quality equipment, but also service expertise necessary to evaluate system problems that occur during initial and long-term operation.

Testing procedures were established for each vendor selected for on-site demonstration of filtration equipment. These procedures were based on the need for testing data to properly complete the final equipment specifications and purchase justification.

The purpose of demonstrating the pilot-scale operations of this filtration equipment was to provide definitive answers to the following design criteria that all would impact the final design:

- removal efficiency of GaAs particles in wastewater
- overall equipment performance
- compatibility with the wastewater corrosion resistance
- purchase, operating and maintenance costs
- health, safety, and environmental concerns

In addition, the following parameters required data for final specification:

Filter Equipment
- filter media type and pore size
- equipment size
- materials of construction
- backwash efficiency and frequency
- optimal differential pressure
- solids loading capability
- optimum operating pressure
- need for automation systems
- utility requirements

Filter Press
- press cake capacity
- overall equipment size
- materials of construction
- closing mechanism design

- maximum operating pressure
- cake dewatering cycle
- cake removal procedure
- need for automation systems
- utility requirements

In order to adequately assess whether the filter manufacturer's filter element was performing in the pilot test apparatus, a comparison of the filter elements performance data was made with published data.

This evaluation of filtration equipment available from these four vendors has resulted in a more realistic engineering design for GaAs solids removal equipment. The final specifications for the equipment will allow future manufacturing process changes to occur without seriously impacting the system and represents an economically and environmentally sound method for substantially reducing arsenic waste generation where the particulate form is involved.

This short-term pilot-scale evaluation was accomplished by testing various filtration units under the same set of process conditions. Not only did this work result in a final equipment specification, but revealed significant wafer fabrication process changes. The conclusions of this process design project are as follows [2]:

1. Arsenic waste generation at HP's facility can potentially be reduced by almost 30,000 kg (33 tons) per year by the installation of filtration equipment.
2. The potential cost savings to HP from installation of filtration equipment to remove GaAs solids prior to the waste water entering the HF treatment system are as follows:
   - potential income of $100,000 per year from sale of the HF treatment solids if free from significant GaAs contamination
   - potential savings of at least $35,000 per year in HF treatment solids disposal costs
   - increased revenues of at least $5,000 from GaAs direct recycling
   - reduced maintenance and health and safety management costs of about $15,000 per year from improved handling of hazardous GaAs solids
3. Filtration equipment followed by a filter press is an effective process in separating GaAs solids from the HP arsenic wastewater.
4. Substantial submicron GaAs solids were confirmed in the wafer fabrication waste stream that were not evident during the initial study.
5. Successful removal of GaAs submicron particles using filtration will probably require the use of a chemical pretreatment.
6. Significant increases in total liquid flow requiring filtration were revealed that were not evident during the initial study.
7. The technology of arsenic waste reduction via filtration for removal of particulate arsenic from wastewater streams should be easily transferable to other semiconductor firms as well as to any other industry where heavy metal solids are produced.

FULL-SCALE INSTALLATION CRITERIA

The first step in the full-scale system installation should be preparation of detailed process diagrams indicating all process piping and control systems and their arrangement. From these process diagrams, construction drawings can be prepared showing exact layout, equipment dimensions, and equipment utility details. While these construction drawings are being prepared, a detailed equipment list including specifications for each component should be assembled. This data serves to verify that equipment purchased does not deviate from original drawings and specification lists.

Once the equipment systems are delivered, they should be thoroughly checked, then installed. This installation should follow the ancillary piping system construction that is to be completed concurrently with equipment manufacturing and construction.

Process system startup and system adjustment should proceed once operational and safety procedures are completed. This will allow fine-tuning of the waste reduction process system to the manufacturing process flow streams for maximum operational efficiency.

Once the systems have been evaluated on a short-term basis, a final evaluation of the system should be performed for more long-term operation verification. At this point, proper training of system operators must be completed so they will be able to operate and maintain the equipment systems according to the manufacturer's and the facility's specifications [3,6].

Detailed Process Diagram Preparation

This is the necessary first step in specification and ordering waste recycling equipment. This step not only utilizes data obtained from previous studies, but illustrates each recycling process step in detail including each system's control and alarm equipment.

Construction Drawings Preparation

Detailed construction drawings should be prepared for installation contractor bidding purposes and for equipment manufacturing and fabrication contractors. These construction drawings form the engineering standard for evaluating final equipment design and conformance to specified construction dimensions. These drawings should be modified only after full evaluation of the design change impacts on the project equipment's expected performance and economics. Thus, these drawings serve as "as-built" drawings once construction and installation is complete.

Detailed Equipment Specifications

In conjunction with the construction drawing preparation, equipment specifications should be detailed for purchasing purposes. Utilizing chemical engineering standards, these specifications maintain control of project construction timing and costs. For example, if on-site equipment dimensions, material, or performance is deemed incorrect or substandard, the manufacturer or contractors should be required to make modifications at their cost, as is standard practice.

Equipment specifications should include the following:

- equipment operation diagrams
- minimum recovery efficiency
- average and peak throughput flow rates
- system turndown capability
- control equipment requirements
- installation diagrams
- metallurgy and elastomer requirements
- electrical and other utility requirements
- dimensional constraints
- general installation procedures

Equipment specifications should also include a list of equipment performance efficiencies and capacities guaranteed by the vendor. Should startup evaluation indicate these criteria are not being attained, adjustments should be made, and if modifications become necessary, they should occur at the vendor's expense.

Equipment and Materials Purchase

Placing equipment orders and letting construction contracts should commence once firm schedules and costs have been established. It is the project engineer's responsibility to assure that delivery dates are met and that system installation cost is controlled.

Manufacturing QA/QC Monitoring

Quality control monitoring is important to assure that equipment delivered and the system installed are as specified. If omissions and/or inadequacies are discovered, necessary action should be taken. Monitoring of equipment dimensions, materials, and control system operation may be carried out at the manufacturer's factory prior to shipment, if deemed necessary.

Equipment Construction and Installation

Waste reduction process equipment should be fabricated and constructed while piping systems are being installed. These two activities can be coordinated such that equipment arrival and installation schedules are maintained. Every effort should be made to maximize the time available for system startup and evaluation.

Process Startup and System Adjustment

The process startup and system adjustment phase probably represents the most critical step in the system installation. During system startup, technical representatives may be required to provide guidance to assure the system is performing as guaranteed. The purpose of these system adjustments should include the following:

- maximize waste recovery efficiency
- minimize utility requirements
- maximize system adaptability to changes in liquid throughput capacity
- adjust alarm set points in the equipment's control systems
- test backup systems for safe operation at maximum flow

Operator Training and Maintenance Schedule Preparation

With the aid of the equipment manufacturer and plant health and safety personnel, operators can be trained to ensure successful economic operation of the waste reduction system. Particular attention should be given to health and safety concerns for proper waste handling. Maintenance schedules should also be prepared in conformance with the equipment manufacturer's recommendations and recommendations of the plant facility's engineers.

REFERENCES

1. Envirosphere Company, "The Reduction of Arsenic Wastes in the Electronics Industry, Final Report." Prepared for the California Department of Health Services Toxic Substances Control Division, Alternative Technology Section, Grant No. 86-T0178, June, 1987.

2. Envirosphere Company, "Process Design to Reduce Arsenic Wastes in the Electronics Industry, Final Report." Prepared for the California Department of Health Services Toxic Substances Control Division, Alternative Technology Section, Grant No. 86-T0113, June, 1988.

3. Perry, R. H. and G. H. Chilton. *Fifth Edition, Chemical Engineers' Handbook*, Section 19. New York:McGraw-Hill Book Company (1973).

4. Wade, R., M. Williams, T. Mitchell, J. Wong and B. Tuse. *Semiconductor Industry Study: Task Force on the Electronics Industry*. State of California Department of Industrial Relations, Division of Occupational Safety and Health, 1981.

5. Sax, N. I. *Dangerous Properties of Industrial Materials—Sixth Edition*. New York:Van Nostrand Reinhold Ginhold Company (1984).

6. Treybal, R. E. *Mass Transfer Operations*, Third Edition. New York:McGraw-Hill Book Company (1980).

7. Schweitzer, P. A. *Handbook of Separation Techniques for Chemical Engineers*. New York:McGraw-Hill Book Company (1979).

UV-Catalyzed Hydrogen Peroxide Chemical Oxidation of Organic Contaminants in Water

D. G. Hager[1]

INTRODUCTION

Chemical oxidation of organic contaminants in water is an alternative technology that can be applied to a broad spectrum of contaminated water, industrial effluents, and concentrated aqueous wastes. The process converts hydrocarbon contaminants to carbon dioxide and water. Any halogens present in the organic molecule are converted to the corresponding inorganic halides.

The ultraviolet light-catalyzed hydrogen peroxide chemical oxidation process (UV/H_2O_2) is applicable for the destruction of most types of organic contaminants whether the concentration is a few micrograms per liter or several thousand milligrams per liter. The UV/H_2O_2 process is effective over a wide pH range. The process creates no waste by-products or air emissions. This process can be used for complete destruction of the organic contaminants or for detoxification of wastewater which then allows subsequent treatments, such as biological processes, to function reliably.

Extensive experience with industrial effluents and contaminated groundwater has demonstrated that the UV/H_2O_2 process is nearly always effective. The cost of treatment varies with type and concentration of contaminant as well as the treatment objective. Fortunately, special laboratory procedures can quickly and inexpensively determine the cost and effectiveness of the process for the particular problem.

THE UV/H_2O_2 PROCESS

Ultraviolet light (UV) catalyzes the chemical oxidation of organic contaminants in water by its combined effect upon the organic contaminant and its reaction with hydrogen peroxide. Many organic contaminants absorb UV light and may undergo a change in their chemical structure or may simply become more reactive to chemical oxidants.

More importantly, UV light, at less than 400 nanometers wavelength, reacts with hydrogen peroxide molecules to form hydroxyl radicals. These very powerful chemical oxidants then react with the organic contaminants in the water. If carried to completion, the end products of hydrocarbon oxidation with the UV/H_2O_2 process are carbon dioxide and water.

The reaction of formic acid with UV-catalyzed hydrogen peroxide illustrates this photochemical oxidation process:

$$H_2O_2 \xrightarrow{UV (<400 \text{ nanometers})} 2 \cdot OH$$

$$HCOOH + \cdot OH \rightarrow H_2O + HCOO\cdot$$
$$HCOO\cdot + \cdot OH \rightarrow H2O + CO_2$$

Like most other chemical oxidations, the UV/H_2O_2 process is dependent upon a number of reaction conditions that can affect both performance and cost. Some process variables are inherent to the properties of the contaminated water while other process variables can be controlled by the treatment system design and operation. Some of the more important process variables are summarized in Table 1.

SELECTION OF OXIDANT

The oxidation potentials for common oxidants are listed in Table 2. As shown, the hydroxyl radical is second only to fluorine in oxidative power.

---

[1] Peroxidation Systems, Inc., 4400 E. Broadway, Ste. 602, Tucson, AZ 85711

Table 1. UV/H₂O₂ Process Variables

1. Variables related to the contaminated water:
   - type and concentration of organic contaminant
   - light transmittance of the water
   - type and concentration of dissolved solids
2. Variables related to treatment process design and operation:
   - UV and H_2O_2
   - pH and temperature conditions
   - mixing efficiency
   - use of catalysts

Table 2. Oxidation Potential of Oxidants

Relative Oxidation Power	Species	Oxidative Potential (Volts)
2.23	Fluorine	3.03
2.06	Hydroxyl Radical	2.80
1.78	Atomic Oxygen	2.42
1.52	Ozone	2.07
1.31	Hydrogen Peroxide	1.78
1.25	Perhydroxyl Radical	1.70
1.24	Permanganate	1.68
1.17	Hypobromous Acid	1.59
1.15	Chlorine Dioxide	1.57
1.10	Hypochlorous Acid	1.49
1.07	Hypoiodous Acid	1.45
1.00	Chlorine	1.36
0.80	Bromine	1.09
0.39	Iodine	0.54

Table 3. Capital Cost and Power Comparison Hydrogen Peroxide vs Ozone

	Hydrogen Peroxide	Ozone
Up to 34 kilograms/day		
Capital		
—Tank and Feed System	$8,000	
—Generator		$135,000
—Power		720 kWh/day
Up to 340 kilograms/day		
Capital		
—Tank and Feed System	$40,000	
—Generator	—	$500,000
—Power		8250 kWh/day

Ozone also forms hydroxyl radicals under UV light catalysis. Neither hydrogen peroxide nor ozone contain metals or halogens that can lead to undesirable by-products during the organic oxidation process. However, hydrogen peroxide is a more cost-effective reactant because each molecule of hydrogen peroxide forms two hydroxyl radicals. Furthermore, hydrogen peroxide has other inherent advantages as the preferred oxidant over ozone. Hydrogen peroxide is supplied commercially as an easily handled liquid (30–50 percent) which has infinite solubility in water. Ozone is a toxic gas with limited water solubility. The water solubility of hydrogen peroxide greatly simplifies the reactor design, in terms of oxidant addition, mixing of the reactants, and elimination of fugitive toxic gases. In addition, hydrogen peroxide storage and feed systems are relatively inexpensive compared to ozone generation and feed equipment (see Tables 3 and 4).

SELECTION OF EQUIPMENT

The UV/H_2O_2 process has been successfully applied to environmental problems by the use of **perox-pure**™ chemical oxidation equipment. Modular, skid-mounted systems are being manufactured and installed in groundwater and industrial effluent applications. The volume of the oxidation chamber and the UV dosage are defined in a series of laboratory tests. The **perox-pure**™ equipment utilizes medium- to high-pressure mercury vapor lamps with proprietary emission specifications. The selection of the specific lamp and the quantity of lamps is based upon the results of special laboratory tests. The type and amount of contaminant dictate which combination of UV energy, hydrogen peroxide, and time of oxidation is most cost-effective.

The rate of chemical oxidation of the contaminant tends to be limited by the rate of formation of hydroxyl radicals. Sufficient UV light at 254 nanometers must be provided to form the hydroxyl

Table 4. Hydrogen Peroxide Advantages as a Chemical Oxidant

- a safe, readily available chemical
- no toxic fumes or gases
- easily stored and pumped
- infinite solubility in water
  — no mass transfer problems associated with gases
  — unlimited dosing capability
- contains no halogens or metals
- degradation products are O_2 and H_2O
- minimal capital investment
- the most cost-effective source of hydroxyl radicals

radicals needed for oxidation of the contaminant in the available time. UV light will also be directly absorbed by many contaminants. This absorbed energy will serve to directly break down the contaminant or make it more receptive to oxidation by the hydroxyl radicals. In addition to providing sufficient energy at 254 nanometers, UV lamps can be designed to emit UV light at other specified wavelengths. This can make the lamp more effective when used on a contaminant that absorbs these wavelengths.

The amount of energy provided to the reactants is the product of intensity and oxidation time. Greater UV intensity requires less oxidation time as the reaction rate will be faster, provided sufficient oxidant is present.

The use of medium- to high-pressure lamps greatly reduces the number of lamps to be assembled and maintained. Furthermore, the UV density obtainable with this type of UV system greatly accelerates the chemical oxidation. Another benefit to the more powerful lamps is that chemical intermediate formation is brief and undesirable by-product formation has not been observed in full-scale **perox-pure**[tm] equipment. These oxidation chambers containing the UV lamps are totally enclosed to obviate any possibility of atmospheric emissions. The design of the oxidation chambers and the selection of UV lamps should be such that the following optimum UV conditions are obtained:

- wavelengths that will be absorbed by the oxidant and the contaminant
- intensity that will penetrate the water at energy levels sufficient to initiate chemical oxidation
- exposure time sufficient to complete reactions with the oxidant and the contaminant

Decontamination occurs in a series of processing steps: the contaminated water is pumped to the **perox-pure**[tm] module; hydrogen peroxide is added to the contaminated water prior to the oxidation chamber; the mixture flows past the UV lamps housed in quartz tubes; the purified water is then discharged or reused. There are no solids, sludges, or air emissions resulting from the process. Figure 1 depicts the **perox-pure**[tm] system schematically.

Water containing low concentrations of organic contaminants can be treated with a single pass through the oxidation chamber. Concentrated aqueous wastes may require operation in a recycle mode. Most systems fall into one of three basic flow configurations:

- In the "flow-through" mode, contaminated water is processed continuously, or during predetermined hours.
- In the "flow-through with recycle" mode, contaminated water is processed continuously with a recycle loop to increase oxidation time and mixing.
- In the "batch-recycle" mode, contaminated water is accumulated in a storage tank. Catalyst, if used, is then added, and the unit is operated until analysis of the water shows that the contaminants have been destroyed.

CONTAMINATED GROUNDWATER PURIFICATION

Groundwater purification by the UV/H_2O_2 process is an application in which chemical oxidation technology has received rapid acceptance. The dissolved organic contaminants are often solvents, fuels, pesticides or other dissolved organic chemicals that are frequently toxic or carcinogenic and occur in low concentrations. The typical approach for groundwater remedial action is extraction of the contam-

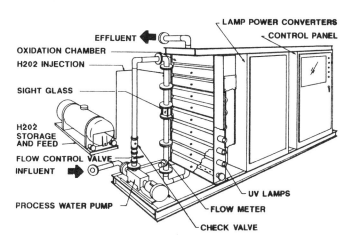

FIGURE 1. The **perox-pure**[TM] system arrangement.

inated water, treatment for organic removal, and reinjection of the purified water. Very low organic concentrations are required to meet most state and federal discharge standards. These standards are frequently the same as those used for potable water quality.

Common treatment technologies include granular activated carbon adsorption (GAC); air stripping (AS), with and without emission treatment; and UV/H$_2$O$_2$ chemical oxidation. UV/H$_2$O$_2$ treatment destroys the contaminant in situ without air emission or waste by-product generation. Both GAC and AS transfer the contaminant to a second media where it requires subsequent attention either as an air pollutant or as a hazardous waste by-product. Both air emissions and hauling of spent adsorbents create additional public liabilities, additional regulatory procedures, such as permitting and monitoring, and extended installation schedules caused by public notification requirements.

Other noneconomic selection factors are presented in Table 5 for alternative groundwater treatment technologies.

The **perox-pure**™ UV/H$_2$O$_2$ process has been selected and installed in a number of groundwater purification projects that have varied widely in water

Table 5. Groundwater Treatment Alternatives: Noneconomic Selection Criteria

	Process Applicability			
	UV/H$_2$O$_2$	Granular Activated Carbon	Air Stripping	Air Stripping + Emission Control
Raw Water Quality Characteristics				
Organic Treatability				
Aromatics	B	B	L	L
Solvents	B	L	V	V
Chlorinated Solvents	B	L	L	L
Pesticides	B	B	NA	NA
Toxic Chemicals	B	L	NA	NA
Molecular Weight				
Low	B	L	L	L
Medium High	B	B	NA	NA
Very High	B	NA	NA	NA
Other Considerations				
Scaling Inorganics	Extra Maintenance	Damage	Damage	Damage
Bacteria Content	Destroys	Augments	Maintenance	Maintenance
Temperature	Insensitive	Sensitive	Sensitive	Sensitive
Variable Flow Rate	Routine Adjustment	Reduced Performance	Reduced Performance	Reduced Performance
Equipment Parameters				
Area	Small	Medium	Small	Medium
Elevation	Low	Medium	High	High
Noise	None	None	High	High
Maintenance	Moderate	Moderate	High	High
Operator Attention	Periodic	Periodic	Periodic	Regular
Energy	High	Low	Low	Moderate
Winterization	Minimal	Extensive	Extensive	Extensive
Permits/Monitoring Required				
Water Discharge	Yes	Yes	Yes	Yes
Air Discharge	No	No	Yes	Yes
Solid Waste Disposal	No	Yes	No	Yes

B = broad range; L = limited to specific contaminants; V = volatile only; NA = not applicable.

Table 6. Groundwater Purification Installations Using the **perox-pure**[tm] Process

Case	Industry	Contaminants	Concentration (μg/L)	Flow Rate (m³/s)	Operating Costs ($ per 1000 cubic meters)		
					UV/H_2O_2	GAC	AS/GAC
A	Food	Trichloroethylene (TCE)	500–1000	0.0126	$250	$500	ND
B	Chemical	Benzene	50,000	0.0016	$135	$170	$225
C	Electronics	Trichloroethane (TCA)	175				
		Dichloroethylene (DCE)	122				
		TCE	122				
		Other	70				
		Total	489	0.0063	$135	ND	$160
D	Fuel	Toluene	7,000				
		Xylene	3,500				
		Benzene	5,000				
		Other	3,900				
		Total	19,400	0.0079			
E	Aerospace	TCE	70				
		Perchloroethylene (PCE)	30				
		Total	100	0.0063	$75	ND	ND
F	Office Machines	DCE	16,000				
		TCE	45,000				
		Toluene	1,000				
		Vinyl chloride (VC)	500				
		Other	1,000				
		Total	63,500	0.0009	$75	ND	ND
G	Military	TCE	100				
		PCE	50				
		TCA	20				
		Other	30				
		Total	200	0.0208	$225	$250	ND

ND = not determined.

volume and contaminant type and concentration. The size of the equipment and therefore capital cost is most heavily dependent upon the volume of water to be treated. The operating cost is directly related to the type and concentration of contaminant. Table 6 summarizes a representative sampling of groundwater installations and provides pertinent cost information as it relates to flow rates and contaminants. In each case the treatment objective was the potable water standard applicable at each site for each contaminant. Cost data are also shown for other treatment processes when they were also evaluated prior to the selection of UV/H_2O_2 chemical oxidation.

INDUSTRIAL WASTEWATER PURIFICATION

Industrial wastewaters vary dramatically in both volume and quality within each Standard Industrial Classification (SIC-Department of Commerce) and among the broad range of industries. Generalization of wastewater treatment capability by any process is subject to misunderstanding and error. Each industrial site must be carefully evaluated, and each treatment option tested.

Most industrial effluents are comprised of the sum of a variety of smaller individual wastewater streams from the manufacturing facility. Good water management practice will separate dissimilar contamination problems (i.e., inorganic verses organic) within each facility and will isolate specific sources of wastewater which are better addressed individually rather than being mixed with other wastewaters (i.e., small volume toxic streams with large volume biodegradable wastewaters).

Pretreatment for Biological Processes

Historically, the relatively low cost of biological treatment has resulted in the selection of biological "end-of-pipe treatment" for many industrial manufacturing facilities. This practice has generally improved the quality of water being discharged to public treatment works or to surface waters. However, in biological processes, specific organic contaminants are transferred from the wastewater to biological sludges that are by-products of the treatment process. Many industrial sludges were disposed into landfills that have become a source of groundwater contamination. Landfills, as such, are being regulated out of existence and therefore are unavailable as a disposal option. Some toxic contaminants are neither destroyed by biological treatment nor transferred to the biological sludge but are found in the industrial discharge.

New regulations (40 CFR parts 414 and 416) were promulgated November 5, 1987 and will become effective June 1, 1990 addressing the organic chemical, plastics, and synthetic fiber industries (OCPSF). These regulations differentiate between those plants with and without end-of-pipe biological plants and between those discharging to a public treatment works or to open waterways. Approximately fifty organic chemicals are regulated.

The UV/H_2O_2 process offers the OCPSF and other chemical utilizing industries an option to combine existing biological treatment with UV/H_2O_2 pretreatment on selected wastewater streams. In this application the UV/H_2O_2 treatment objective would be detoxification as opposed to complete oxidation. This combination will improve the final effluent quality without creating toxic sludges. Treatment technologies commonly used for volatile and dissolved organic contaminants removal from industrial effluents are listed and compared in Table 7.

The UV/H_2O_2 Process Applied to Specific Contaminants

Performance data for the application of the UV/H_2O_2 chemical oxidation process is specific to the equipment as well as to the variety of operating variables described earlier. Commercially available UV/H_2O_2 equipment varies in type and amount of UV energy applied to the wastewater. The importance of this fundamental variable is illustrated in Figure 2 that compares the destruction of phenol by the UV/H_2O_2 process by varying the UV dosage from 75 to 1000 watts per liter. As shown, the reaction time for phenol destruction is inversely proportional to the intensity of UV energy being applied.

By combining higher UV dosage with the infinite solubility of hydrogen peroxide, the **perox-pure**tm process design has opted for smaller equipment with higher energy utilization per volume of wastewater being treated.

During the commercial development of the UV/H_2O_2 chemical oxidation process, hundreds of contaminated industrial wastewater samples were tested under a variety of process conditions.

Table 7. A Comparison of Pretreatment Options for Biological Treatment

	UV/H_2O_2	Ozone	Steam Stripping	Granular Activated Carbon
Raw Water Quality				
Organic Treatability				
Volatile	Yes	Limited	Yes	Yes
Dissolved	Yes	Yes	No	Yes
Toxic	Yes	Yes	Yes	Yes
Other Considerations				
Scaling Inorganics	Extra Maintenance	Minimal	Minimal	Damage
Bacteria	Destroys	Destroys	Extra Maintenance	Augments
Equipment Parameters				
Area	Small	Medium	Small	Medium
Elevation	Low	Medium	High	High
Noise	Low	Low	Moderate	Low
Maintenance	Moderate	High	Moderate	Low
Operator Attention	Periodic	Regular	Regular	Periodic
Energy	High	High	High	Low
Winterization	Minimal	Moderate	Extensive	Extensive
Permits/Monitoring Required				
Water Discharge	Yes	Yes	Yes	Yes
Air Discharge	No	Yes	Yes	No
Solid Waste Disposal	No	No	No	Yes
Operating Area	No	Yes	Yes	No

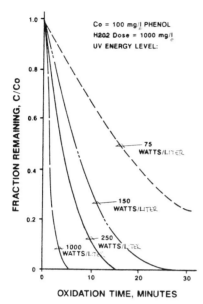

FIGURE 2. Photochemical oxidation rate vs. UV energy.

In order to compare and collate the data gathered from diverse industrial wastewater sources and developed under a variety of process operating conditions, first-order reaction rate constants were determined. The rate constants were then scaled-up at a constant UV dosage to given final values representing reaction rate constants in full-scale **perox-pure**[tm] equipment. Figure 3 illustrates the differing reaction rates for trichloroethylene (TCE) vinyl chloride, dichlorophenol, benzene, toluene, xylene, methylene chloride, and acetone.

Table 8 lists numerous chemical contaminants that have been successfully treated by the **perox-pure**[tm] Process. Two sets of data are shown: Column A provides a range of reaction rate constants; and Column B provides a range of hydrogen peroxide dosages.

A range of values is given for each contaminant to allow for variations in water color, suspended solids,

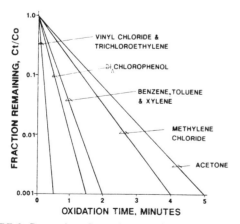

FIGURE 3. Destruction of organic contaminants using the **perox-pure**[tm] process.

total volatile organic compounds (VOC) or chemical oxygen demand (COD), and so forth. Higher values of the constant A reflect better quality water. Higher values of the Constant B reflect poorer water quality. These constants are applicable to contaminant concentrations up to several thousand milligrams per liter. The estimated H_2O_2 dosage should not normally exceed 500 milligrams per liter.

The equations below can be used in preparing preliminary estimates of **perox-pure**[tm] operating and capital costs for the destruction of organic contaminants in aqueous wastestreams. The full-scale oxidation time [Equation (1)] can be used along with the wastestream flow rate to calculate equipment size, electrical power costs, etc. The hydrogen peroxide dosage [Equation (2)] can be used to determine the quantity of hydrogen peroxide needed and the hydrogen peroxide cost.

$$t = (1/A) \ln (C_i/C_e) \quad (1)$$

where

t = full-scale oxidation time (minutes)
A = constant from Table 8
C_i = influent contaminant concentration (mg/L)
C_e = effluent contaminant concentration (mg/L)

$$D = 1 \, BC \quad (2)$$

where

D = H_2O_2 dosage (kilograms per 1000 cubic meters)
B = constant from attached table
C = total influent VOC concentration (mg/L)

Using the oxidation time derived in Equation (1) and the flow rate of a given application, Figure 4 can be used to select the appropriate **perox-pure**[tm] models (see Figures 5–7).

Destruction of Concentrated Organic Aqueous Wastes

Concentrated aqueous hydrocarbon waste solutions result from a variety of chemical processes. The disposal of these concentrated solutions without air emissions or without the use of landfills is a formidable challenge for many industries. UV/H_2O_2 chemical oxidation offers an alternative solution for the on-site destruction of numerous concentrated organic wastes.

Both thermal and chemical oxidation converts hydrocarbons to carbon dioxide and water. However,

Table 8. **Perox-pure**™ Reaction Rate Constants for Chemical Contaminants

Contaminant	A	B	Contaminant	A	B
Acenaphthene	2.0–4.0	1.5–3.0	Formaldehyde	0.8–2.0	1.0–2.0
Acetic acid	0.2–0.6	0.4–0.8	Formic acid	1.2–2.5	0.8–2.0
Acetone	0.4–0.8	1.5–3.0	Freon-TF (trichlorofluoromethane)	1.5–4.0	1.0–3.0
Acetonitrile	0.3–0.8	1.0–2.5	Fructose	0.6–1.6	1.0–2.5
Acrylonitrile	1.5–3.5	0.4–0.8	Glucose	0.6–1.6	1.0–2.5
Aldecarb	2.0–4.0	0.2–0.6	Glycerin	0.3–0.8	0.8–2.0
Aniline	0.8–1.5	1.0–3.0	Hexane	0.8–2.0	1.0–3.0
Benzene	2.0–6.0	0.5–1.5	Hydrazine	0.5–3.0	0.8–2.0
Benzoic acid	1.5–3.0	1.0–2.5	Hydroquinone	0.8–2.0	1.5–2.5
Bromoform	1.0–3.5	1.5–3.5	Isopropanol	0.8–1.6	1.5–2.5
n-Butylamine	0.3–0.6	1.5–2.5	Methanol	0.5–1.5	1.0–2.5
Butyric acid	1.0–2.0	0.6–1.5	Methylcellulose	0.2–0.8	1.0–2.0
Captan	1.0–2.5	1.5–3.5	Methylene chloride	0.4–3.0	1.0–2.5
Carbon tetrachloride	1.2–3.8	1.0–2.5	Methylethylketone	1.0–2.0	1.5–3.0
Catechol	1.5–2.5	0.8–2.0	Methylisobutylketone	0.5–1.5	1.5–3.0
Chloroacetic acid	1.0–2.0	1.0–2.5	Monomethyl hydrazine	0.8–2.5	0.8–2.0
Chloroaniline	1.0–2.5	1.5–3.0	Morpholine	0.3–0.8	1.0–2.5
Chlorobenzene	2.0–4.0	0.5–1.5	Naphthalene	1.5–4.0	0.5–1.5
Chloroform	1.0–3.5	1.5–3.5	Nitrobenzene	2.0–4.0	0.5–1.5
Chlorophenols	3.0–8.0	1.5–3.0	Nitroguanidine	1.0–3.0	1.0–2.5
meta-Cresol	0.3–0.6	0.5–1.5	Nitrophenols	3.0–8.0	1.0–2.5
Cyclohexane	0.6–1.4	0.8–2.0	Nitrotoluene	1.5–3.5	0.5–1.5
Cyclohexanone	0.4–1.0	1.0–2.0	PCBs	1.5–4.0	0.5–2.0
2,4-Dichlorophenoxyacetic acid			Pentachloronitrobenzene	1.0–2.0	1.5–2.5
(2,4-D)	0.8–2.0	1.5–2.5	Pentachlorophenol	3.0–8.0	1.0–2.0
DIMP (Disomethylpropane)	0.8–2.0	1.0–2.5	Pentane	0.8–1.6	1.0–2.5
Diaminodichlorobiphenyl	2.5–4.5	1.5–3.0	Phenol	2.0–10.0	0.5–2.0
Dibromodichloropropane	0.3–0.8	1.5–2.5	Polyvinyl alcohol	0.3–0.8	1.0–2.5
Dichlorobenzidene	2.0–5.0	1.5–2.5	Propionic acid	0.5–1.5	0.5–2.0
Dichloroethane	0.8–3.0	1.0–3.0	Resorcinol	1.5–4.0	0.8–2.0
Dichloronitroaniline	0.6–2.0	1.5–3.0	Sodium Thiocyanate	0.4–1.0	1.0–2.0
Dichloropentadiene	1.5–3.0	1.0–2.5	Tannic acid	1.5–3.0	1.0–2.0
Dichlorophenoxyacetic acid	0.8–1.6	0.8–2.0	Tartaric acid	1.5–3.0	1.0–2.0
Dichloropropane	1.0–2.5	1.0–2.0	Tetrachloroethane	0.5–2.0	0.5–1.5
Dichlorotrifluoroethane	0.6–1.5	0.5–1.5	Tetrachloroethene	3.0–8.0	0.8–1.5
Dimethyl sulfoxide	2.0–5.0	0.6–2.0	Tetrahydrofuran	1.5–3.0	0.5–1.5
Dinitrophenol	4.0–6.0	1.0–2.5	Toluene	1.5–6.0	0.5–1.3
Dioxane	1.0–2.0	0.8–1.6	Trichloroethane	0.5–2.0	0.4–1.0
Ethanol	0.5–1.5	1.0–2.5	Trichloroethene	5.0–10.0	0.4–1.2
Ethyl Acetate	0.8–1.6	1.0–2.0	Trichloropropane	2.0–5.0	1.0–2.0
Ethyl Benzene	2.0–5.0	0.5–1.5	Trichlorophenol	3.0–8.0	1.5–3.0
Ethyl Diamine	0.4–0.8	1.0–2.5	Unsymmetrical dimethylhydrazine		
Ethylene Glycol	2.0–4.0	1.5–2.5	(UDMH)	0.8–2.0	0.8–2.0
Ethylenediaminetetraacetic acid			Vinyl chloride	3.0–6.0	0.6–1.5
(EDTA)	0.4–1.0	2.0–3.0	Xylene	1.0–5.0	0.5–1.5

there are a number of factors which should be considered in addition to the basic economics of each process.

An important advantage of the chemical oxidation process is that it can be used as a pretreatment to detoxify aqueous wastes. With thermal oxidation it is necessary to operate at very high temperatures (1200°C) to minimize formation of toxic compounds. In so doing, much fuel is consumed evaporating the water.

On the contrary, UV/H_2O_2 chemical oxidation may be easily operated at less than full oxidation by control of the amount of oxidant and power. If intermediate reaction products are not objectionable, considerable savings in oxidant and power can be attained by this procedure. Chemical oxidation is particularly applicable to a detoxification step prior to biological treatment. Phenolic aqueous wastes, for example, can be oxidized to biodegradable intermediate chemicals at a fraction of the cost of thermal oxidation.

Halogenated or sulfonated hydrocarbons are not desirable for combustion in incinerators. These compounds generally have poor fuel value and more im-

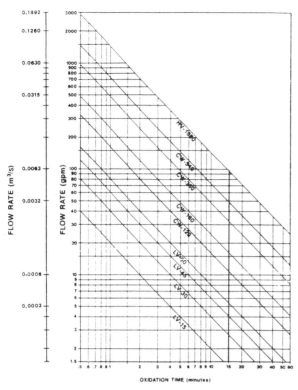

FIGURE 4. Perox-pure™ equipment selection.

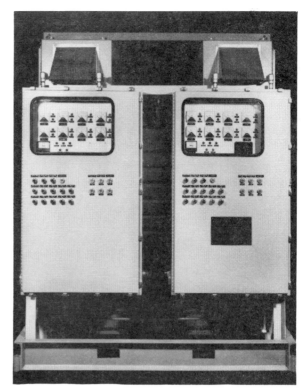

Figure 6. Perox-pure™ model CW 360.

portantly, produce acid gas discharges that are corrosive and illegal to discharge to the atmosphere. Expensive alloy construction and gas scrubbers are required to burn these compounds. There are no gaseous emissions with chemical oxidation at economical concentrations, and the lower temperature liquid (less than 83°C) discharge stream is less corrosive and more easily neutralized than the exhaust gases from incineration.

In UV/H_2O_2 chemical oxidation, the oxygen requirement is supplied by the oxidizer, not from air as in thermal oxidation. For example, only 3.2 kilograms of hydrogen peroxide are required for chemical oxidation of 1 kilogram of methanol. This is less than 1 liter of a 50 percent hydrogen peroxide solution required for chemical oxidation of 0.45 kilograms of methanol, regardless of the amount of water with which it is associated. The end products of this reaction are 0.6 kilograms of carbon dioxide per kilogram of methanol plus water. The solubility

Figure 5. Perox-pure™ model CW 540.

Figure 7. Perox-pure™ model LV 60.

Table 9. Chemical Oxidation vs Thermal Oxidation Selection Factors

Item	Chemical Oxidation	Thermal Oxidation
Nonaqueous Solutions	Not Economical	Suitable
Free Oils	Not Suitable	Suitable
Suspended Solids	Not Suitable	Suitable
Halogenated Hydrocarbons	Suitable	Not Suitable
Sulfonated Hydrocarbons	Suitable	Not Suitable
Air Discharges	None	Permit Required
Solid Residue	None	Ash
Liquid Discharges	Equals Influent	None
Economical Size	Small to Large	Large
Noise Level	Low	High
Safety Considerations	Few	Many
Economic Concentrations	<10,000 ± mg/L	>10,000 ± mg/L
Permitting Requirements	Limited	Extensive
Heat-up Time	Nil	1–2 hours
Full-Time Operator	None	Required
Time for Repair	Short	Long
Maintenance Costs	Low	High

of carbon dioxide is 168 kilograms per cubic meter at standard temperature and pressure. This is much greater than 0.6 kilograms of carbon dioxide produced. Therefore, there will be no emissions from the chemical oxidation of 0.12 kilograms of methanol in one cubic meter of water, as compared to almost 311 cubic meters of combustion products produced by incineration of the same solution.

An additional factor in favor of chemical oxidation of these compounds is that the halogen exerts no demand for oxidant while greatly increasing the weight of the contaminating molecule. The net result is that chemical oxidation tends to be less expensive than thermal oxidation when applied to chlorinated compounds. Because of these factors, incineration of halogenated or sulfonated hydrocarbons will not usually be an economical choice. Table 9 summarizes additional selection factors when considering thermal or chemical oxidation.

UV/H_2O_2 chemical oxidation equipment varies widely in size, design, and capacity. A given unit can vary over a 30:1 range in capacity depending upon the resistance of the contaminant to chemical oxidation. Capital cost will then, likewise, vary as widely in terms of cost per cubic meter to be treated.

Chemical oxidation equipment will also vary in cost per unit of capacity, with larger units costing less. A balance will need to be struck between many units scattered at point of use to minimize collection costs and larger units to minimize equipment costs.

In Table 10, UV/H_2O_2 equipment costs are compared to the costs of an incinerator for two volumes of phenolic wastewater.

The primary operating costs for an on-site UV/H_2O_2 chemical oxidation system are those for the oxidizing agent and for electrical power to generate the UV light.

Table 11 summarizes hydrogen peroxide and power costs for the destruction of toluene, TCE, methylene chloride, phenol and methanol using a UV/H_2O_2 **perox-pure**tm system. Values are given for concentrations ranging from 300 to 10,000 mg/L.

Table 10. Alternate Costs of Concentrated Phenolic Waste Destruction

Case	Volume of Waste Water	Capital Costs ($000)		Operating Costs (per cubic meter)	
		UV/H_2O_2	Incineration	UV/H_2O_2	On-Site Incineration
A	3.785 cubic meters per day	$15.33	$13.21	$34.61	$58.12
B	37.7854 cubic meters per day	$18.49	$34.35	$12.42	$28.27

Table 11. UV/H₂O₂ Destruction Costs vs Concentration of Aqueous Concentrations

Concentration (mg/L)	Operating Costs ($ per cubic meter)				
	Toluene	TCE	Methylene Chloride	Phenol	Methanol
300	5.28	0.79	5.03	4.23	7.13
1,000	—	1.85	5.81	11.36	11.10
5,000	—	—	12.15	52.84	36.99
10,000	—	—	19.55	95.11	60.77

These data can be compared to thermal oxidation by using an operating cost of $15.85 per cubic meter for fuel costs ($0.50 per 10⁵ Joules and an efficiency of 75 percent). It can be seen that UV/H$_2$O$_2$ is an attractive alternative for destruction of hydrocarbon wastes with concentrations below 1000 mg/L or as high as 5000 mg/L as in the case of methylene chloride.

REFERENCES

1. Froelich, E. M. "UV/Hydrogen Peroxide Treatment—A New Chemical Oxidation Process for Toxic Organics," Presented at California Water Pollution Control Association, Anaheim, CA, May 10, 1985.
2. Hager, D. G. "The UV/Hydrogen Peroxide Process: An Emerging Technology for Groundwater Treatment," Presented at HazMat West 85, Long Beach, CA, December 3-5, 1985.
3. Hager, D. G. and C. E. Smith. "The Destruction of Organic Contaminants in Water by Chemical Oxidation," Presented at Haztech International, Denver, CO, August 13-15, 1986.
4. Hager, D. G. and Loven, C. G. "On-Site Destruction of Organic Contaminants in Water," Presented at HWHM National Conference and Exhibition, Washington, DC, March 16-18, 1987.
5. Hager, D. G., C. G. Loven and C. L. Giggy. "Chemical Oxidation Destruction of Organic Contaminants in Groundwater," Presented at HMCRI National Conference and Exhibition, Washington, DC, November 16-18, 1987.
6. Cheuvront, D. A. and G. H. Swett. "Innovative Groundwater Treatment Technologies with Zero Air Emissions," Presented at Hazmacon 88, Anaheim, CA, April 5-7, 1988.
7. Peterson, R. L. "Chemical Oxidation of Aqueous Hydrocarbon Solutions as an Alternate to Thermal Oxidation."
8. Hager, D. G., C. G. Loven and C. L. Giggy. "On-Site Chemical Oxidation of Organic Contaminants in Groundwater Using UV Catalyzed Hydrogen Peroxide," Presented at AWWA Annual Conference and Exposition, Orlando, Florida, June 19-23, 1988.
9. Cheuvront, D. A., C. L. Giggy, C. G. Loven and G. H. Swett. "Groundwater Treatment with Zero Air Emissions," Presented at The Petroleum Hydrocarbons and Organic Chemicals in Groundwater Conference, Houston, TX, November 8-11, 1988.

Polysilicate Heavy Metals Mitigation Technology

George J. Trezek[1]

ABSTRACT

A versatile cost-effective system for treating heavy metals has been developed and successfully applied to a host of waste streams. These include such materials as auto shredder residue, arc furnace dust, incinerator ash, various types of filter press cake, contaminated soils, and waste water treatment plant and other sludges resulting from a variety of manufacturing activities. Basically, the treatment involves the application of the polysilicate blend to the waste stream followed by the addition of a cementitious material such as lime, pozzalime, or cement and curing for a one- to two-day period. It is believed that the efficacy of the treatment is the transformation of a metal or more commonly the metal oxide into a metasilicate such as lead oxide to a form of lead metasilicate. Typically, the amount of cementitious material needed to complete the treatment is on the order of 10 percent; consequently, the treatment results in a relatively small volume increase. In certain cases some processing of the waste stream must be performed in order to adjust the size distribution prior to treatment.

The development of the technology has progressed through a series of laboratory-scale studies, pilot field tests and full-scale in-plant applications. The leachability of the treated waste has been evaluated through the determination of soluble threshold limit concentration (STLC) values by both the California (CAM) and EPA (EP tox) wet leachate tests. The CAM test using citric acid is about one to two orders of magnitude more aggressive in leaching metals. A treatment data base encompassing several hundred tests on the previously mentioned wastes has been assembled. Thus far, the most extensive experience has been with treating auto shredder residue and heavy metal contaminated soil. Five shredding installations generating between 50 to 150 tons per day of residue currently have the technology for the on-line treatment of this waste stream. Typically, the STLC values of the principal metals of concern i.e. lead, cadmium and zinc, as measured by the CAM test, can be reduced by as much as an order of magnitude after treatment. The mobile system units have been used for the treatment of heavy metal contaminated soil. Two such units having processing capacities in the range of 7 to 8 tons per hour and 60 to 70 tons per hour are available. The equipment is versatile and easily adapted to the treatment of other types of materials. For example, the smaller unit has been used to treat bag house dust while incinerator ash was treated with the larger system.

INTRODUCTION

The polysilicate treatment process is a technology that deals with the mitigation of heavy metals. Although the silicate technology has been applied to reducing elevated levels of metals in both liquids and solids, this particular process, which has become known in the industry as the Trezek Process, is applicable to solids such as heavy metals contained in soil, ash, bag house dust, etc.

The development of the process began in earnest in early 1985 as a project to ascertain the technical and economic feasibility of treating the heavy metals contained in auto shredder residue. The California Department of Health Services (DHS) required that this material be managed as a hazardous waste because it failed to pass the California wet extraction test (CAM test) for certain metals such as lead, cadmium, and zinc. After the successful development of a treatment protocol on the laboratory scale, a pilot system was installed as part of the shredding plant process line. A full-scale permanent on-line treatment system was installed toward the end of 1985.

Following the initial success with auto shredder residue, the treatment was applied to mitigating heavy metals in other types of materials. These have included heavy metal contaminated soil, bag house dust, electric arc furnace dust, incinerator ash, filter press cake, foundry sand, waste water treatment sludges, and sludges from a variety of other manu-

---

[1] Professor, Department of Mechanical Engineering, University of California, Berkeley, CA 94720

facturing operations. Developments in the technology have also proceeded on the equipment side—that is, in the delivery of the treatment beyond the laboratory scale. In addition to the fixed in-line concept, equipment has been developed which allows the treatment to be placed in the field with a mobile system.

Consideration will be given to the nature of the technology, the types of delivery equipment used to implement the treatment, and a discussion of laboratory and field results.

NATURE OF THE TECHNOLOGY

The polysilicate treatment technology is a chemical treatment that utilizes commercially available soluble silicate solutions and various cementitious materials such as cement, lime, pozzalime, and fly ash. As is the case with many emerging technologies, the development of this treatment has been to a large extent applications driven, caused by the growing and often urgent need to mitigate hazardous wastes. Thus, the quantification of the underlying mechanisms is not complete so that the treatment is applied on a semiempirical basis. The underlying mechanism is believed to be the formation of insoluble or low solubility metasilicates. For example, lead oxide would be transformed to lead metasilicate. The mechanisms governing the process are thought to occur as a series of three reactions or steps of which the first, formation of the metasilicate, is the most important.

The treatment proceeds according to the following: The material is thoroughly wetted by the polysilicate blend. The silicate can be in the form of a potassium silicate, sodium silicate, or a blend of potassium and sodium silicate. The polysilicate is mixed with water and is typically sprayed into the material as it enters the mixer. The amount of water is determined by the initial moisture content of the material requiring treatment. The next aspect of the treatment involves the addition of the cementitious material. The combination of the cementitious material and water will produce a pozzalonic reaction which typically results in a stiff or solid matrix after the treated material has cured. The curing or drying step is considered the final phase of the treatment.

Each waste stream is characterized and subjected to a range of treatments in order to determine an optimum protocol. There also appears to be a preferred method of implementing the treatment. For example, the blended silicate solution is first added to the waste prior to mixing in the cementitious material. The treated material must be allowed to cure. Depending upon the moisture content, this final step of the treatment may require one to two days.

Potassium silicates are a family of chemicals with a wide range of physical and chemical properties. Their various characteristics permit their use in a host of diverse industrial applications such as protective and decorative coatings, soaps and detergents, binders, and flux coating for welding rods. They are clear, highly viscous liquids having a pH in the range of 11.3 to 11.7. The viscosity is affected by the SiO_2/K_2O ratio, concentration, and temperature. For example, the lower the ratio, the more alkaline at a given solids content, the lower the viscosity. When the silicate is mixed with water it quickly changes to a solution whose viscosity approaches that of water. Also, relatively small temperature increases, on the order of 10°C, can cause a five-fold decrease in viscosity.

Thus far, the majority of experience and test results have been obtained using a commercial potassium silicate blend known as K_2O, manufactured by Lopat, Inc. of Wanamassa, New Jersey. The Lopat material is made in two parts: part A is a blend of three different viscosity potassium silicates and part B contains a catalyzer and dispersing agent. The silicate (part A) and catalyzer (part B) can be made in various ratios. For mixing in the field, the simplest approach is to blend the initial material so that equal amounts of parts A and B are used. After parts A and B are mixed, the resulting solution is diluted with an appropriate amount of water.

This treatment technology is sensitive to the type and concentration of metals in the waste, the nature of the waste material, i.e., soil, sludge, etc., the size distribution, the amount and type of silicates, and the amount and type of cementitious material. Only small amounts, on the order of 5 to 15 percent, of cementitious materials are used in the treatment. For certain materials the additions of polysilicate and cementitious material are on the order of one-third to one-half gallon per ton and 7 to 10 percent, respectively. This differs significantly from other silicate-based treatment techniques which require as much as a 100 percent increase in volume for an effective mitigation. Thus, although the concepts embodied in this technology appear to be similar to others, the underlying mechanism is chemical treatment which differs from other solidification/fixation/stabilization systems. It is also important to note that the laboratory evaluation for determining the solubility requires milling the treated material to pass a No. 10 (2 millimeter) standard sieve. Consequently, a mechanism based solely on encapsulation would fail after milling. Some limited scanning electron microscopic observations of the treated material suggest the formation of the metasilicate structure along with other polymer matrix structures.

Some important features of this technology are summarized as follows: (a) the silicates can be custom blended; (b) the silicate blend in combination with the cementitious material can be optimized in order to achieve a cost-effective treatment methodology; (c) typically only small volume increases result in the treated material; (d) treated materials are in compliance with regulatory agency standards for nonhazardous wastes; (e) treated materials such as soils can be backfilled after curing, thereby eliminating further disposal; (f) the possibility exists that some treated residues could be made into usable aggregates; (g) the constituents of the treatment, i.e., silicates, pozzalime etc., are nonhazardous and used in other commercial applications; and (h) the use of the technology is straightforward and amenable to a mobile system.

TREATMENT SYSTEM EQUIPMENT

For the most part, the equipment needed to deliver the polysilicate treatment is composed of conventional components. These include a pug mill or screw-type mixer, feed hopper and spray chamber arrangement, chemical metering and delivery pumps, cementitious material hopper and feeder, and an appropriate feed and discharge system. In certain situations, the material requiring processing must be preconditioned before entering the treatment system. Typically, this involves screening. For example, in the case of soil, large nonhazardous items such as rocks, stones, and pieces of wood not requiring treatment are removed as an initial processing step. This is also the case for certain types of residues, i.e., auto shredder residue, where items such as large pieces of seat cushions, rubber, and ferrous and nonferrous metals not requiring treatment are removed through screening. Because it has been determined through numerous testing that the heavy metals reside in the finer sizes, screening is almost always part of the overall treatment process.

The basic unit operations comprising the treatment process can be assembled either as (a) an inline system which becomes part of the plant generating the material requiring treatment or (b) a mobile system which can be taken to the site. It should be noted that a different set of permit regulations is applicable to each of the two cases. In the first instance, inserting the treatment in the process line creates a situation where waste materials are in effect not generated. The modification of the process line is viewed as a classic source reduction approach and is outside of regulatory purview by both the Environmental Protection Agency and state programs. Basically, with the in-line approach, the material being generated is not yet a waste, and therefore no hazardous waste treatment permit is required. On the other hand, the mobile system approach qualifies for a transportable treatment unit (TTU) permit. In California this type of treatment qualifies for a TTU permit by rule operation under new regulations promulgated by the DHS.

Consistent with the above permitting situation, in-line treatment systems have been installed in five auto shredding plants in California. Although each plant shreds car bodies, the process lines vary depending upon the particular approach taken toward recovery of ferrous and nonferrous materials. Thus, the placement and configuration of the polysilicate treatment equipment varies between the plants. These plants process on the order of 50 to 150 tons of residue per day. The first unit operation is either wet or dry screening. The oversize material has an aluminum fraction which is further removed while the undersize material is subjected to polysilicate treatment. Additional ferrous recovery occurs from the treated material prior to final curing. The undersize fraction enters the feed hopper–spray chamber which is mounted at the beginning of a screw type mixer. This 25-foot unit is powered by a 40 horsepower hydraulic motor and gear box arrangement. Cementitious material is stored in a silo and automatically conveyed to a small hopper and screw feeder above the mixer. A chemical delivery system premixes the polysilicates and water for pumping to the spray chamber.

Two types of mobile systems have been developed to deliver the treatment on an on-site basis. Both of these units are extensions of the permanent or in-line treatment systems that have been installed in the auto shredding plants. The smaller of the two units is operated by the Trezek Group Inc. and was partially funded by a grant from the DHS. As shown in Figure 1, the treatment system is contained on two trailers. Basically, the function of the two trailers is as follows: The white trailer contains the pug mill mixer, small air compressor for operating the automatic sampler, the chemical delivery system and a small feed hopper for regulating the flow of cementitious material. The system is designed to process soil-like material at a rate of seven to eight tons per hour. The mixer is powered by a 15-horsepower direct drive hydraulic motor. The polysilicates are stored in a 100-gallon stainless steel tank, metered into a 25-gallon water mixing tank and pumped to a series of spray nozzles located at the base of the feed hopper. An automatic sampler, attached to the discharge

FIGURE 1. Polysilicate treatment mobile system units—processing rate 7–8 TPH.

FIGURE 2. STS large scale (60–70 TPH) polysilicate treatment mobile system.

opening, can be operated at preset intervals for the collection of representative samples. The gray trailer, shown in its upright assembled form, is used as the storage and delivery system for cementitious material. This unit is a modified end dump trailer having a cover, pneumatic loading pipe, fixed end gate with a discharge cone and a small bag-house-like dust separator. When the system is in service, a screw conveyor is attached to a valve in the bottom cone and delivers cementitious material to the feed hopper on the blender. This storage trailer can be raised into its upright position prior to loading by means of a wet kit mounted on the mixer trailer which is operated by a power takeoff from the hydraulic unit. The system is designed so that it can be brought to the site and set up within several hours. Appropriate water and power utilities are required for operation. The power can be either in the form of a connection to existing on-site utilities or by means of an auxiliary portable diesel engine generator unit.

The other mobile system has a processing capacity in the range of 65 tons per hour. It is operated by Solid Treatment Systems Inc. which is a wholly-owned division of the BKK Corporation. Although it embodies the basic elements of the polysilicate treatment technology, it has features that differ from the smaller mobile system. As seen in Figure 2, the overall treatment apparatus is also contained on two trailers. The basic treatment unit trailer consists of twin feed hoppers, a twin screw pug mill, cementitious material storage silo, and a discharge conveyor. The storage silo is hydraulically elevated (Figure 3) after the unit arrives on the site. A diesel engine generator set mounted at the rear of the trailer allows the unit to be self-contained. Direct loading into the feed hopper can be accomplished with a front-end loader. The erected unit is shown in Figure 4 processing heavy metal contaminated soil at a rate of approximately thirty tons per hour. The unit contains a certified belt scale on the feed conveyor which allows a continuous monitoring of the feed rate and cumulative tons of material processed. The polysilicate de-

FIGURE 3. STS system mobilization.

livery system is arranged in a separate trailer shown in the background of Figure 2. A 2000-gallon storage tank allows the polysilicate and water blend to be delivered to spray nozzle system at the point of material entry into the pug mill. The flow system can be operated in either a batch or continuous mode. For continuous operation, the polysilicates are injected into the holding or buffer tank by means of a calibrated metering pump from two, 250-gallon tanks. The 55-gallon polysilicate drums can be pumped directly into these tanks from outside the trailer. Irrespective of the water line hookup, the treatment unit can be put into service within four or five hours after arrival at the site.

TREATMENT APPLICATION RESULTS

A procedure has been established for implementing treatment that typically consists of obtaining representative samples of the material, usually a 5-gallon size, and performing a laboratory-scale treatment. A number of treatment protocols are conducted on a series of 500-gram samples, and an optimized version is then selected for a full-scale field implementation. This also allows the treatment costs to be established consistent with the type and concentration of heavy metals in the material requiring treatment. Basically, the laboratory results serve as a good starting point. Experience has shown that some further adjustments are often necessary in the field when the treatment begins to operate at the multiton-per-hour range. For example, before treating a 5000-ton pile of soil, an initial run would be made on a 100 to 200-ton sample with an appropriate laboratory analysis verification.

Since 1985, when the polysilicate treatment began being applied to various classes of wastes, several hundred test results have been obtained and assembled into a treatment data base. Instead of dealing with laboratory results, attention will be focused on results obtained from the field. These will consist of results from both the fixed in-line systems and the mobile systems that have been operated on a demonstration and commercial basis. Examples will include auto shredder residue, contaminated soil, arc furnace dust, and incinerator ash.

Auto Shredder Residue

Auto shredder residue is the by-product of recycling scrap or junk automobiles through a shredder or hammermill and removing the ferrous components. This method of dealing with scrap automobiles started around 1962. Prior to this time, the common practice was to remove the radiator and battery, cut the engine, transmission, and frame from the light steel body, and open air burn the hull to remove the combustibles. The metal pieces were cut and sold as heavy melting scrap and the body was baled. As the method of steel making shifted from the open hearth to the oxygen furnace, the level of copper, zinc, lead, aluminum, nickel, and chromium impurities in the bundled scrap caused its usage to be reduced to the point where the so-called old style method of recycling automobiles was no longer viable. Thus, the shredding method with subsequent recovery of ferrous and nonferrous components became the accepted practice. The remaining materials such as the glass, plastics, rubber, foam seat cushions, and some nonferrous components constitute a material known as auto shredder residue.

The composition of the residue has changed since the implementation of the shredding method. For example, between the 1960s to the 1980s the total fer-

FIGURE 4. STS unit processing material at 30 TPH.

rous metals have decreased from about 86 to 76 percent, the nonferrous metals such as aluminum, copper, and zinc die cast have increased from 4 to 6 percent and the nonmetals have increased from 10 to 18 percent. In the nonmetals category, plastics have registered the largest increase, i.e. from less than 1 to nearly 6 percent. On the whole, the residue accounts for between 20 to 25 percent of the infeed to the mill.

Around 1984, the DHS declared that auto shredder residue must be managed as a hazardous waste because it failed to pass the CAM standards. This was primarily due to elevated soluble threshold limit concentration levels (STLC levels) of lead, zinc, cadmium, and to a lesser extent copper chromium, and nickel. Depending upon the nature of the scrap being processed, polychlorinated biphenyls (PCBs), particularly aroclor 1254 and 1260, can also exceed the standards. It is interesting to note that in terms of the heavy metals, the residue will usually pass the EP tox standards so that the material may be considered as nonhazardous in regions outside of California. However, PCBs have been a problem with auto shredder residue in other parts of the country, particularly on the east coast. The PCB problem is thought to arise from simultaneous shredding of white goods, transformers, etc.

The in-line processing system has been installed in five of the eight auto shredding plants in California. Extensive testing indicated that the heavy metals problem resides in the fine-size materials. Consequently, a screening system is included as part of the overall treatment system. In order to reduce the levels of heavy metals in the residue, the California auto shredders in concert with DHS require that tail pipes and mufflers be removed. They also require that gas tanks be removed. With the treatment, the tail pipe and muffler restriction may be rescinded.

A set of typical treatment results are summarized in Table 1. Data taken over a three-year period at a plant shredding only auto bodies indicates the following STLC CAM levels are applicable to the heavy metals in the residue: lead, 100 to 200 mg/L; zinc, 400 to 600 mg/L; cadmium, 2 to 4 mg/L; copper, 1 to 5 mg/L; nickel, 5 to 15 mg/L; chromium, 5 to 10 mg/L. A comparison of the data from the three facilities shows a consistency in the feed material to the treatment system. The polysilicate treatment successfully reduces all levels to below the standard with reductions in some cases by more than an order of magnitude. The variation in reduction levels is a function of the manner of treatment application; that is, the quantity of polysilicate and cementitious material used in the treatment. In general, higher levels would represent a treatment protocol to be optimized for cost-effectiveness consistent with meeting the standard.

Contaminated Soil

The polysilicate treatment technology has been used to treat heavy metal contaminated soil at two sites in California. The first project involved a demonstration of the technology on 100 tons of soil

Table 1. STLC Results of Polysilicate Treatment for Auto Shredder Residue

Auto Shredder		Lead (mg/L)		Cadmium (mg/L)		Zinc (mg/L)		Nickel (mg/L)		Chromium (mg/L)		Copper (mg/L)	
Unit	Sample Series	Initial	Final	Initial	Final	Initial	Final	Initial	Final	Initial	Final	Initial	Final
I	1	82.8	17.0	2.53	0.22	665	90.1	3.73	0.26	3.98	2.10	0.15	0.98
	2	60.5	19.0	1.60	0.16	485	81.5	2.3	0.54	2.95	2.60	0.21	0.80
	3	49.5	25.0	1.28	0.67	170	74.5	1.88	0.59	2.40	1.70	0.25	0.29
	4	27.5	15.3	0.63	0.43	205	76.0	0.55	0.34	1.90	1.08	0.18	0.51
	5	30.5	4.95	0.69	0.04	345	9.8	1.45	0.71	2.05	1.90	0.19	1.30
	6	38.5	12.4	1.30	0.18	535	55.5	2.5	0.54	2.95	2.90	0.11	0.85
II	1	168	6.65	4.0	0.21	482	4.80	6.80	1.70	8.2	3.2	1.5	7.4
	2	109	1.80	3.3	0.10	315	1.50	5.3	1.10	6.3	3.5	0.45	4.1
	3	156	4.95	3.8	0.23	418	5.3	7.6	1.75	6.5	2.7	2.60	5.8
III	1	32	0.34	2.8	0.13	820	210	7.0	3.4	7.6	4.9	ND	ND
	2	110	ND	2.3	ND	670	170	5.1	2.1	6.2	3.2	ND	ND
	3	130	0.54	3.5	0.23	500	240	3.8	3.0	3.7	3.4	0.03	0.34
STLC CAM Standard		5 mg/L Regulation 50 Auto Shredders		1 mg/L		250 mg/L		20 mg/L		560 mg/L		25 mg/L	

at the port of Los Angeles using the small mobile system. Following this successful demonstration, the process was applied on a commercial basis using the large unit to the treatment of 5000 tons of soil at a steel manufacturing facility.

The soil treated in the demonstration project was taken from a site that was formerly used for scrap metal operations. Because of these prior operations, the principal metals of concern were lead, zinc, cadmium, copper, and nickel. A size distribution analysis of the soil prior to treatment revealed a wide range in size covering at least two orders of magnitude, i.e., from 0.1 to greater than 10 millimeters. Large or oversize material, that is, stones, rocks, and pieces of metal remaining from past operations, required removal prior to the actual polysilicate treatment. As shown in other studies, the heavy metal contamination resided in the fine-size material. As was the case in prior applications, the oversize fractions did not require treatment.

The mobile treatment system was moved to the site and powered with a mobile diesel engine generation set. A water line from a nearby hydrant was installed with appropriate connections to the polysilicate delivery tank. The treatment equipment including a feed hopper and screen is shown in Figure 5 in its assembled or arranged format on the site. The basic steps in the treatment sequence are as follows:

a. The initial phase is material conditioning, which involves the adjustment of the size distribution. As previously mentioned, the heavy metals problem resides in the fine size material. In the case of soil, various components such as rocks, stones, and pieces of wood must be removed prior to entering the treatment unit blender. The soil requiring treatment is loaded into a feed hopper with a front end loader and conveyed to the screen. At the screen, the soil is separated into two fractions, the oversize components are removed, and the fine undersize soil is put into a form for treatment.

b. The next phase of the treatment is concerned with chemical delivery. As shown in Figure 5, the discharge conveyor underneath the screen delivers material requiring treatment to the feed hopper on the blender. As the material enters the blender feed hopper, it becomes wetted by the polysilicate solution. In effect, the feed hopper on the blender is actually a spray chamber where the initial contact is made between the polysilicate and the material. The chemical delivery system is versatile in that the amounts of silicate can be adjusted to the feed rate of

FIGURE 5. Processing heavy metal contaminated soil at Port of Los Angeles.

material independent of the dilution water. High-pressure pumps deliver the diluted polysilicate water blend to the spray nozzles on the blender feed hopper. These pumps, operated independently, can supply different sets of spray nozzles and are individually activated depending upon the feed rate and moisture content of the material entering the blender.

c. The mixing phase consists of material transversing through the pug mill blender. There are three zones in the mixing chamber: the initial zone where the material continues to be wetted, i.e., the polysilicate and material are brought into intimate contact in a thoroughly wetted situation short of becoming a slurry; the mixing zone where the cementitious material is added and mixed into the wetted soil; and the exit zone where the material is discharged through an automatic sampler. As shown in Figure 5, the cementitious material is added at approximately the midpoint of the mixer. This material is transported by screw conveyor from the storage trailer to the hopper feeder on the blender trailer and into the mixer. The final processing occurs as the cementitious material is combined with the silicate wetted soil. A representative sample of the treated material can be collected by automatic sampling of the blender discharge at preset intervals consistent with the material throughput. This treatment was operated at a rate of about 7 to 8 tons per hour so that on the order of a 1-pound sample was removed and dropped into a collection box or 5-gallon container at each sampling interval.

d. The final step in the treatment is the curing or drying of the material. The treated material leaving the blender via a discharge conveyor is basically placed in piles which are turned with a front-end loader until dry. The mean size would be one to two orders of magnitude larger than the original untreated soil.

The total amount of material for treatment was taken from six zones on the site. Approximately 20 tons of material was taken from each zone and treated separately. As previously mentioned, the soil being treated contained elevated levels of lead, zinc, cadmium, copper, and nickel. The results of the treatment on the six separate piles assembled from the sites are summarized in Table 2. With the exception of copper, the reduction in the STLC levels was well over 90 to 100 percent. In general, the treatment was not as effective on copper where the highest reduction was 87 percent.

The commercial treatment of the soil using the large-scale sixty-five-ton per-hour STS unit followed a procedure similar to the previously described work at the port of Los Angeles. The actual equipment arrangement has already been given in Figure 2. This soil that was contaminated with heavy metals, lead, zinc, and cadmium from previous scrap metal handling operations also contained some partially treated bag house dust. As the aftermath of a previously unsuccessful treatment, the dust, mixed with lime and water, hardened into nuggets and was buried in the soil requiring treatment. The nuggets constituted a relatively small percentage of the overall material and were removed along with rocks, debris, etc. by screening prior to treatment.

In order to verify the efficiency of the treatment at the higher rates, which were essentially an order of magnitude higher than previous demonstrations of the technology, an initial run was conducted in which 200 tons of material were treated at a rate of 40 tons per hour. The test consisted of four 50-ton runs. Six individual treatment samples were collected before and after each run. The results of individual results were in good agreement with composite samples for the same runs. Having established the treatment, the remainder of the material was treated at a rate of about 65 tons per hour. A compilation of the overall data showed that the initial concentrations (mg/L), as measured by the CAM test, of lead, zinc, and cadmium were typically on the order of 200, 1500, and 5 to 7 respectively. The concentrations (mg/L) of the treated material were on the average of 2 for lead, 16 for zinc, and 0.35 for cadmium.

Arc Furnace Dust

A series of tests designed to evaluate the feasibility of applying the polysilicate technology to the treatment of bag house or electric arc furnace dust (K061) were conducted at Cascade Steel using the small mobile system. Although the equipment arrangement shown in Figure 6 was similar to the previous studies, no screening system was required. In addition, a special removable feed hopper having a capacity of about 3 tons was constructed so that dust could be collected at the bag house discharge. The hopper was placed over the mobile system spray chamber and metered into the treatment unit through an adjustable

Table 2. Field Test Results for Soil from the Port of Los Angeles

Sample Pile	STLC Concentrations*									
	Lead		Zinc		Cadmium		Copper		Nickel	
	Initial	Final	Initial	Final	Initial	Final	Initial	Final	Initial	Final
1	213	5.01	528	4.3	2.1	0.03	127	24.8	18.2	1.9
2	149	13.0	443	13.0	1.6	0.11	92.5	29.9	10.9	2.1
3	127	0.5	258	0.06	2.1	0.01	70.5	15.9	9.7	0.80
4	86	1.9	225	4.6	0.9	0.04	113	33.1	28.7	2.6
5	152	1.6	454	18.2	2.0	0.02	128	26.7	28.6	1.4
6	107	3.3	242	1.2	1.5	0.1	91.8	12.3	19.2	1.2

*CAM test results.

FIGURE 6. Treatment units and special feed hopper for processing bag house dust.

bottom slide door and vibrator arrangement. The nature of this material is such that manageable treatment rates were on the order of 2 tons per hour. In this material, lead was the heavy metal of concern. The results of the tests, summarized in Table 3, showed that the material could be treated so that the EP tox concentration of lead would be below 5 mg/L. The tests showed a continuously varying concentration of lead in the feed ranging between 580 to 890 mg/L. In the absence of knowledge of the initial concentration, the treatment would have to be adjusted for the worst case situation in order to insure successful treatment. Extensions in the treatment are also being made in order to comply with the recent toxicity characteristic leaching procedure (TCLP) standards promulgated by EPA.

The emerging polysilicate technology has proven to be an effective low-cost method for the mitigation of heavy metal contaminated materials. The economics of rendering materials nonhazardous is favorable compared to conventional forms of disposal. Significant reductions in metal leachability are achieved with minimal increases in volume or weight. Typically, the treatment will shift the size distribution toward a larger mean particle size that will greatly reduce the effect of total metal concentrations on airborne pathways. Continuing efforts are being made to expand the treatment data base and extend the range of applicability in order to achieve lower final leachable metal concentrations.

Incinerator Ash

The polysilicate treatment has been successfully applied to the treatment of incinerator ash generated in a mass burning municipal solid waste to energy facility. The large STS mobile system was used for the treatment which enabled ash to be treated at a rate of 60 tons per hour. This material, containing a combination of the bottom ash and fly ash, was screened prior to treatment so that only material passing a one-inch screen entered the unit. The principal heavy metals of concern were lead, zinc, and cadmium. Initial test results indicated the following STLC CAM level reductions: lead, from 14 to 40 mg/L to 1 to 2 mg/L; zinc, from 350 to 600 mg/L to 1 to 10 mg/L; and cadmium, from 2 to 5 mg/L to 0.2 to 0.3 mg/L.

CONCLUDING REMARKS

The emerging polysilicate technology has power to be an effective low-cost method for the mitigation of heavy metal contaminated materials. The economics of rendering materials nonhazardous is favorable compared to conventional forms of disposal. Significant reductions in metal leachability are achieved with minimal increases in volume or weight. Typically, the treatment will shift the size distribution toward a larger mean particle size that will greatly reduce the effect of total metal concentrations on airborne pathways. Continuing efforts are being made to expand the treatment data base and extend the range of applicability in order to achieve lower final leachable metal concentrations.

Table 3. Electric Arc Furnace Dust Treatment

Test Series	Quantity Treated (lb)	Concentration of Lead (mg/L)*	
		Initial	Final
1	3950	580	26
	3950	580	31
	2300	580	1.7
2	2990	580	0.98
	3450	580	0.35
	4000	890	37
3	4000	890	45
	3500	890	32

*EP tox test.

2.17

Electrochemical Oxidation of Refractory Organics

Brian G. Dixon,[1]* Myles A. Walsh,[1] R. Scott Morris[1]

INTRODUCTION

The contamination of the nation's water resources by organic and inorganic chemicals is a serious and growing problem. As the population continues to expand, development pressures persist and the supply of clean water dwindles. These simple facts make the recycling and purification of all available water of paramount importance. It has been projected that the industrial portion of systems wastewater treatment alone will grow by 11 percent per year at least through the year 1995 [1]. Clearly new treatment processes will be required and developed to supplement established technologies. Many effective techniques to remove organic chemicals from water have been proposed over the years. The best of these technologies are widely employed and include adsorption on activated carbon, thermal or chemical oxidation (incineration, ozonation), air stripping, filtration (membranes, filtration-coagulation), and biological removal. Of the variety of developing processes, catalytic oxidation of pollutants has one of the most promising futures [2]. Thermal, electrochemical, and photochemical catalytic techniques are being developed with the hope that they can significantly reduce the costs of organic chemical destruction to innocuous products. Catalysis offers a number of potential advantages over established oxidative techniques. Some of these advantages are low-temperature operation, faster reaction rates and the possibility of the direct activation of molecular oxygen. This latter advantage could minimize the need for using more expensive and hazardous chemical oxidants such as potassium permanganate, chlorine dioxide, hydrogen peroxide, and ozone. The following report summarizes the results of a research and development program designed to determine the practicality of using a capacitively driven electrochemical cell, in the absence of added electrolyte, to remove organic pollutants from water.

APPLICABLE WASTE STREAMS

The overall approach of the technology lends itself to a variety of potential waste streams. Potential applications lie anywhere that the presence of organic compounds in water represents a pollution problem. The system is designed such that the electrochemical cell construction could be scaled up for larger flow volume waste streams. Since the system is based upon catalytic decomposition of pollutants, care must be taken to assure that catalyst poisoning is relatively slow and that the catalyst can be regenerated. For most applications these constraints will, in all likelihood, limit the technology to relatively low pollutant concentrations. On the other hand, the adsorptive-catalytic nature of the process will allow for the removal and degradation of undesirable compounds at extremely low concentrations. Particular targets of this research were aromatic hydrocarbons, like those found at petroleum spill sites, and chlorinated solvents at concentrations in the parts per million.

PROCESS DESCRIPTION

The primary objective of the research program was to establish the practicality of oxidizing model

---

*Author to whom correspondence should be addressed.

[1]Cape Cod Research, Inc., Box 600/95 Main Street, Buzzards Bay, MA 02532

organic pollutant compounds as a function of an impressed AC electric field [3,4]. A key part of this ambitious undertaking was to demonstrate that this electrooxidation could be accomplished *in the absence* of a dissolved electrolyte. The program was split into two fundamental and more or less equal parts (described in Sections 4 and 5, respectively). In the first part, cyclic voltammetry was used to evaluate various supported electrocatalysts for their effectiveness in initiating the reactions of designated pollutants. The results of these screening tests would then establish the most effective catalyst-support system for a given pollutant. The second part of the program involved the design and construction of experimental electrochemical test cells which, when filled with the appropriate catalyst-support and separator materials, would actually oxidize (or reduce) designated pollutant compounds, hopefully to innocuous decomposition products, in an efficient manner. Following such demonstration on a small scale, larger flow volume apparatus could be designed and built to establish the scaleup capabilities of the process as a whole. More specifically, answers to the following questions were sought:

- Can cyclic voltammetry (CV) be used to identify which electrocatalysts are active for a given pollutant?
- Can the CV screening results be used to predict effective catalytic materials for use in an electrochemical flow cell?
- Can the catalysts be easily immobilized onto various support surfaces?
- Are the supported electrocatalysts effective in removing organics from water in the absence of added electrolytes?
- How does the variation of cell parameters such as support material, flow rate, and surface area change the observed results?

As will be shown, positive answers to these questions have established the feasibility of the overall approach. In addition, these answers lead to insight into the techniques needed for the construction of an effective water treatment electrochemical reactor.

CATALYSTS, POLLUTANTS & ELECTROCHEMISTRY

Catalysts-Preparations & Evaluations

Although the various catalysts were immobilized onto a number of different support materials, the basic procedures used for a given catalyst were much the same. Therefore, a general synthetic procedure for each catalyst will be described. In all cases, the weight of the support before and after catalyst loading was obtained. Nickel-coated diatomaceous earth and 5 percent ruthenium on carbon were used as received from Alfa Chemicals and Engelhard Industries, respectively. The supports evaluated include activated carbons, graphite, diatomaceous earth, ceramics, and various oxides. In addition, a number of forms of these materials were tested including powders, pellets and, where available, fabrics.

Iron (III) Tetraphenylporphyrin (FeTPP)

FeTPP (Stream Chemicals) was first dissolved in tetrahydrofuran (THF). The supports were mixed into small quantities of this solution and the slurry placed in an ultrasonic bath for ten minutes. This mixture was left to sit for one hour at room temperature and then the THF was evaporated off, first by using a rotary evaporator and then by drying the particulate under vacuum for one hour at ~40°C. The resultant supported catalyst was then activated [5] by heating in argon for one hour at 400°C in a sealed quartz container. This procedure polymerizes the monomeric chelate into an insoluble organometallic complex. Such porphyrin catalysts are known to activate oxygen, in the presence of water, to yield hydrogen peroxide by the mechanism shown in Scheme 1.

Immobilized 2-aminoanthraquinone

Covalently bonding anthraquinone onto high surface carbon blacks involved two steps as shown in Scheme 2. First the carbon black was brominated by adding excess liquid bromine to the black in a sealed flask and heating to 255°C for three hours. This brominated material was then mixed into a flask containing 2-aminoanthraquinone (AAQ) and aluminum trichloride dissolved in dry chloroform. After refluxing for eight hours at 65°C, the catalyst was filtered and washed with dry methanol to remove any excess 2-AAQ. The resulting catalyst was vacuum dried for greater than one hour at 40–60°C prior to use in order to remove residual methanol or chloroform.

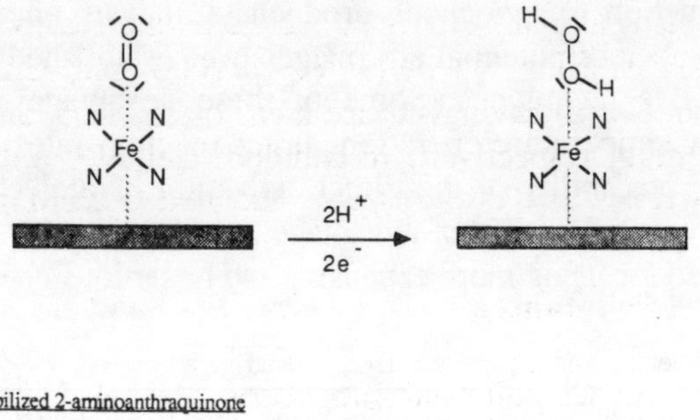

Immobilized 2-aminoanthraquinone

SCHEME 1.

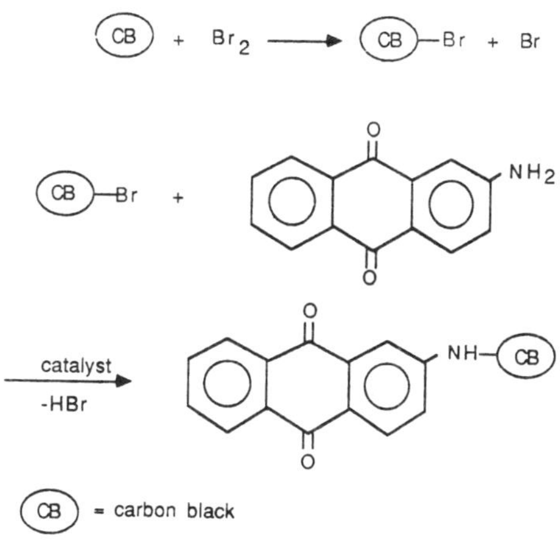

SCHEME 2.

PbO_2, NiO, MnO_2

The preparation of these oxides onto a desired support was carried out by first preparing a suitable solution of the desired metal nitrate in methanol, adding the support to make a slurry and eventually removing the solvent in vacuo. The resultant metal nitrate impregnated support was then heated to pyrolyze the nitrate at 250–500°C to give a supported active metal oxide catalyst.

Ruthenium–Titanium Spinel

$$(TiO_2)_x (RuO_2)_{1-x}$$

The ruthenium–titanium spinel is a mixed metal oxide that was synthesized by first preparing Beer's solution as follows [6]: 3g $RuCl_3 \cdot 3H_2O$ was dissolved in 1.2 mL HCl (37%); 18.6 mL of n-butanol and 9 mL tetra n-butyl titanate were added to form a solution. This mixture was stirred for at least one hour at ambient temperature. The desired surface was then coated with this mixture and heated at 500°C for 2 minutes. This procedure was repeated as necessary to give the desired degree of loading. The support materials evaluated included three carbon blacks, Sterling R, Black Pearl 2000 and CSX (Cabot Corp.) having surface areas of 25, 1475, and 1475 m²/g, respectively. In addition, Celite™, a diatomaceous earth, was used as obtained from Manville Corp.

Model Pollutants

Five model pollutant compounds were evaluated with each catalyst system. These compounds were phenol, toluene, chloroform, trichloroethane, and diethylphthalate. Each pollutant was tested at 50 and 500 parts per million (ppm) in water. The latter was prepurified by passage through an activated carbon column followed by deionization. A noteworthy observation was made in that a yellow discoloration of the counter electrode was observed during the analyses with phenol. This color is probably due to quinonic decomposition products which are formed on the electrode surfaces.

Cyclic Voltammetry: Analytical

To rapidly screen a large number of combinations of substrate, catalyst, test conditions, and model organic compound to be electrooxidized, an analytical rotator (Pine Instruments ASR 2) was adapted to accept graphite disks. These disks were made of spectrographic grade graphite (Ultra Carbon Corp.) and when mounted in the rotator had a disk area of 1.13 cm².

In order to test the effectiveness of these catalyst-supports for electrooxidations of organic materials, they must first be immobilized onto the graphite disks. The catalysts were each ultrasonically mixed with an aqueous Teflon dispersion (ICI AD-1) in a 70 weight percent catalyst:30 weight percent Teflon™ ratio. Each mixture was sprayed onto six identical disks to produce a very thin surface layer which was then dried and heated in argon to 230°C to sinter the Teflon and bind the catalyst to the disk's surface.

Catalyst testing involved placing the catalyzed disk in the analytical rotator, prewetting the catalyst with sulfuric acid, rinsing repeatedly with deionized water, and then lowering the rotating disk into the test cell.

The test cell consisted of a beaker covered with a lid which is penetrated by a double junction saturated calomel electrode (SCE), by a port for injecting precise amounts of model organic compound (phenol, toluene, chloroform, trichloroethane, or diethylphthalate), by a platinum wire connected to a counter electrode (a 1 cm² smooth platinum flag), and by the shaft of the analytical rotator. The test solution was initially at pH = 7 in a phosphate buffer containing 0.1M KCl. To this was added aliquots of buffer containing the organic pollutant to make up final solutions containing 50 to 500 ppm of the organic. Cyclic voltammograms (CV) were generated using an ECO Model 567 function generator which drives an ECO Model 551 potentiostat. The voltammograms were recorded using an ECO Model 862 XY recorder.

Cyclic Voltammetry: Results and Discussion

Tables 1 and 2 and Figure 1 contain the results of

Table 1. Cyclic Voltammetry of Support Activities*

Support	Pollutant	Ppm (H₂O)	Anodic (mA) 100 mV	500 mV	800 mV	Cathodic (mA) 100 mV	500 mV	800 mV
Black Pearl 2000	Phenol	50	−0.19	0.06	0.12	0.19	0.37	**0.62**
		500	−0.19	0.19	**0.50**	−0.06	**0.50**	**1.37**
	Toluene	50	−0.06	−0.31	−0.44	0.25	0.12	−0.31
		500	−0.25	**−0.50**	**−0.62**	0.37	0.25	−0.37
	Chloroform	50	−0.06	−0.13	−0.13	0.13	0.06	0.00
		500	−0.13	0.19	0.25	0.13	0.06	0.00
	Trichloroethane	50	0.00	−0.06	−0.06	0.19	0.13	0.00
		500	−0.06	−0.19	−0.13	0.25	0.19	−0.06
	Diethylphthalate	50	0.13	−0.13	−0.13	0.19	0.13	−0.13
		500	0.06	−0.31	−0.31	0.31	0.19	−0.19
Sterling R	Phenol	50	0.06	0.12	0.37	−0.37	0.12	**0.50**
		500	−0.12	0.19	0.12	−0.31	−0.31	**0.75**
	Toluene	50	−0.06	−0.06	−0.06	0.05	0.00	−0.12
		500	−0.19	−0.25	−0.19	0.25	0.12	0.06
	Chloroform	50	−0.25	−0.38	−0.19	−0.19	−0.19	−0.38
		500	−0.13	−0.38	−0.44	−0.19	−0.25	**−0.50**
	Trichloroethane	50	0.06	−0.06	−0.13	0.13	−0.06	−0.13
		500	0.06	−0.13	−0.19	0.25	0.00	−0.13
	Diethylphthalate	50	0.13	−0.25	−0.25	0.31	−0.13	−0.31
		500	0.13	−0.38	−0.44	0.38	−0.13	−0.38
CSX	Phenol	50	−0.12	0.12	0.19	−0.19	0.25	**0.50**
		500	−0.12	0.31	**0.62**	−0.19	**0.69**	**1.31**
	Toluene	50	−0.12	−0.19	−0.31	0.12	−0.06	−0.25
		500	−0.12	−0.25	−0.44	0.12	−0.06	−0.44
	Chloroform	50	−0.06	−0.13	0.25	0.19	−0.06	−0.06
		500	−0.19	0.19	0.19	−0.13	−0.25	−0.38
	Trichloroethane	50	−0.06	−0.25	**−0.50**	0.38	0.06	−0.19
		500	−0.06	−0.44	**−0.83**	0.56	0.06	−0.38
	Diethylphthalate	50	0.00	−0.06	0.00	−0.06	−0.19	0.06
		500	0.00	−0.06	−0.19	0.06	−0.13	−0.19
Diatomaceous Earth	Phenol	50	0.00	0.37	**0.56**	−0.25	0.25	0.31
		500	0.06	0.75	**1.19**	−0.50	0.50	**1.31**
	Toluene	50	−0.19	−0.06	−0.19	0.19	0.06	−0.12
		500	0.00	−0.31	**−0.56**	0.37	0.18	−0.25
	Chloroform	50	0.13	−0.06	−0.06	0.19	−0.13	−0.13
		500	0.19	−0.13	−0.13	0.13	−0.19	−0.25
	Trichloroethane	50	−0.13	−0.06	−0.06	0.13	0.00	0.00
		500	−0.13	−0.06	−0.13	0.13	−0.06	−0.06
	Diethylphthalate	50	0.31	−0.06	−0.19	0.31	−0.31	−0.25
		500	**0.50**	−0.13	0.25	0.31	−0.38	−0.49

*Conditions and supports as described in Table 2.

Table 2. Cyclic Voltammetry of Pollutant Catalyst Activity*

Catalyst**	Pollutant	Ppm (H₂O)	Anodic (mA) 100 mV	500 mV	800 mV	Cathodic (mA) 100 mV	500 mV	800 mV
FeMTPP/bp	Phenol	500	**0.62**	**0.50**	**0.56**	−0.12	0.31	**0.56**
	Toluene	500	0.38	**−0.75**	**−0.94**	**1.19**	0.31	**−0.62**
	Chloroform	500	−0.06	−0.19	**−0.50**	0.12	−0.12	−0.38
	Trichloroethane	500	−0.37	−0.31	−0.37	0.44	0.12	−0.06
	Diethylphthalate	500	0.19	0.00	−0.12	−0.12	−0.06	0.25
FeMTPP/sterling	Phenol	500	**0.62**	−0.31	**−0.50**	**1.06**	**1.12**	**0.81**
	Toluene	500	**−1.44**	0.06	**0.62**	−0.44	0.12	**0.75**
	Chloroform	500	**0.75**	**0.87**	0.06	−0.25	**−0.75**	−0.37
	Trichloroethane	500	−0.12	**−0.62**	**−1.00**	0.56	0.25	−0.37
	Diethylphthalate	500	−0.19	−0.37	−0.44	0.00	−0.25	−0.44
FeMTPP/csx	Phenol	50	0.25	0.06	0.12	0.37	0.31	0.19
		500	0.25	−0.19	−0.12	**0.94**	**0.87**	0.25

Table 2. (continued)

Catalyst**	Pollutant	Ppm (H₂O)	Anodic (mA)			Cathodic (mA)		
			100 mV	500 mV	800 mV	100 mV	500 mV	800 mV
	Toluene	50	0.00	−0.06	−0.12	0.12	0.06	0.00
		500	0.00	−0.25	−0.19	0.12	0.00	−0.12
	Chloroform	50	−0.12	−0.25	−0.12	0.12	0.19	−0.06
		500	−0.06	−0.37	−0.25	0.25	0.25	−0.19
	Trichloroethane	50	0.12	−0.25	−0.31	0.31	0.19	−0.06
		500	0.19	−0.25	−0.44	0.37	0.12	−0.25
	Diethylphthalate	50	0.37	−0.06	−0.25	0.12	−0.12	−0.25
		500	**0.50**	−0.12	−0.37	0.37	−0.19	**−0.50**
FeMTPP/celite	Phenol	50	0.12	0.19	0.31	−0.13	0.19	0.25
		500	0.12	0.37	0.75	−0.06	**0.50**	1.06
	Toluene	50	−0.12	−0.06	−0.05	0.12	0.12	0.00
		500	−0.19	−0.12	−0.12	0.25	0.18	0.00
	Chloroform	50	0.00	−0.06	−0.06	0.06	0.06	0.00
		500	0.00	−0.06	−0.06	0.13	0.06	0.00
	Trichloroethane	50	0.00	−0.13	−0.06	0.13	0.13	−0.06
		500	0.06	−0.19	−0.25	0.31	0.25	−0.06
	Diethylphthalate	50	0.06	−0.25	−0.31	0.25	0.00	−0.19
		500	0.13	−0.31	−0.44	0.44	0.13	−0.13
NiO/bp	Phenol	500	0.12	−0.25	0.06	0.12	**0.75**	**0.81**
	Toluene	500	0.12	−0.25	−0.37	0.25	0.00	−0.37
	Chloroform	500	0.25	0.00	**−0.79**	0.12	0.06	−0.25
	Trichloroethane	500	−0.06	−0.06	−0.19	0.12	0.00	−0.19
	Diethylphthalate	500	0.00	−0.31	−0.25	0.25	−0.12	−0.06
NiO/sterling	Phenol	50	−0.12	0.25	0.25	−0.31	0.12	**0.50**
		500	0.00	0.19	0.37	−0.12	0.25	**0.87**
	Toluene	50	0.12	−0.19	−0.25	0.12	0.06	−0.12
		500	0.25	−0.44	**−0.56**	0.50	0.12	−0.31
	Chloroform	500	−0.06	−0.12	−0.25	0.12	0.00	−0.19
	Trichloroethane	50	0.12	−0.37	**−0.50**	0.31	0.12	−0.25
		500	0.19	**−0.50**	**−0.75**	0.50	0.19	−0.44
	Diethylphthalate	50	−0.06	−0.12	−0.19	0.06	0.00	−0.12
		500	0.00	−0.12	−0.25	0.12	0.00	−0.19
NiO/csx	Phenol	500	−0.06	0.00	**0.69**	0.31	**0.81**	**1.50**
	Toluene	500	0.00	−0.12	−0.12	0.00	0.00	−0.06
	Chloroform	500	0.00	−0.12	−0.19	0.06	−0.06	−0.25
	Trichloroethane	500	0.12	0.12	0.00	0.00	−0.06	−0.12
	Diethylphthalate	500	0.12	**−0.75**	−0.12	−0.12	0.06	−0.31
Ni/celite	Phenol	500	0.31	0.19	**0.94**	0.19	**0.69**	**1.37**
	Toluene	500	0.12	−0.19	−0.19	0.25	0.12	−0.12
	Chloroform	500	0.06	−0.11	−0.25	0.25	0.00	−0.37
	Trichloroethane	500	−0.06	−0.37	**−0.50**	0.25	0.00	−0.37
	Diethylphthalate	500	0.25	−0.31	**−0.50**	**0.75**	0.31	−0.12
2AAQ/bp	Phenol	500	0.12	**0.56**	**2.00**	**−0.56**	**0.62**	**1.88**
	Toluene	500	−0.06	0.06	0.06	0.19	0.19	0.12
	Chloroform	500	0.12	−0.19	−0.25	0.44	0.19	0.00
	Trichloroethane	500	0.06	−0.12	−0.19	0.12	0.00	−0.37
	Diethylphthalate	500	0.37	−0.06	−0.19	**0.69**	0.44	0.19
2AAQ/sterling	Phenol	50	−0.08	0.06	0.20	0.13	0.13	−0.06
		500	0.06	**0.81**	**0.94**	−0.31	0.38	**1.63**
	Chloroform	50	0.06	−0.13	−0.13	0.13	0.06	−0.13
		500	0.19	−0.13	−0.13	0.31	0.25	−0.06
	Trichloroethane	50	−0.19	−0.38	−0.38	0.44	0.31	−0.13
		500	−0.19	−0.44	**−0.51**	**0.56**	0.38	−0.20
	Diethylphthalate	50	−0.26	−0.25	−0.38	0.38	**0.50**	0.07
		500	−0.32	−0.38	**−0.50**	**0.56**	**0.63**	−0.06
2AAQ/csx	Phenol	500	0.37	0.37	**1.19**	0.31	**0.69**	**1.37**
	Toluene	500	0.06	−0.19	−0.25	0.19	0.06	0.19
	Chloroform	500	−0.06	−0.25	−0.37	0.06	−0.12	−0.31
	Trichloroethane	500	0.00	0.25	0.31	−0.37	−0.25	0.00
	Diethylphthalate	500	−0.19	−0.12	−0.12	−0.19	−0.25	−0.31

(continued)

Table 2. (continued)

Catalyst**	Pollutant	Ppm (H_2O)	Anodic (mA)			Cathodic (mA)		
			100 mV	500 mV	800 mV	100 mV	500 mV	800 mV
2AAQ/celite	Phenol	50	0.00	−0.16	−0.16	−0.47	−0.47	−0.31
		500	0.31	0.16	**0.62**	0.16	**0.62**	**0.94**
	Toluene	50	0.06	−0.31	**−0.50**	0.44	0.19	−0.12
		500	0.12	−0.19	−0.44	0.37	0.12	−0.12
	Chloroform	50	0.12	−0.06	0.25	−0.06	−0.25	−0.31
		500	0.25	−0.19	−0.31	−0.06	−0.31	−0.40
	Trichloroethane	50	0.12	−0.19	−0.37	0.00	−0.19	−0.25
		500	0.25	−0.31	**−0.56**	0.25	−0.19	**−0.56**
	Diethylphthalate	50	0.00	−0.25	−0.44	−0.06	−0.37	**−0.56**
		500	0.00	−0.25	−0.44	−0.06	−0.37	**−0.56**
MnO_2 on CSX	Phenol	500	−0.25	0.19	**0.62**	−0.25	0.31	**0.81**
	Toluene	50	0.06	−0.25	−0.31	−0.31	0.12	−0.06
		500	−0.19	−0.37	−0.16	0.25	0.12	0.00
	Chloroform	500	−0.19	−0.44	−0.44	0.19	0.12	−0.06
	Trichloroethane	500	**−0.50**	−0.06	−0.12	0.00	−0.06	−0.31
	Diethylphthalate	500	0.19	−0.31	**−0.50**	0.19	0.00	−0.25
PbO_2 on CSX	Phenol	50	−0.06	0.12	−0.25	−0.12	0.00	0.25
		500	−0.06	−0.19	0.06	0.06	0.25	0.44
	Toluene	500	−0.19	0.37	**0.62**	−0.44	0.06	**0.50**
	Chloroform	500	0.00	−0.25	−0.25	0.06	0.06	−0.31
	Trichloroethane	500	0.00	0.31	−0.31	0.06	−0.06	−0.25
	Diethylphthalate	500	0.06	−0.19	−0.31	0.06	−0.06	−0.25
5% Ru on C	Phenol	500	0.06	0.06	0.19	0.12	0.12	**0.62**
	Toluene	500	−0.12	−0.12	−0.12	0.19	0.06	−0.06
	Chloroform	50	0.00	−0.25	0.12	0.12	0.06	−0.12
		500	0.19	−0.12	−0.12	0.37	0.19	−0.12
	Trichloroethane	50	−0.06	−0.37	−0.44	0.25	0.00	−0.25
		500	−0.06	−0.37	−0.44	0.19	−0.06	−0.25
	Diethylphthalate	50	0.00	−0.19	−0.25	0.19	−0.06	−0.25
		500	0.00	−0.25	−0.31	0.12	−0.12	−0.31
Ru-Ti Complex	Phenol	500	0.47	**0.62**	**0.94**	0.31	0.47	**0.94**
	Toluene	500	0.48	0.16	−0.31	0.16	−0.31	**−0.62**
	Chloroform	50	**−0.62**	−0.16	0.16	**−0.62**	−0.16	0.16
		500	−0.31	0.31	**0.78**	−0.47	0.47	**1.09**
	Trichloroethane	50	**0.69**	**0.81**	**0.50**	**0.62**	**0.88**	**0.94**
		500	**0.56**	**0.63**	**0.75**	**0.63**	**0.81**	**0.81**
	Diethylphthalate	50	0.31	−0.25	**−0.63**	0.25	−0.31	**−0.88**
		500	**−0.62**	−0.31	**−0.50**	0.06	0.25	**−0.56**

*All CVs were run at 50 mV/sec between −100 to 900 mV. All were run in 0.1 M KCl/ph7 phosphate buffer solution.
**Catalyst Codes: Carbon Blacks: bp: Black Pearl 2000 (Cabot Corporation), CSX: (Cabot Corporation), Sterling R: (Cabot Corporation), Catalyst Loadings: 30 mg catalyst/250 mg support; FeMTPP = iron (III) mesotetraphenylporphyrin; NiO = nickel (II) oxide; Ni = nickel metal; 2AAQ = 2-aminoanthraquinone; MnO_2 = manganese (IV) oxide; PbO_2 = lead (IV) oxide; Ru = ruthenium metal; Ru-Ti = mixed metal oxides, a spinel of ruthenium–titanium.

the parametric screening studies on the various combinations of four supports with and without the eight different catalysts and four different pollutants. As expected, a wide range of reactivity response is observed. It is of interest that the four virgin supports show low levels of activity (Table 1). This activity is not unexpected since the carbon blacks, in particular, are known to have electrochemically active groups, such as quinonyl, carboxyl, and hydroxyl groups, on their surfaces. The kind and number of these groups depend upon the specific manufacturing technique used in making each of the blacks. The other support, Celite™, also has reactive surface functional groups, such as metal oxides, which may function as catalysts as well. In these tables, a negative value indicates reduction and a positive value oxidation. The first column identifies the catalyst on a given support. Columns two and three indicate the pollutant identity and concentration, respectively. The three anodic columns contain the data, in milliamps, of the difference between the anodic current sweep at a given voltage (e.g., 100mV) in the presence and absence of the pollutant. The same is true for the cathodic wave. This is probably made clearer by examination of Figure 1, which represents a typical set of scans for 12 weight percent MnO_2 deposited upon

Black Pearl 2000 carbon black. This figure shows that the electrochemical response is larger in the presence of phenol at 500 ppm, indicating the activity of the immobilized catalyst. The data contained in Table 2 show that the electrochemical response of a given pollutant is very catalyst dependent. For the purposes of the following discussion, mV difference readings of plus or minus 0.5 or greater (bold numbers in Tables 1 and 2) will be considered significant, although this is a somewhat arbitrary cutoff point. Phenol, an easily oxidized substrate, shows a significant response with all of the catalysts, albeit with some only at the higher 500 ppm concentration. As expected, the other pollutants are not nearly as reactive but all show reactivity with at least one catalyst. Surprisingly, the ruthenium–titanium spinel appears to be the most generally effective catalyst, with all of the pollutants giving a significant response. In addition, the FeTPP is effective for toluene, chloroform, and trichloroethane, while 2-amino-anthraquinone showed a significant activity only for diethylphthalate. The latter was found to be the most recalcitrant compound which is not surprising since it is already in a relatively high state of oxidation.

Conclusions: CV Analyses

The CV analyses accomplished what they were designed to do. Correlations between electroactive catalysts and pollutant type were established. These correlations were then used to choose the catalysts for the electrochemical test cell to be described in the next section. It is important to remember that these screening test results are not necessarily directly applicable to the cell studies since the former work was done under very controlled conditions and in the presence of electrolyte. Nevertheless, an overall analysis of the data leads to a number of interesting conclusions:

1. All of the organic pollutant compounds respond electrochemically to more than one catalyst.
2. The Ru–Ti spinel showed remarkable activity with all of the catalysts.
3. Iron (III) tetraphenylporphyrin (FeTPP) also showed significant catalytic activity for all of the compounds evaluated.
4. PbO_2, MnO_2, and Ru catalysts exhibited catalytic activity only for phenol.
5. Phenol and diethylphthalate gave the greatest and least electrochemical responses of the pollutants studied, respectively.

The catalysts used in the flow cell studies about to be described were primarily chosen based upon their electrochemical response and, secondarily, by ease of their immobilization onto a given support material.

ELECTROCHEMICAL FLOW CELL STUDIES

The objective of this second part of the study was to translate the results of the parametric screening tests into an actual electrochemical test cell. This laboratory-scale apparatus would accomplish an effective degradation of organic pollutants in water *in the absence* of any added electrolyte. Such a process had not previously been achieved in a practical fashion.

Cell Design and Construction

The electrochemical cell used for the laboratory tests is shown in Figure 2. Here the driving electrodes were made of either Monel or titanium and were connected to AC power via a rheostat to vary the input power. The solution was pumped into the cell through the inlet at the bottom and out the top. The driving electrodes were connected by bolts insulated with ceramic washers. Between the driving electrodes was a hollow polypropylene (PP) or polymethylmethacrylate (PMMA) chamber that was filled

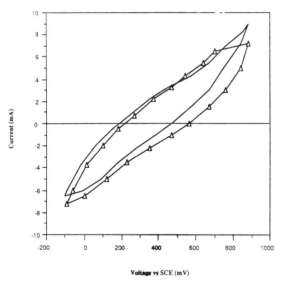

FIGURE 1. Cyclic voltammetry of manganese dioxide catalyzed reaction of phenol.

Conditions: Current Range: 10 mA, Y setting: 0.5 V/in., X setting: 0.2 V/in. Rate: mV/sec., Scan Range: −100 to 900 mV
  12% (wt) MnO_2 on Black Pearl 2000 carbon black
  —△— = 0.1 KCl/Potassium hydrogen phosphate buffer (pH = 7)
  —— = Buffer plus 500 ppm Phenol

FIGURE 2. Electrochemical test cell.

with catalytic or insulator materials as desired. Packed cells held a volume of 100–500 mL of solution. The photograph of Figure 2 shows the transparent PMMA chamber filled with surface catalyzed alumina pellets.

Experimental Variables: Description and Control

This second part of the study involved a great number of experimental variables. These included:

Catalyst: FeTPP, MnO_2, and anthraquinone were the workhorse catalysts. The combination of catalyst and support will hereafter be referred to as the auxiliary electrodes.

Catalyst Support: A total of sixteen support materials were evaluated. They can be split into three basic categories: carbon based (conductive), synthetic polymer, and ceramic (nonconductive). Pellets, spheres, and woven or fabric materials were evaluated.

Pollutant: Phenol, toluene, chloroform, trichloroethane and diethylphthalate (gold label, Aldrich Chemical Co.) were used at levels of 1 to 500 ppm in high purity water (Burdick and Jackson).

Flow Rate: One liter of oxygen saturated solution containing the pollutant at the desired concentration was used for each test. This solution was recirculated through the cell at a known flow rate, usually at 40–60 mL/minute. In all cases it was established that this concentration had stabilized prior to the application of any power.

Separator Materials: These electrically insulating materials were used to separate one auxiliary electrode layer (electroactive catalyzed support) from another and/or to separate the driving electrodes from the auxiliary electrodes. They included a variety of cloth materials such as organic polymer fabrics, woven ceramics, commercial battery separators, alumina pellets, and cellulosic sheet.

Cell Packing Geometry: This variable involves how the various packing materials, auxiliary electrodes, and separators, are arranged within the cell cavity. The variously catalyzed auxiliary electrodes were tested in the presence and absence of separator materials. In the cases where pollutant disappearance was observed, uncatalyzed support control experiments were carried out to show that the pollutant loss was correlated with the catalyst. The number of catalyzed layers varied between one and thirty depending upon the material types used.

Surface Area: Intuitively, this must be an important parameter. It is well known in heterogeneous thermal oxidation catalysis that the adsorption of the reactants onto the catalytic surface is a crucial, often determinate, step in achieving a satisfactory reaction rate [7]. To a first approximation, the higher the active surface area, the more effective the reaction. Surface area is also a difficult parameter to study analytically. In addition, the quantity of adsorption of a subject pollutant onto a support such as activated carbon is quite large and must be carefully monitored so that it is not confused with the electrochemical decomposition phenomena.

Chemical Analyses

The reaction progress in all cases was followed directly using gas chromatography. Samples of cell effluent were collected as a function of time, temperature, and applied power. Analyses were carried out on a Hewlett-Packard Model 5890 GC equipped with a flame ionization detector, split-splitless injector and a Supelco SPB-5 bonded column. This column has a polydimethyldiphenylsiloxane active phase with an i.d. = 0.25 mm and a length of 30 meters. One microliter injections were made with at least four repeat injections per sample. Data analysis was accomplished on a Hewlett-Packard Model 3393 Integrator computer. In all cases a baseline pollutant concentration was well established prior to any power application.

Electrical Measurements

A large number of electrical measurements of various kinds were made in an attempt to define the electrical parameters controlling the capacitance driven catalytic decompositions. Specifically, wattage readings across the driving electrodes, cell capacitances and resistances were measured as a function of the different cell packing materials and geometries, pollutant presence and identity, flow rate, and temperature. Typical resistance values were between 50 and 10^6 ohms with power readings between 2 and 140 watts. The values measured were found to be somewhat variable and sensitive to changes in cell packing configuration and the flow of liquids.

Cell Studies: Results and Discussion

The CV data presented in Section 4 suggests that cyclic voltammetry is a useful screening test for evaluating the activity of potential electroactive catalysts with chosen pollutants. It is of particular interest that the organometallic FeTPP and the Ru–Ti mixed metal oxide spinel were the only catalysts found to be active for all of the pollutants.

Interestingly, the catalyst MnO_2, immobilized onto conductive carbon cloth, has been found to give the most promising results. Figure 3 exemplifies the results and is a plot of phenol concentration as a function of time in the presence and absence of an applied AC field. The experiments were carried out as follows: preweighed carbon cloth (Stackpole Fibers Co., Panex #SWB8; 32 mils thick, 8 Harness weave) was cut into pieces in a dimension to fit in the perpendicular direction within the flow cell. These pieces were then immersed in a solution of $Mn(NO_3)_2$ in CH_3OH for a period of one hour at ambient temperature. Subsequent removal of the CH_3OH *in vacuo* resulted in the carbon cloth being impregnated with $Mn(NO_3)_2$ at the desired level. Pyrolysis of this cloth was carried out by gradual heating to a maximum of 400°C until gas evolution ceased. This procedure resulted in a catalytic loading of activated MnO_2 on carbon cloth after the pyrolysis. The flow cell was then packed with these catalyzed pieces in horizontal layers alternating with layers of insulating ceramic cloth (Al_2O_3; Nextel B-14, Fiber Materials, Inc.). In addition, each driving electrode was insulated from the carbon fabric ends by multiple pieces of ceramic cloth placed between the former and the latter. One liter of 500 ppm phenol in H_2O was pumped through the cell at 50 mL/min in the absence of an applied field to assure a stable phenol peak area reading was attained.

Reaction progress was followed by gas chromatography as described on page 000. As Figure 3 shows, approximately 30 percent of the phenol disappeared in a little over an hour upon application of 80V AC across the driving electrodes. Removal of the field for two and a half hours resulted in an immediate leveling off of the phenol concentration. Reapplication of the 80V field caused a resumption of phenol loss until a total of 50 percent of the phenol was gone. In both cases the solution temperature was closely monitored to assure that thermal oxidation was not occurring. Neither solution underwent a temperature change of more than 5°C.

No products of phenol decomposition were found in the GC analyses. This is not surprising since some of the expected products are only poorly water soluble (e.g., muconic acid). In addition, it is likely that such products may only very slowly diffuse away (or not at all) from the site where they formed as is well known in heterogeneous catalysis [8]. This possibility is supported by Figure 3 showing a cumulative 50 percent drop in phenol concentration followed by no further disappearance. The suggestion here is that reaction products slowly poison the electroactive sites on the cloths. Quite possibly these successful experiments indicate the presence of only a very small area of a very effective catalytic surface that rapidly poisoned. The study of the surfaces of heterogeneous catalysts is, at best, difficult, and the nec-

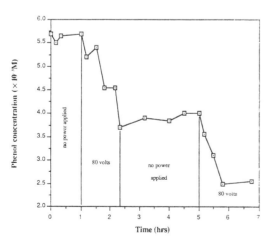

FIGURE 3. Disappearance of 500 ppm Phenol in Water. Conditions: 25 wt % manganese oxide on carbon fabric; ceramic fabric separators 22–27C: flow rate = 50–60 mL/min.; 1L solution.

essary equipment was not available to study the phenomena in more depth during this work. Therefore no attempt was made to identify the active catalytic site. Figure 4 contains the results of a similar electrooxidation experiment but one using toluene as the test pollutant rather than phenol. In this instance also, a decrease in toluene is observed as a function of the applied electric field. In this case 50 volts were applied after the test solution was equilibrated. As can be seen, the applied voltage initiated a significant drop in the toluene level when compared to the control sample. Experiments run under virtually identical conditions but substituting chloroform, trichloroethane, or diethylphthalate for phenol yielded no comfirmable, electrically induced decomposition results.

A large number of experiments were carried out using FeTPP and anthraquinone as catalysts in place of MnO_2 but under the same conditions described above for phenol–MnO_2. Here, too, no confirmable instance of electrically induced decomposition was observed.

Of the other catalyst–support combinations, all were found to have significant shortcomings as compared to the carbon cloth. The ceramic materials are all insulators and it appears a conductive catalyst–support combination is required for the auxiliary electrode structure. The carbon blacks are very high surface area and highly conductive, but are very difficult both to work with and to immobilize catalysts upon. They and the carbon pellets also adsorb tremendous amounts of organic materials making experimental data interpretation time consuming and difficult. It is worth noting that the Ru–Ti spinel could not be placed upon the various carbon-based supports because its preparation includes a calcination step at 500°C in air that oxidizes and destroys any carbonaceous supports.

ADVANTAGES OVER CONVENTIONAL TREATMENT SYSTEMS

The development of an electrochemical system that can efficiently degrade organic pollutants at low temperature using AC current represents a significant advance over existing technologies and one that possesses many potential applications. The advantages of such a system would include:

- use of ubiquitous AC, as opposed to DC, current
- the ability to electrochemically clean the electrodes should they poison
- the capability of tailoring the catalyst, catalyst support, and cell design parameters to fit specific applications
- the ability to scale up the design and construction of the apparatus using electrical similarity parameters
- low-temperature operation

An especially attractive application for this technology is at small waste generation sites. For these applications, the electrochemical cell could be designed to fit the volume of solution to be handled and easily retrofitted. The continued development of this technology promises to yield a novel technique for the removal and destruction of organic pollutants from water. This technique will not replace but rather compliment the state-of-the-art procedures.

STATE OF DEVELOPMENT

The research and development work thus far carried out has identified a promising new electrochemical technique for removing organics from waste streams or polluted water. Work is currently underway that will identify the crucial electrical parameters of the cell. In addition, new cell packing configurations and catalyst combinations are being evaluated. These studies will establish promising variable combinations that can then be packaged into a practical system. One approach will be to scale up the design and evaluate the performance of the parallel plate electrochemical cell shown in Figure 1.

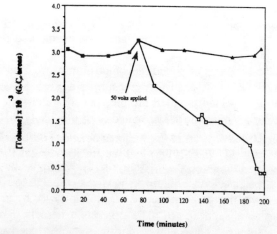

FIGURE 4. Disappearance of toluene (saturated) in water. Conditions: 25 wt % manganese oxide on carbon fabric; ceramic fabric separators 22–27°C: flow rate = 50–60 mL/min.; 1L solution.
□ sample to which voltage was applied
▲ control samples

Swiss Roll: High Flow Rate Electrochemical Cell

In anticipation of scaleup work, a second generation apparatus called a Swiss or jelly roll cell has been designed and constructed [9]. This apparatus is shown by Figure 5. This cell has the driving electrodes running down the center of the cell. The auxiliary electrodes and separator materials are attached to the driving electrodes and wound into a spiral in alternate layers around the center axle. Encasing this whole setup is a PVC pipe with fittings that allow the solution to flow through and gases to be injected if necessary. Figure 5 shows an end-on view of the pipe containing the central driving electrodes with a spirally wound titanium foil attached. On the surface of this foil, a layer of the Ru–Ti spinel catalyst has been immobilized using a procedure as described previously. Flow rate studies carried with this cell have given a maximum flow rate of between 50 and 2000 ml/min. This compares with a maximum rate of 70 ml/min for the parallel plate cell.

CONCLUSIONS

A number of significant conclusions can be drawn from this study. The first part of the research demonstrated that cyclic voltammetry can be an effective analytical tool in the evaluation of the electrochemical reactivity of pollutant compounds with a variety of catalysts. Of the catalysts evaluated, iron (III) tetraphenylporphyrin, 2-aminoanthraquinone and the ruthenium–titanium mixed metal oxide spinel showed the most significant levels of activity with the widest range of pollutants. Of these, the ruthenium–titanium is of the greatest interest. This catalyst's remarkable activity with all of the model pollutants is worthy of further study, especially since it is already highly oxidized and thus should be stable under the reaction conditions. The second part of the study was by design much more exploratory in nature but nevertheless suggested a direction for further development work. It was found possible to electrochemically induce the disappearance of phenol or toluene from a dilute solution in the absence of added electrolyte using an AC current. This result is, to our knowledge, an unprecedented finding. It is also promising in that the most widespread organic pollution problem is gasoline spills for which toluene was specifically chosen as a model. This suggests that the technology, once fully developed, may be useful at the site of petroleum spills. Clearly, much work remains to be done before a practical electrochemical cell can be marketed.

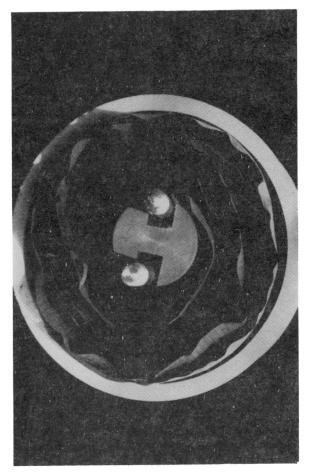

FIGURE 5. Swiss roll electrochemical cell.

Current investigations involve the examination of new catalyst–support combinations and a careful study of the effects of the rate of alteration of the cell current upon pollutant degradation rates and products.

ACKNOWLEDGEMENTS

The authors would like to acknowledge the generous support of the National Science Foundation (Grant #ECE-8420105) and the Environmental Protection Agency (Contract #68-02-4428) for support of this research.

REFERENCES

1. *Chemical Marketing Reporter*. New York, NY: ©Schnell Publishing Company Incorporated (August 20, 1987).

2. Glaze, W. H. Chapter 5, "Oxidation of Organic

Substances in Drinking Water," in *Control of Organic Substances in Water and Wastewater*. B. B. Berger, ed. Park Ridge, New Jersey:Noyes Data Corporation (1987).

3. Dixon, B. G., K. L. Gagnon, S. McGookin and M. A. Walsh. "Removal of Primary Organic Pollutants by Electrooxidation," NSF Final Report. Grant No. ISI-8420105 (November, 1987).

4. Dixon, B. G., S. McGookin and K. L. Gagnon. "Electrooxidative Purification of Waste Water," EPA Final Report. Contract No. 68-02-4428 (April, 1987).

5. Jahnke, H., M. Schornborn and G. Zimmerman. "Organic Dyestuffs as Catalysts for Fuel Cells," *Topics in Current Chemistry*, 61:133 (1976).

6. Beer, H. B., U.S. Patent 3,632,498 (1972).

7. Golodets, G. I. *Heterogeneous Reactions Involving Molecular Oxygen*. New York, NY:Elsevier Science Publishers B.V. (1983).

8. Deviney, M. L. and J. L. Gland. *Catalytic Characterization Science: Surface and Solid State Chemistry*, American Chemical Society Symposium Series 288 (1985).

9. Robertson, P. M., P. Berg, H. Reiman, K. Schleich and P. Seiler. "Application of the Swiss Roll Electrolysis Cell in Vitamin-C Production," *J. Electrochem. Soc.*, 130:591 (1983).

2.18

PACT® Systems for Industrial Wastewater Treatment

John A. Meidl[1]

INTRODUCTION

Recent public awareness of harmful environmental effects of toxic and hazardous waste is creating a demand for adequate treatment of those wastes. Public sentiment has been instrumental in moving Congress to action which in turn has resulted in Environmental Protection Agency (EPA) regulations that are beginning to affect all facets of wastewater treatment and disposal, impacting both public and private entities. Liquid toxic waste, for example, cannot now be treated by landfilling or burying—treatment via destruction is the preferred disposal method. Biological treatment may be possible for dilute levels of toxic organics, incineration or wet oxidation for very concentrated wastes.

CONVENTIONAL TREATMENT METHODS

In evaluating the merits and pitfalls of any treatment program, one must consider not only the wastewater treated, but also the residuals generated and the quality of the air discharged as a result of the treatment effected. To illustrate these factors, the treatment of benzene, toluene, and xylene (BTX) will be briefly discussed.

Benzene, toluene, and xylene are organic compounds, liquids at 20°C, having the chemical and physical characteristics at 20°C as listed on p. 178.

Granular activated carbon (GAC) adsorption and biological treatments are often considered for removing BTX from industrial wastewaters.

GAC Adsorption

Carbon has been used to absorb and treat the general class of organics as defined in Table 1 [1] and has been used to treat dilute concentrations of toxic and metal substances to tertiary levels. However, in treating such a mixture, the use of a physical treatment step alone is often not adequate since desorption phenomena occur (chromatographic effect) under transient loading conditions, thus causing the release of previously adsorbed materials/toxic organics.

The Freundlich isotherms for BTX are shown in Figure 1. As shown, benzene is the least sorbable of the three compounds. This is expected since benzene is a 6-carbon molecule, toluene and xylene are 7- and 8-carbon atoms, respectively. A comparison of carbon requirements to reduce about 20 mg/L of each compound by 50, 75, and 90 percent is shown in Table 2. As shown, significantly more carbon is needed to remove benzene than xylene. Treating wastes, then, when a whole host of organics other than BTX are involved becomes a complex task.

Abnormal breakthrough behavior can occur in GAC. Also, change in waste characteristics from testing to start-up can dramatically change design and predicted carbon use.

The GAC system also adsorbs organic materials that can be easily treated biologically. Such biodegradable material takes up valuable space on the surface of the carbon that otherwise could have adsorbed nonbiodegradable compounds. This characteristic results in inefficient use of the GAC. A perspective on realizable savings is best described by work at the Stringfellow Quarry site, which will be discussed later.

---

[1] Zimpro/Passavant Inc., Rothschild, WI 54474

Compound	Chemical Formula	Boiling Point °C	Density, g/ml	Vapor Pressure, atm	Henry's Constant	Solubility mg/L
Benzene	C_6H_6	80	0.879	0.125	5.49	1,750
Toluene	C_7H_8	111	0.866	0.037	6.66	534
Xylene	C_8H_{10}	140	*	0.009	5.27	175

Biological Treatment

Conventional activated sludge (AS) treatment will generally be able to treat BTX-contaminated wastewaters. As Geating [2] shows, benzene can be treated to low concentrations. Similarly, toluene and xylene can be removed.

An important factor to consider in biological treatment, however, is how other compounds that may be included as a part of the BTX wastestream respond to this treatment. Work by Weber [3] as shown in Table 3, for instance, notes that compounds like 1,2-dichlorobenzene and 1,2,4-trichlorobenzene are not well treated at all. True treatment—the summation of removals occurring as a result of biodegradation and floc sorption—is quite poor. Much of the "treatment" is stripping that occurs in the aeration basin of the AS process. One must question whether air quality standards will be exceeded and, if so, how to best treat the gas phase.

Weber's work shows that actual treatment (the result of both AS sorption and biodegradation) accounts for about 84 percent removal of benzene and toluene, 75 percent removal of xylene, and only 35 and 1 percent removal of dichlorobenzene and trichlorobenzene, respectively.

Table 1. Organic Compounds Amenable to Activated Carbon Adsorption

Aromatic hydrocarbons (benzene, toluene, xylene)

Polynuclear aromatics (napthalene, anthracene, biphenyls)

Chlorinated aromatics (chlorobenzene, polychlorinated biphenyls, aldrin, endrin, toxaphene, DDT)

Phenolics (phenol, cresol, resorcinol and polyphenyls)

Chlorinated aliphatic hydrocarbons (carbon tetrachloride, perchloroethylene)

High molecular weight aliphatic acids and aromatic acids (tar acids, 2,4-dichlorobenzoic acid, sulfonated lignins, benzoic acid)

High molecular weight aliphatic amines and aromatic amines (aniline, toluene, diamine)

High molecular weight ketones, esters, ethers, and alcohols (dextran, ethylene glycol)

Surfactants (alkyl benzene sulfonates, linear alcohol sulfates)

Soluble organic dyes (methylene blue, indigo carmine)

Recognizing that biological treatment of some compounds may be a problem and that stripping of volatile organic compounds (VOCs) occurs, it would be prudent to ascertain the chemical composition and stripping tendency of VOCs prior to the design of a biological waste treatment system because it may mean additional treatment steps, covered basins, and an off-gas emissions control system.

It must also be recognized that, as a result of biotreatment, a biological sludge is generated that must be controlled and disposed of.

Combined Activated Sludge—Powdered Activated Carbon (PAC) Treatment

A treatment system that takes advantage of both the biodegradation characteristics of organics as well as their adsorption characteristics is a powdered carbon enhanced activated sludge system as shown in Figure 2. In such a system, PAC is carried in sufficient concentrations in the activated sludge aeration system to ensure capture of complex and volatile organic substances that would otherwise create treatment problems in separate AS or GAC systems. Such a system, in fact, has been shown to be more efficient in treatment than even a combined AS + GAC system [4,5]. Greater detail on the process, known as Zimpro/Passavant's PACT® system, is shown in Figure 2.

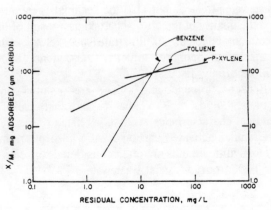

FIGURE 1. Carbon adsorption isotherms.

Table 2. Treatment of BTX by Carbon Adsorption (Starting Concentration = 20 mg/L)

Effluent Conc., mg/L	10	5	1
% Removal	50	75	95
Compound		Carbon Dose, mg/L	
Benzene	190	1,000	18,000
Toluene	100	250	700
Xylene	60	120	200

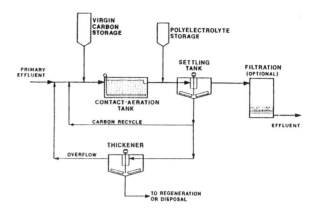

FIGURE 2. PACT® general process diagram.

THE BASIC POWDERED ACTIVATED CARBON TREATMENT APPROACH: THE PACT® SYSTEM

Generally, granular activated carbon should be used to treat only very dilute concentrations of organics (ppb range) because, with more concentrated organics, the use of such a physical treatment step alone is often not adequate to compensate for desorption phenomena (chromatographic effect) occurring under transient loading conditions. The powdered carbon enhanced activated sludge system, however, not only removes toxic organics from wastewater, but also reduces the tendency of upsets of the active biological population, thus allowing both physical adsorption and bio-oxidation/stabilization to occur simultaneously.

The System

The powdered carbon process as it is designed today will effectively treat conventional and nonconventional pollutants. The process, called the PACT® wastewater treatment system, is shown in Figure 2.

General Process Description

Wastewater is pumped to the contact aeration tank of the PACT® system. Powdered activated carbon is added to the flow to provide the desired wastewater carbon dose rate or, when spent carbon regeneration is included, to provide makeup carbon to replace treatment system losses. The carbon dose rate may range from as low as 50 mg/L for dilute wastes to several hundred mg/L for high-strength, difficult-to-treat wastewaters.

The mixed-liquor suspended solids (MLSS) concentration in the PACT® system exceeds that of conventional activated sludge systems and may reach 15,000 mg/L when higher carbon dose rates and spent carbon regeneration are employed. At higher MLSS levels, the mixed liquor typically comprises equal parts of volatile active carbon, biomass, and suspended ash. For high-strength, difficult-to-treat wastewaters, the volatile carbon concentration may exceed 50 percent of the total mixed-liquor suspended solids. A positive dissolved oxygen concentration is maintained by the aeration system to ensure efficient biological treatment.

A small dose of liquid cationic polyelectrolyte is added just ahead of the final clarifier to ensure maximum overflow clarity. Dose rates typically range from 0.5 to 1.5 mg/L. Clarifier hydraulic loading

Table 3. AS Treatment of Toxic Organics (% of Influent)

Compound	Effluent	AS Sorption	Biodegraded	Off-Gas (Stripping)
Nitrobenzene	2	0	98	1
Benzene	1	0	84	16
Toluene	1	0	83	17
Xylene	1	0	75	25
Ethylbenzene	1	0	78	22
Chlorobenzene	1	0	80	20
1,2-Dichlorobenzene	6	0	35	59
1,2,4-Trichlorobenzene	10	1	0	90
Lindane	96	4	0	0

rates for a PACT® system range from 400–800 gallons per day (gpd)/ft². While the hydraulic loading rates are comparable to conventional activated sludge systems, the solids loading rates are much higher, typically 100–125 lb/ft² at average flow.

Following clarification, effluent filtration is employed if low-suspended-solids requirements must be met. Filtration is usually recommended if less than 20 mg/L suspended solids must be consistently obtained.

Settled clarifier solids are continuously recycled to the aeration basin with a portion wasted to regeneration or solids disposal.

Since a portion of the biological solids and spent powdered activated carbon is wasted daily from the PACT® system, virgin powdered activated carbon addition is therefore required to maintain the desired aeration basin carbon concentration. The addition rate of virgin carbon is determined by treatment requirements, waste load, wastewater characteristics, and carbon type.

The PACT® system involves the addition of powdered carbon to the aeration basin of the activated sludge process, providing treatment in a single process step that is more effective than a simple combination of carbon adsorption and biotreatment.

Process Control Parameters

Operation of the PACT® system is controlled by adjustment of specific process parameters. There are six process parameters that can be encountered in the operation of the PACT® system. These parameters are as follows:

1. Hydraulic detention time (HDT)
2. Solids residence time (SRT)
3. Mixed-liquor carbon suspended solids ($MLCSS$)
4. Carbon dose (CD)
5. Carbon type
6. Timed placement of carbon

The hydraulic detention time (HDT) is calculated by dividing the volume of aeration basin V by the influent flow rate Q such that

$$HDT = V/Q$$

and is a measure of the length of time that the wastewater is in contact with the powdered carbon and biological solids. The solids residence time (SRT) is a measure of the average time that the solids, both powdered activated carbon and biological solids, are retained in the PACT® system. The SRT is usually calculated by dividing the total mass of solids in the PACT® treatment system by the amount of solids removed W daily from the system. In calculating the SRT, the total solids in the PACT® system are assumed to be in the aeration basin. Therefore,

$$SRT = \frac{MLSS \times V}{W}$$

where $MLSS$ is the total mixed liquor suspended solids concentration that comprises mixed-liquor carbon suspended solids ($MLCSS$), biological solids (biomass), and inert suspended solids (ash).

The carbon dose (CD) is defined as the amount of virgin powdered activated carbon (and regenerated carbon, if applicable) that is added to the influent stream and is calculated by dividing the mass of carbon added per unit time by the influent flow rate, i.e., CD = (mass of carbon added per unit time)/Q; $MLCSS$ is dependent on the operating parameters CD, HDT, and SRT, through the relationship

$$MLCSS = \frac{CD \times SRT}{HDT}$$

The carbon type influences the operation of the system due to differences in ash content, adsorptive capacity, and settling characteristics for a given carbon.

Timed placement of the carbon deals with the point in the PACT® system where the virgin or regenerated carbon is added or, in the case of the batch PACT® system, when in the operating cycle the carbon is added.

Advantages

Carbon adsorption in the PACT® system provides a mechanism for completely removing toxic or inhibitory substances as a result of the system's biooxidation potential. That is, organics are retained in the system for the period of time approaching the solids residence time of the system rather than the hydraulic detention time as would occur in a conventional biological treatment process.

Maintaining a large quantity of powdered activated carbon adsorbent in the PACT® system provides stability against shock and variable organic loadings. McShane et al. [6] also report similar results in treating leachates from a hazardous site in California. Major priority pollutants present were acetone, methyl ethyl ketone (MEK), tetrahydrofuran, and 1,2-dichloroethylene.

That PACT® is able to reduce toxic organics to low levels is well documented in pilot- and full-scale operations. Rollins' et al. [7] research demonstrated in

Table 4. Treatment of Organic Chemicals Production Wastewater: Activated Sludge vs. PACT® Performance
(Five Month Data Average; Equivalent Operating Conditions)

	Influent	Activated Sludge Effluent	PACT® Effluent
Biochemical Oxygen Demand (BOD$_5$), mg/L	4,035	17	11
Chemical Oxygen Demand (COD), mg/L	10,230	296	102
Total Organic Carbon (TOC), mg/L	2,965	65	25
R—Cl, mg/L	5.08	0.91	0.10
Phenol, mg/L	8.1	0.22	0.01
Color, APHA Units		820	94

parallel-operated PACT® and activated sludge studies that chlorinated aromatic hydrocarbons (R—Cl) were far lower in concentration in the PACT® effluent than in the activated sludge effluent (see Table 4) even though R—Cl levels to PACT® were spiked to as high as 106 mg/L.

Extensive testing at DuPont's Chambers Works Plant [8] as shown in Table 5 produced similar results.

Full-scale operation at the Chambers Works facility [9] shows trends for organic priority pollutant removal by the PACT® system arranged from the easiest to the most difficult to remove (see Table 6). Volatiles are generally well removed, with only 1,1,1-trichloroethylene and chloroform below 90 percent removal. Acid extractables are also generally well removed. Phenolics removability decreases with increased ring chlorination. Aromatic compounds are well removed. Di- and trichlorobenzene and dinitrotoluene are the hardest compounds to remove, yet 70 percent of these are removed at the powdered activated carbon dosages used (125 mg/L).

Table 5. Priority Pollutant Removal by the PACT® System and Activated Sludge Treatment of Chambers Works Wastewater

		% Removal	
Pollutant	Feed (ppb)	Activated Sludge	PACT® System
Base-Neutral Extractables			
1,2-dichlorobenzene	18	90.6*	>99.0
2,4-dinitrotoluene	1000	31.0	90.0
2,6-dinitrotoluene	1100	14.0	95.0
Nitrobenzene	330	94.5*	>99.9
1,2,4-trichlorobenzene	210	>99.9*	>99.9
Acid Extractables			
1,4-dichlorophenol	19	0	93.0
2,4-dinitrophenol	140	39.0	>99.0
4-nitrophenol	1100	25	97.0

*Aeration stripping a factor.

Phenolics removal as reported by Churton and Skrylov [10] showed that PACT® treatment of synfuels wastewater containing 227 mg/L of phenol was possible to levels below 1 mg/L.

Priority pollutant removal during PACT® treatment of a domestic and pharmaceutical wastewater was reported by Cormack et al. [11]. Spikes of various priority pollutants showed removal by PACT® to be significantly better than could be expected by the conventional activated sludge system. Furthermore, Simms [12] reports that the 53-million-gallons per day (mgd) PACT® facility in Kalamazoo, Michigan, is removing heavy metals (cadmium, chromium, copper, lead, silver, zinc) and priority pollutants (chloroform, methylene chloride, 1,2-dichloroethane, 1,1,1-trichloroethane, tetrachloroethane, chlorobenzene, benzene, and toluene) that would not be achievable by conventional treatment processes.

Canney et al. [13] report PACT®'s ability to effectively treat concentrated waste and concentrated toxic organics. These contaminants were measured as extractable organic chlorine (EOCl), and the results of the study are reported in Table 7.

Although concentration analyses of components in waste discharges are important, a significant consideration is the discharges' impact on the existing biota at the receiving estuary/stream. It has been demonstrated that PACT® is able to effectively reduce toxicity of wastewater treatment plant discharges. For example, Eckenfelder's [4] test results in Table 8 show that the lethal concentration, LC$_{50}$ (the percent of effluent that produces death or immobilization in half (50%) of the test species) for PACT® are far better than for conventional activated sludge and, should improved performance be desired, only a simple adjustment of carbon dose is necessary. Similar results detailing PACT®'s advantages over AS are reported by Randall [15].

As previously mentioned, the ability to improve performance in the PACT® system is directly related

Table 6. Relative Removability of Compounds by the Chambers Works PACT® System

Removal (%)	Average Feed Conc., ppb	Average PACT® Effluent Conc., ppb	Compound	Class
(>99)	1,770	nil	Methylchloride	Volatile
(>99)	33	nil	Naphthalene	Base-neutral extractable
(>99)	454	2.1	Nitrobenzene	Base-neutral extractable
(>99)	28	nil	N-Nitrosodiphenylamine	Base-neutral extractable
(99)	519	4.7	Toluene	Volatile
(99)	19	nil	1,2-Dichloroethane	Volatile
(99)	3.6	nil	1,2-trans-Dichloroethylene	Volatile
(98)	105	0.85	Benzene	Volatile
(98)	1,720	30	Chlorobenzene	Volatile
(98)	161	5.0	2,4-Dinitrophenol	Acid extractable
(97)	1,020	10	4-Nitrophenol	Acid extractable
(>95)	18	nil	N-Nitrosodi-n-propylamine	Base-neutral extractable
(>95)	3.6	nil	Methyl bromide	Volatile
(95)	11.4	0.6	2-Chlorophenol	Acid extractable
(95)	94	1.4	Carbon tetrachloride	Volatile
(95)	155	3.0	Trichlorofluoromethane	Volatile
(94.9)	174	6.3	BOD_5	
(94)	611	38	Phenol	Acid extractable
(94)	192	13	2-Nitrophenol	Acid extractable
(94)	41	1.7	Ethylbenzene	Volatile
(94)	41	1.9	Trichloroethylene	Volatile
(94)	280	12.3	Chloroethane	Volatile
(93)	24	1.7	Tetrachloroethylene	Volatile
(>90)	2	nil	2,4-Dimethylphenol	Acid extractable
(>90)	1.6	nil	Acenaphthalene	Base-neutral extractable
(>90)	0.6	nil	Anthracene	Base-neutral extractable
(>90)	1	nil	Fluoranthene	Base-neutral extractable
>90)	0.8	nil	Phenanthrene	Base-neutral extractable
(89)	13	0.6	1,1,1-Trichloroethane	Volatile
(81)	2.1	0.4	Pentachlorophenol	Acid extractable
(81)	201	20.5	Chloroform	Volatile
(80.8)	174	32.9	Soluble TOC	
(73)	370	100	1,3- and 1,4-Dichlorobenzene	Base-neutral extractable
(67)	0.3	0.1	2,4,6-Trichlorophenol	Base-neutral extractable
(66)	523	169	1,2,4-Trichlorobenzene	Base-neutral extractable
(65)	1,900	243	2,4-Dinitrotoluene	Base-neutral extractable
(64)	1,640	575	2,6-Dinitrotoluene	Base-neutral extractable
(63.7)	1,440*	484*	Color	
(44)	214	120	1,2-Dichlorobenzene	Base-neutral extractable

*APHA color units.

to carbon dose and/or carbon levels carried in the mixed liquor. Grieves [16] research in treating refinery wastewater, the results of which are presented in Table 9, shows improved organics removal when carbon mixed liquor is increased. Similarly, Hutton [17] reports improved removal of toxic organics (Table 10) as the carbon dose is increased.

Attention thus far has focused on effluent quality without consideration of aeration off-gas quality and sludge residuals impact, two other areas where toxic organics can escape treatment.

Off-Gas Quality

A concern regarding off-gas quality would be in the tendency to strip volatiles while aerating and thus not truly affect treatment of the organic species. In this regard, PACT® has been shown to be able to control organic emissions much more effectively than conventional biological processes.

In parallel testing of PACT® and activated sludge systems at Baltimore [18], aeration off-gas quality has shown total hydrocarbon levels using PACT® and activated sludge to be 78 ppm and 208 ppm, respectively.

Weber and Jones [19] conclude that "the addition of activated carbon to activated sludge systems can be an effective technique for providing enhanced removal of nonbiodegradable and poorly degradable compounds." The study showed that stripping vola-

Table 7. Performance Summary: PACT® Treatment of Oxidized Wastewater Mixture

Operating Conditions			
Solids Residence Time, days	12.4[1]	16[2]	11.6
Hydraulic Detention Time, days	6.4	6.0	6.0
Mixed-Liquor Volatile Carbon, mg/L	5,100	5,830	5,260
Mixed-Liquor Volatile Biomass, mg/L	5,300	7,010	5,940
Performance			
EOCl, mg/L			
Feed	75	113–150	300
Effluent	<0.05	0.06	0.16
% Removal	>99	>99	>99
COD, mg/L			
Feed	18,800	19,200	16,700
Effluent	563	739	904
% Removal	97	96	95
BOD_5, mg/L			
Feed	9,880	9,010	—
Effluent	11	10	12
% Removal	>99	>99	>99

[1]SRT ranged from 2.7 to 24 days.
[2]SRT adjusted from 25 to 12 days during this phase.

Table 8. Treatment of a Chemical Manufacturing Wastewater

	BOD (mg/L)	TOC (mg/L)	Color (APHA units)	Cu (mg/L)	Cr (mg/L)	Ni (mg/L)	LC_{50} (%)
Influent	320	245	5365	0.41	0.09	0.52	—
Activated Sludge Effluent	3	81	3830	0.36	0.06	0.35	11
PACT® Effluent @ Carbon Dose							
@ 100 mg/L	3	53	1650	0.18	0.04	0.27	33
@ 250 mg/L	2	29	323	0.07	0.02	0.24	>75
@ 500 mg/L	2	17	125	0.04	<0.02	0.23	>87

Table 9. Effect of Mixed-Liquor Carbon on PACT® Effluent Organics

	Effluent Concentration, mg/L		
Mixed-Liquor Carbon Concentration, g/L	0*	4	10
Soluble Organic Carbon	21	10	8
Soluble COD	80	40	33
Phenol	0.048	0.03	0.028

*Control activated sludge.

Table 10. Extractable Organic Removal PACT® vs. Activated Sludge

	Effluent Concentration, ppb		
	Activated Sludge	PACT®	PACT®
Carbon Dose, mg/L	0	100	300
Base Neutral Extractables			
1,2-Dichlorobenzene	1.7	0.4	<0.1
2,4-Dinitrotoluene	950	70	53
Nitrobenzene	18	1.7	<0.1
Acid Extractables			
2,4-Dichlorophenol	22	3.1	1.3
2,4-Dinitrophenol	86	3.9	<0.1
4-Nitrophenol	830	100	29

Table 11. Fate of Toxic Organics (% of Influent)

Compound	Activated Sludge		PACT® @ 100 mg/L Carbon Dose	
	Effluent	Off-gas	Effluent	Off-gas
Benzene	<1	16	<1	14
Toluene	<1	17	<1	0
o-Xylene	<1	25	<1	0
1,2-Dichlorobenzene	6	59	<1	6
1,2,4-Trichlorobenzene	10	90	<1	6
Lindane	>95	0	<1	0

% Treatment = 100 − (effluent + off-gas).

tiles is highly likely out of activated sludge, however, as noted in Table 11, addition of powdered carbon to the aerator increases volatiles retentivity and nearly eliminates volatiles stripping.

Work by Copa [20] at a Superfund site showed that PACT® did not strip organics out of the contaminated water. Organics in the water consisted of benzene, chlorobenzene, dichloromethane, chloroform, ethylbenzene, toluene, 1,2-dichloroethylene, trichloroethylene, and tetrachloroethylene. Hoffman [21] also showed that when treating a Superfund leachate spiked with trichloroethylene, 2,4-dichlorobenzene, and 2-chlorophenol to levels as high as 100 mg/L, no detectable traces of those organics were found in the effluent, and losses due to their stripping out of the PACT® aeration basin ranged from 0.011 to 0.098 µg/L of total chlorinated organic compounds.

Residuals Management

Proper residuals (waste solids) management is an important aspect for any system because residuals could carry the concentrated form of the toxic organic extracted in the wastewater treatment step. Such residuals from the toxic organics treatment must be disposed of in a secure landfill or subjected to oxidation/incineration such as wet air oxidation (WAO) or thermal (open flake) combustion for destruction.

As with any wastewater treatment process, spent residuals (carbon) also exist from PACT®—the spent residual being composed of biosolids, spent carbon, and adsorbed organics. Landfilling of filter pressed spent residual is possible. The advantage of PACT® over activated sludge, however, is that the resultant spent materials can be over 50 percent solids for PACT® pressed residuals versus only 15 percent for activated sludge pressed residuals.

With PACT®, however, comes a lucrative option: being able to economically recover a commodity (powdered activated carbon) while destroying associated contaminants (conventional toxic organics), and being able to do so while in a slurry of less than 10 percent solids. That option uses wet air oxidation [22]—an aqueous phase oxidation of organics plus inorganic materials—and a simultaneous regeneration of powdered activated carbon. The operation is conducted at elevated temperatures (less than 500°F) and pressures (less than 1100 psig) as shown in the general flow scheme of Figure 3.

In wet air oxidation, the bulk of the organic waste stream is converted to carbon dioxide and water. Because of the aqueous phase oxidation, no oxides of nitrogen or sulfur exist in the off-gases. Also, oxidation of the spent slurry is generally self-supporting (autogenous), requiring no outside source of fuel energy to maintain operating temperatures.

Wet air oxidation is being used, commercially, to successfully destroy and detoxify high-strength and/or toxic wastes. Early work by Randall and Knopp [23] and Canney and Schaefer [24,25], describe the wet air oxidation of a range of compounds, showing high destructive efficiencies thereby, many exceeding 99 percent. Randall [26] also cites the effectiveness of wet air oxidation for the simultaneous regeneration of powdered carbon and destruction of toxic organics.

If wet air oxidation regeneration were to be used with PACT®, the only residuals for disposal would consist of a blowdown ash. Although the ash may

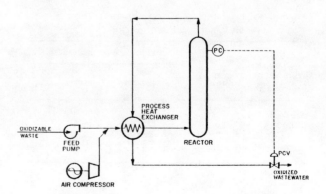

FIGURE 3. Wet air oxidation general flow diagram.

contain the oxide forms of heavy metals, the ash will be stable, sterile, and nonleaching [27,28]. The ash can also be dewatered to solids concentrations as high as 70 percent, with normal ranges being 40 to 60 percent.

There are a number of examples of PACT, and PACT + wet air oxidation regeneration being successfully used; some of these installations are noted on the PACT® Users List (see Table 12).

Examples of how PACT®, and PACT® + wet air oxidation can be used to treat toxic organics are best described in later sections.

PACT® SYSTEMS TREATMENT EXAMPLES

Treatment at a Superfund Site

As part of the Stringfellow Quarry site cleanup effort under the Comprehensive Environmental Response, Compensation, Liability Act (CERCLA), and under contract with the State of California Department of Health Services, demonstration treatment of the quarry's leachate was undertaken by Zimpro/Passavant Inc. using the PACT® system [29].

The PACT® process equipment used to treat the Stringfellow Quarry's contaminated groundwater was housed in a mobile trailer (see Figure 4). The process treated 2.4 gpm and operated at a hydraulic detention time of 16 hours, a solids residence time of 15 days, and a powdered carbon dose of 667 mg/L. In doing so, the single-stage PACT® system accomplished greater than 70 percent removal of COD and DOC (see Table 13).

Organic priority pollutant analyses showed that PACT®'s effluent was essentially free of all organic priority pollutants, even though benzene, chlorobenzene, dichlormethane, chloroform, 1,2-dichloroethylene, trichlorethylene, tetrachlorethylene, ethyl benzene, and toluene were found in the influent. Also, to test the capability of PACT® to control stripping of organic compounds, aeration tank off-gas analyses were done. Off-gas total hydrocarbon (*THC*) analyses from the aeration tank ranged between 2.5 to 11 ppm (volume/volume) as methane—comparable to background *THC* levels at the site—verifying that PACT® does limit volatilization of organics from the aeration basin.

In contrast to this study, an engineering study was done in 1984 using a granular activated carbon analysis to predict the treatment necessary to achieve the same COD level accomplished by the PACT® process [30]. The process comparison between PACT® and GAC treatment of an influent with a 500 mg/L COD level at a flow rate of 60 gpm is shown in Table 14.

Table 12. PACT® System Users

Municipal	Industrial
Vernon, Connecticut*	DuPont (2)
Medina, Ohio*	General Electric
Mt. Holly, New Jersey*	Bofors-Nobel*
Burlington, North Carolina*	Exxon
Bedford Heights, Ohio*	Tenneco
Kalamazoo, Michigan*	Upjohn
El Paso, Texas*	Alcoa
North Olmstead, Ohio*	Crompton Knowles
Jessup, Maryland*	Moore Business Forms (3)
	Huron Valley Hospital
Kimitsu, Japan*	Ciba-Geigy
Oga, Japan*	Powell-Duffryn
Senroku, Japan*	ICI United States
Oizumi, Japan*	Koppers
Ibargi, Japan*	Rollins Environmental (2)
	BKK Landfill
	Southern Yeast
	Tosco
	Bethlehem Steel
	Unocal*
	BPCL Refinery*
	Nalco
	Domtar
	Central Services
	Safety Kleen
	Aldrich Chemical

*Wet air regeneration system.

The GAC replacement estimate is based on adsorption studies conducted on the waste in question.

The comparative cost analysis that used the 1984 engineering study, the on-site PACT® work, and the GAC adsorption work as bases is shown in Table 15. Though PACT® was higher in capital costs, a fivefold savings in operation and maintenance costs favors PACT®. Total costs for granular carbon, in fact, are

Table 13. PACT® Performance Summary Stringfellow Quarry

Operating Conditions	
Feed Rate, gpm	2.43
Hydraulic Residence Time, day	0.74
Solids Residence Time, days	15
Mixed Liquor, mg/L	
Suspended Solids (SS)	20,780
Carbon	9,700
Biomass	3,240

Performance	Influent	Effluent
COD, mg/L	1,788	467
BOD_5, mg/L	55	5
DOC¹, mg/L	550	154
SS, mg/L	103	18

¹Dissolved organic carbon.

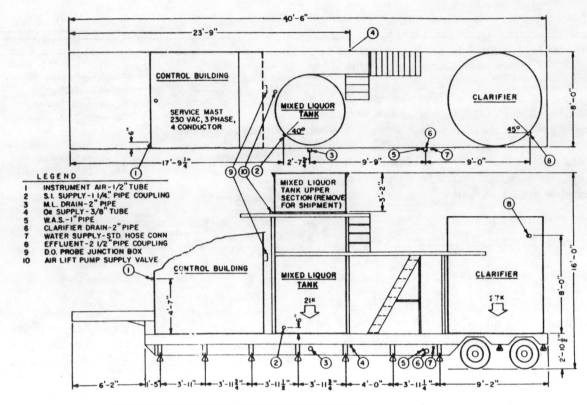

FIGURE 4. Mobile plant.

found to be at least twenty times more than the cost of powdered carbon. Annual costs taken over a two-year period at 10 percent interest shows PACT® to be one-half the cost of a GAC system. The difference between PACT® and GAC would be even greater if annual costs were determined for a period greater than two years.

Reductions in GAC requirements via pre-biotreatment to reduce GAC costs would not be favorable to the GAC treatment alternate because BOD_5 from the lime precipitation step is already very low. Even if biotreatment were possible, substantial capital and operating costs would occur that would still make the complex biotreatment + GAC system more costly in both capital and operating areas than PACT®.

Table 14. 60-gpm Process Comparison: PACT® vs. GAC Stringfellow Quarry, California

	PACT®	GAC
Flow, gpm	60	60
Chemical Pretreatment (Metals Removal)	Lime + Caustic	Lime + Caustic
Sand Filtration, gpm/ft²	2.5*	2.5
Carbon Treatment		
Detention Time, Hours	18	0.5
SRT, days	15	—
MLSS, g/L	20	—
Carbon Dose, mg/L	670	7000
Solids Dewatering	Press	Press

*Included in evaluation even though not necessary to meet the 500 mg/L effluent COD goal.

Table 15. 60-gpm Comparative Cost Analysis: PACT® vs. GAC Stringfellow Quarry, California

	PACT®	GAC
Capital Cost (1984), $	$1,700,000	$1,200,000
Annual Operating Costs, $		
Chemicals	80,000	75,000
PAC @ 40¢/lb	70,000	—
GAC @ 95¢/lb	—	1,800,000
Power @ 9¢/kWh	65,000	30,000
O&M @ $20/h	220,000	150,000
TOTAL	$435,000	$2,055,000
Annual Costs (2 yrs, 10%)		
Amortized Capital	980,000	691,000
O&M	435,000	2,055,000
TOTAL	$1,415,000	$2,746,000

Analysis based on same effluent quality.

Table 16. Partial List of Permitted Compounds: Bofors-Nobel Inc., Muskegon, Michigan

Acetone	Chloroaniline	Isophorone
Aliphatic amine	Chlorobenzene	Methylene chloride
Allyl alcohol	Chlorophenol	Methylpyridine
Ammonium dithiocarbamate	Cresol	Nitrocresol
Ammonium thiocyanate	Dichlorobenzene	Nitrophthalic acid
Aniline	Dichlorobenzidene	Perchloroethylene
β-Chloroaniline	Dimethylaminexylanol	Phenol
β-Naphthylamine	Dinitrotoluene	Phenoxybiphenyl
Benzene	Di-N-Propylformamide	Phenylnaphthalene
Benzidine	Diphenylether	Phthalic Acid
Benzoic acid	1,2-Dichloroethane	2-Propanol
Biphenyl-OL	Chlorobiphenyls	Sodium acetate
Bipyridene	Ethyl acetate	Tetrachloroethylene
Bis(ethyl, hexyl)phthalate	Ethyl benzene	Toluene
	Formaldehyde	

Treatment of RCRA Wastewater and CERCLA Groundwater

Since March 1983, Bofors-Nobel Inc., a Michigan toll manufacturer of herbicides and organic chemicals, has used an innovative wet air oxidation + PACT® + wet air regeneration systems approach to treat both Resource Conservation and Recovery Act (RCRA) waste as well as CERCLA contaminated groundwater [31,32].

Bofors-Nobel, in acquiring Lakeway Chemicals Inc. of Muskegon, Michigan, also inherited a site on which 370 million pounds of contaminated sludges had been discarded. The sludges, contaminated with dichlorobenzidine (DCB) at concentrations as high as 5,000 mg/L presented a formidable problem to Bofors-Nobel because the contamination had entered the groundwater table. Barrier wells were selected to contain the contaminated water, requiring the pumping of more than 1 mgd from the groundwater to on-site treatment. In an effort to stem the pollution created by past Lakeway practices and to meet pretreatment requirements of the Muskegon County wastewater system, Bofors-Nobel evaluated various technologies for the treatment of the contaminated groundwater. Extensive pilot plant testing of biological, adsorption, and PACT® systems showed the PACT® system to be not only more reliable, but far more economical than other alternatives.

Also, Bofors-Nobel's expanded organic chemicals manufacturing operation produced a variety of RCRA wastes, many too toxic for even PACT® to handle, but too dilute to economically incinerate (see Table 16). In this case, wet air oxidation was selected to destroy RCRA waste toxicity, acting as a pretreatment step ahead of PACT®. The wet air oxidation process could also be used to regenerate spent carbon from the PACT® process while destroying any adsorbed toxic organic on the carbon.

The treatment system designed to accommodate the needs of Bofors-Nobel is shown in Figure 5. Two package wet air oxidation units were supplied: one unit dedicated solely to detoxification, the other used mainly as a carbon regeneration process but having the ability to go on-line as a "detox" unit. As a result of this planning, Bofors-Nobel received the Presidential Council on Environment Quality, Hazardous Waste Management Award in 1983.

Currently, more than 1.5 mgd of contaminated groundwater and RCRA process wastewater is being treated by the systems shown in Figure 5. Performance of the wet air oxidation detox unit and the PACT® system is shown in Table 17. (OCA and DCB are listed since these compounds are regularly monitored.) Overall, more than 99.5 percent COD removal is being obtained. Wet air oxidation has been destroying an average of 99.8 percent of the toxic organics fed to it; its discharge is nontoxic and easily treated in the PACT® system.

The overall annual operating costs for the PACT® + wet air regeneration system, including solids disposal costs and county wastewater treat-

Table 17. Waste Treatment Performance: Bofors-Nobel Inc.

Wet Oxidation Detoxification		
	Influent	Effluent
Toxic organics, mg/L	600–1200	ND–5
COD, g/L	70–80	30–40
BOD_5, g/L	—	>15
PACT® Wastewater Treatment		
	Influent	Effluent
Orthochloroanaline, µg/L	53,000	<10
Dichlorobenzidene, µg/L	12,000	<2
Ammonia Nitrogen, mg/L	150–200	<10

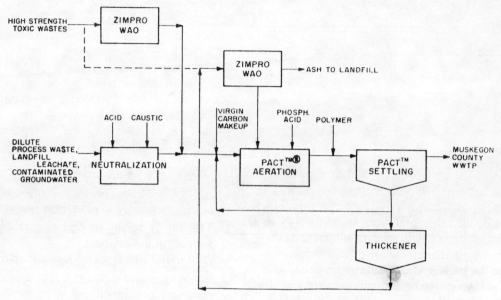

FIGURE 5. Wastewater treatment facility—Bofors-Nobel, Inc.

ment charges, is about $1 million, or just under $2.00/1,000 gallons. This amounts to less than 10 cents per pound of COD treated.

In considering wet air regeneration cost-effectiveness, the annual cost for regeneration, virgin carbon makeup, and solids disposal is less than $300,000. Without regeneration, those same annual costs would exceed $1 million and the problem of contaminated solids and its latent liability would not be eliminated.

Surface Runoff Waters

Powell Duffryn Terminals (PDT), Bayonne, New Jersey has installed a PACT® system to treat surface runoff water prior to discharging it to New York harbor [33].

The Bayonne terminal covers thirty acres and is one of the major terminals serving the industrial northeastern United States. It is capable of taking the world's largest ocean-going chemical parcel tankers, and storing a full range of chemical products. Capacity is over 1 million barrels.

Alternatively, the facility is equipped for direct transfer between ship and truck or rail.

In addition, Powell Duffryn offers, through its TEPCO division, contract blending and packaging for such products as antifreeze. Surface water is collected and contained by a system of concrete pads and dikes the company installed to prevent groundwater contamination. The terminal faced increasingly stringent requirements for its effluent discharge, including a bioassay test for toxicity reduction.

To comply, PDT processes up to 55,000 gallons of surface runoff a day. PDT runoff contains a "high" COD and a number of dissolved organics that must be reduced before the effluent is discharged into the harbor.

In order to meet its stringent National Pollutant Discharge Elimination System (NPDES) permit (Table 18), PDT piloted the PACT® system in 1985 and determined it would reliably achieve treatment objectives. Testing (Table 19) at two different carbon doses showed that the dosages used were very con-

Table 18. Powell-Duffryn Terminals: NPDES Permit (mg/L unless specified otherwise)

Flow, mgd	0.055
pH	6–9
Oil and Grease (O&G)	15 maximum
COD	150
TOC	Monitor
BOD	50
Total Suspended Solids (TSS)	50
Methylene chloride	0.055
Perchloroethylene	0.065
Trichloroethylene	0.065
Chloroform	0.040
Carbon tetrachloride	0.030
1,1-Dichloroethane	Monitor
1,2-Dichloroethane	0.085
Styrene	Monitor
Butyl acetate	Monitor
Ethylene diamine	Monitor
Vinyl acetate	Monitor
Vinyl chloride	0.065
Phenol	0.035
Toxicity (Bioassay)	96 hour LC_{50} 50%
Priority Pollutants	Monitor

Table 19. PACT® System Pilot Testing:
Powell-Duffryn Terminals

	Phase I	Phase II
Operating Conditions:		
HDT, days	1.1	1.1
SRT, days	8.3	8.3
Carbon Dose, mg/L	1,790	940
Mixed-Liquor Carbon, mg/L	13,500	7,100
MLSS, mg/L	16,900	10,000
Performance Results:		
Biogrowth Rate, g bio/g COD$_r$[1]		0.09
COD in (mg/L)	636	643
out	19	14
BOD$_5$ in (mg/L)	230	182
out	<6	<6
DOC in (mg/L)	117	118
out	<4	<4
Color in (APHA units)	98	40
out	8	6
O&G in (mg/L)	40	30
out	<1	<1

[1]COD$_r$—Chemical oxygen demand removed.

Table 20. PACT® System Performance
Powell-Duffryn Terminals

	Typical Values Runoff	PACT® Effluent
COD, mg/L	>600	44
BOD$_5$, mg/L	up to 1,800	18
SS, mg/L		18
O&G, mg/L		2.7
Priority Pollutants, mg/L		<0.005
Toxicity, 96 hour LC$_{50}$		>100%

servative and enabled the PACT® system to easily meet treatment objectives.

Stress testing was also conducted and showed PACT® to be extremely stable.

As matter of interest, waste sludge from the PACT® system was tested for its metals-leaching characteristics. No appreciable leaching was observed from the spent carbon.

Based on this testing, a PACT® system package wastewater plant furnished by Zimpro/Passavant Inc. was installed consisting of equalization, pH control, two-stage prefab PACT® system, sludge storage, virgin carbon silo, and associated ancillary equipment. Full-scale performance of the system is shown in Table 20.

Anaerobic PACT® System—Leachate Treatment

An anaerobic powdered carbon system has been shown to have advantages over other conventional biological anaerobic systems when treating high-strength wastes such as a 10.5 g/L COD, 4.5 g/L BOD$_5$, 3.2 g/L DOC leachate.

The anaerobic PACT system: 1) will yield better performance than conventional suspended contact systems or packed bed systems; 2) will require no internal support media and no special liquid-gas separation device(s); 3) will not foam within the digester reactor, a problem that is a particular problem in noncarbon suspended contact units; 4) will be able to be operated with controlled sludge wasting and, thus, at controllable solids residence times; and 5) will be able to control toxic organics.

A comparison of biological packed-bed anaerobic units with the powdered carbon unit is shown in

Table 21. Anaerobic Comparison: Anaerobic PACT® System vs. Packed Bed System

	Packed Bed		Suspended Contact	
Media	Rock	Saddles	PAC	PAC
Operating Conditions				
COD Loading, lb/ft³/d	0.8	0.4	0.64	0.60
SRT, days	—	—	79	21
Carbon Dose, mg/L	0	0	165	166
Performance				
Solids Yield, lb bio/lb COD$_r$	0.04	0.07	0.03	0.06
Gas Production, ft³/lb COD$_r$	8.0	6.9	6.9	7.2
COD				
Effluent, g/L	5.1	3.8	2.4	3.0
% Removal	52	63	76	70
BOD				
Effluent, mg/L	1,100	510	200	400
% Removal	74	89	95	90
DOC				
Effluent, mg/L	—	—	0.7	0.9
% Removal	—	—	77	72

Table 21. The data indicate that the anaerobic PACT® system yields a lower strength effluent BOD_5 and COD than the packed-bed system. Solids yields from all anaerobic systems were generally low.

As expected, biological solids yield for the 70-day *SRT* unit was much lower than the 21-day *SRT* unit. A slightly better effluent quality was also achieved in the 70-day unit because of the higher level of carbon carried in the mixed liquor of that unit, duplicating the relationship seen in aerobic PACT® systems.

In this case, discharge from the anaerobic PACT® system could be directly sewered without further treatment. If required, however, as in the case of a NPDES discharge permit, a second stage PACT® system could be added. In this case, virgin carbon would be added to the second stage (aerobic) PACT® system, with its waste sludge of partially spent carbon routed to the first stage (anaerobic) PACT® system.

REFERENCES

1. *Wastewater Treatment System Manual, Series 200*, published by Zimpro Inc., pp. 200–207 (1984).
2. Geating, John. "Literature Study of the Biodegradability of Chemicals in Water," EPA Grant No. 12806699-01 (September 1981).
3. Weber, Walter, Jr. and Bruce Jones. "Toxic Substance Removal in Activated Sludge and PACT® Treatment Systems," *WPCF Conference*, New Orleans, Louisiana (October 1984).
4. O'Brien, G. et al. "Effect of New Regulations on a Commercial PACT® Wastewater Treatment Plant," *Purdue Conference* (May 1989).
5. Hutton, D. "Priority Pollutant Removal-Comparison of the PACT® Process with Activated Sludge Followed by GAC Columns," *Symposium on Application of Adsorption to Wastewater Treatment*, Vanderbilt University (February 1981).
6. Mcshane, S., A. Lebel, T. Pollock and B. Stirrat. "Biophysical Treatment of Landfill Leachate Containing Organic Compounds," *41st Industrial Waste Conference*, Purdue University (May 1986).
7. Rollins, R., C. E. Ellis and C. L. Berndt. "PACT®/Wet Air Regeneration of an Organic Chemical Waste," *37th Industrial Waste Conference*, Purdue University (May 1982).
8. Hutton, D. and S. Temple. "Priority Pollutant Removal: Comparison of PACT® Process and Activated Sludge," *52nd Annual WPCF Conference*, Houston, TX (October 1979).
9. Hutton, D. "Removal of Priority Pollutants with a Combined Powdered Carbon—Activated Sludge Process," *ACS* (March 1980).
10. Churton, B. and V. Skrylov. "Studies to Treat Process Wastewater from a Coal Liquefaction Plant," *Summer National Meeting of the American Institute of Chemical Engineers*, Cleveland, OH (1982).
11. Cormack, J., D. Y. Hsu and R. G. Simms. "A Pilot Study for the Removal of Priority Pollutants by the PACT® Process," *38th Industrial Wastes Conference*, Purdue University (May 1983).
12. Simms, R. "Successful Removal of Toxics at Kalamazoo Using the PACT® Process," *59th Annual WPCF Conference*, Los Angeles, CA (October 1986).
13. Canney, P., C. Modorski, R. Rollins and C. Berndt. "PACT® Wastewater Treatment for Toxic Waste Cleanup," *39th Industrial Waste Conference*, Purdue University (May 1983).
14. Eckenfelder, W. "Technologies for Toxicity Reduction in Industrial Wastewaters," *Toxicity Reduction Seminar*, Hyatt Regency, New Brunswick, New Jersey (November 20 1985).
15. Randall, T. "Wet Oxidation of PACT® Process Carbon Loaded with Toxic Compounds," *38th Industrial Waste Conference*, Purdue University (May 1983).
16. Grieves, C. G. "Powdered Carbon Enhancement Versus Granular Carbon Adsorption for Oil Refinery Batea Wastewater Treatment," *51st Annual Conference, Water Pollution Control Federation*, Anaheim, CA (October 1978).
17. Hutton, D. and S. Temple (previously cited).
18. Zimpro Inc. Report, "Pilot Activated Sludge and Powdered Carbon Enhanced Activated Sludge Study at Baltimore, Maryland's Back River WWTP," (January 1986).
19. Weber, Walter, Jr. and B. Jones (previously cited).
20. Copa, W., T. Vollstedt and P. Canney. "Demonstration of Unit Scale Powdered Activated Carbon Treatment of Hazardous Wastes," Zimpro Inc. Report, Contract No. 83-82053, Dept. of Health Services, CA (November 1984).
21. Hoffman, M. Internal Zimpro Inc. Report; "PACT® Treatability of Mid-State Landfill Leachate Spiked with Trichlorethylene, 1,4-Dichlorobenzene, and 2-Chlorophenol," (September 1986).
22. Zimpro Inc. *Wastewater Treatment System Manual, Series 200*, pp. 200–263 to 104 (previously cited).
23. Randall, T. and P. Knopp. "Detoxification of Specific Organic Substances by Wet Oxidation," *J. Water Poll. Control Fed.* 52, 8:2117 (1980).
24. Canney, P. and P. Schaefer. "Detoxification of Hazardous Industrial Wastewaters by Wet Air Oxidation," *1983 Spring National AIChE Meeting*, Houston, TX (1983).
25. Canney, P. and P. Schaefer. "Wet Oxidation of Toxics: A New Application of Existing Technology," *Toxic and Hazardous Waste Proceedings, 15th Mid-Atlantic Industrial Waste Conference*. M. D. Lagrega and L. K. Hendrian, eds. Ann Arbor Science Publ. (1983).
26. Randall, T. (previously cited).
27. Zimpro Inc. *Wastewater Treatment System Manual, Series 200*, pp. 200–266 to 68.
28. Internal Zimpro Inc. Communication, "Wet Air Regeneration Unit Metals Analysis" (October 18, 1983).
29. Copa et al. (previously cited).
30. "Fast-Track Remedial Investigation Feasibility Study," Stringfellow Site, Riverside, CA (April 1984).
31. Zadonick, L. "Comprehensive Site Cleanup at Bofors-

Nobel, Inc." *16th Mid-Atlantic Industrial Waste Conference* (June 1984).

32. Peterson, Ron and John Meidl. "Treatment of Contaminated Groundwater and RCRA Wastewater at Bofors-Nobel, Inc." *4th National RCRA Conference on Hazardous Waste and Hazardous Materials* (March 1987).

33. Meidl, John. "Treatment of Benzene, Toluene and Xylene and Other Pollutants in Surface Runoff," *Seventh Annual ILTA National Operating Conference* (June 1987).

Destruction of Cyanides in Electroplating Wastewaters Using Wet Air Oxidation

H. Paul Warner[1]

ABSTRACT

Many of the technologies normally applied for the destruction of cyanide in waste streams containing cyanides and metals result in the generation of sludges that contain high concentrations of cyanide (200 to 1000 mg/kg), most of which is strongly complexed with constituent metals. In order to reduce the total cyanide content of these sludges, we investigated the possibility of treating the original waste stream by a technology that would destroy the cyanide, both free and complexed, prior to precipitation of the metals. This paper presents the results of the application of wet air oxidation for cyanide destruction prior to sludge generation. Experience and engineering judgment strongly suggest that this technology could also be applied to liquids and sludges generated by other technologies that contain high concentrations of cyanides.

INTRODUCTION

Pursuant to Section 3004 (m) of the Resource Conservation and Recovery Act (RCRA), enacted as part of the Hazardous and Solid Wastes Amendments (HSWA) on November 8, 1984, the Environmental Protection Agency (EPA) is investigating alternatives for treating cyanide-containing electroplating and heat-treating wastes prior to placement in landfills. The agency has previously established treatment standards for metals in the sludges from these wastes with the First Third Listed Hazardous wastes (53 FR 31137, August 17, 1988). Treatment standards for the wastewater from these wastes were "soft hammered" with the First Third rule. The agency is now developing standards for all of these wastes and will, with additional regulations, set standards for cyanide. Cyanide standards were reserved by the agency in the First Thirds rule. This paper will discuss the application of wet air oxidation as one of the technologies that can be applied for treatment of cyanide-containing electroplating wastes, specifically, spent plating bath (F007).

For the discussion that follows, it must be pointed out that data related to the raw waste selected for treatment have been claimed as Confidential Business Information (CBI) by the generator and cannot be used in this presentation. However, by using constituent concentrations of a "typical" F007 waste, a relatively accurate evaluation of this treatment process can be made. The constituent concentrations of this "typical" F007 waste will be the average of other F007 wastes as found in the agency's proposed background document for the regulation of cyanides [1].

It should also be pointed out that prior to the pilot-scale work, a bench-scale run was made which, due to the corrosivity of the raw waste, mandated the use of titanium as the construction material for the treatment system. Only the smallest scale system was constructed of titanium and, therefore, was chosen for this project.

Approach

Alkaline chlorination is one of the most commonly applied treatment technologies for F007 wastes, and is usually followed by precipitation, clarification, and finally dewatering of the generated sludge. In a well-operated system, the final liquid discharge meeting effluent discharge concentration regulations may be directly discharged to a surface stream or publicly owned treatment works (POTW). However, the dewatered sludge will normally contain, along with varying concentrations of regulated metals, high concentrations of total cyanide (200 to 1000 mg/kg).

---

[1]U.S. Environmental Protection Agency, Cincinnati, OH 45268

These cyanide concentrations would more than likely restrict continued land disposal of the dewatered sludge without additional treatment. Taking into consideration the necessity of disposal, alternative treatment techniques were evaluated. Wet air oxidation of F007 wastes has been successfully demonstrated by Zimpro/Passavant in Casmalia, California. Their technology was selected for further evaluation and for the generation of data in support of the agency's regulatory program.

WET AIR OXIDATION PROCESS [2]

Wet air oxidation is the liquid-phase oxidation of organics or oxidizable inorganic components at elevated temperatures and pressures. Oxidation is brought about by combining the wastewater with a gaseous source of oxygen (usually air) at temperatures and pressures in the range of about 175° to 327°C (360°–620°F) and 2069 to 20,690 kPa (300–3,000 psig), respectively. The solubility of oxygen in aqueous solutions is enhanced at elevated pressures, and the elevated temperatures provide a strong driving force for oxidation.

Wet air oxidation has been demonstrated at bench-scale, pilot-scale, and full-scale as a technology capable of breaking down hazardous compounds to carbon dioxide and other innocuous end products. Cyanide in electroplating wastes is converted to carbonate and ammonium ions when oxidized as shown by the reactions:

$$2\ NaCN + O_2 + 4\ H_2O = Na_2CO_3 + (NH_4)_2CO_3$$

The major processing steps in the wet air oxidation process are wastewater pressurization/air compression, preheat, reaction, cooling, depressurization, and liquid-gas separation. Figure 1 is a flow diagram of the wet air oxidation process.

The wastewater or slurry is brought to system pressure by a high-pressure pump. Air from a compressor may be added directly to the waste or to dilution water and preheated to raise the temperature of the mixture at the reactor base such that the exothermic heat of reaction will increase the mixture temperature to the desired maximum. Preheating can be accomplished by an external heat source as shown in Figure 1 or by the reactor effluent. Start-up energy is provided by the external heat source to the preheater or to an auxiliary heater. Residence time for the oxidation reaction is provided by the reactor; the temperature of the wastewater-air mixture rises as the reaction occurs. The reactor effluent is cooled with cooling water (as shown in Figure 1) or with the wastewater-air mixture. Cooling is usually to about

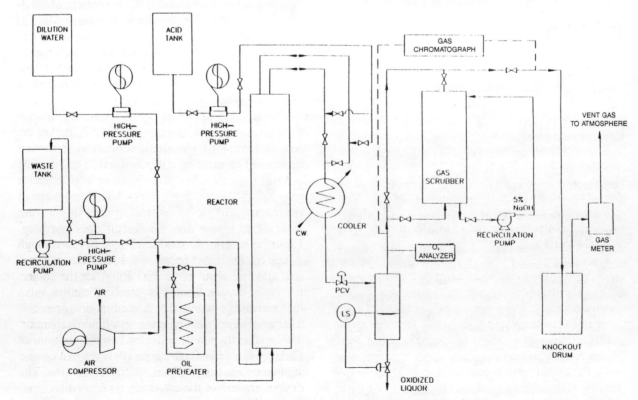

FIGURE 1. Schematic of Zimpro/Passavant Wet Air Oxidation Process.

35–57°C (95° to 135°F). A control valve reduces the pressure of the oxidized liquor spent air mixture. The gas phase is disengaged from the liquid phase in the separator vessel. Off-gas from wet oxidation systems is usually treated to reduce the concentration of hydrocarbons. Wet scrubbing, which is commonly used to cool the gas stream, results in some reduction of hydrocarbons. Adsorption columns and afterburning provide additional organic emissions reduction.

The overall F007 treatment system at Zimpro/Passavant may be described in three operations: (1) a blending step to control feed parameters, (2) the wet air oxidation process, and (3) the treatment of oxidized liquor.

Feed Blending Operation

Four 55-gallon drums of F007 waste were mixed in a water-heated stainless steel tank to ensure a homogenous feed composition. The waste required heating to about 43–54°C (110°–130°F) to maintain its liquid state. Below that temperature range, the waste crystallized because of the high concentration of sodium carbonate, which made handling very difficult. After the waste was thoroughly mixed, the drums were refilled and placed in a hot water bath, maintained at 43°C (110°F). As waste was needed in the performance of the test run, a drum of waste was removed from the water bath, thoroughly agitated with a portable mechanical mixer, and then pumped into the treatment system feed tank. The feed tank is equipped with a heating coil, mechanical mixer, and a recycle line to ensure the feed is maintained at the proper temperature and is homogenous.

Wet Air Oxidation Treatment

The steady-state operating conditions of the Zimpro/Passavant wet air oxidation process are shown in Table 1. All the parameters listed in Table 1 are key operating parameters; however, the single most important parameter used to determine steady state was maintaining the reactor outlet temperature of about 232°C (450°F).

Operating phases of the wet air oxidation process include: warm-up (with tap water followed by waste feed), stabilization, steady state, and cool down (with tap water). The warm-up period with tap water typically requires about four hours followed by an additional two hour warm-up with the waste feed. Waste feed warm-up continues until the operating conditions were within approximately 5 percent of the desired conditions, which signals the beginning of the stabilization period. During the stabilization period, the operating conditions were continually fine-tuned. Typically, 1 to 3 residence times (i.e., about 1 to 3 hours) were allowed before steady state is begun. The steady-state period continues as long as needed to collect all the required samples or until the system falls outside steady-state conditions specified in Table 1. Following steady state, the system is switched to tap water and cooled down over a period of about 6 to 8 hours.

In the wet air oxidation treatment process dilution, water (i.e., tap water) is pumped at a rate of 2.72 gal/h (1745 psig) and combined with 1.1 ft³/min of compressed air prior to passing through an oil preheater (Figure 1). The preheater heats the tap water/air stream to about 271°C (520°F) as it enters the base of the pilot-scale titanium reactor at 1710 psig. Approximately 2 feet from the bottom of the 3-inch (i.d.) 15-foot-long reactor, the waste feed is pumped into the reactor at a rate of 2.75 gal/h. The total influent flow rate to the reactor is 5.47 gal/h. Heat tapes, spaced evenly along the length of the reactor at 15-inch intervals, along with the heat of reaction maintain the temperature at about 232°C (450°F). The oxidized liquor exits the reactor through one of two exit ports. One port is used as the reactor outlet for the oxidized liquor, and 18 percent nitric acid is pumped at a rate of 0.55 gal/h (1735 psig) through the other port to remove any carbonate plugging at the exit port. Every four hours through the test run, the valves controlling the exit ports are reversed to ensure that the reactor does not plug. From the reactor, the oxidized liquor passes through a tube-in-tube water cooler that brings the temperature down to about 41°C (106°F). After the cooler, the liquor passes through a pressure control valve that returns the wastewater to atmospheric pressure. The wastewater then enters a gas-liquid separator. The oxidized liquor is collected from the bottom of the separator and the off-gas passes through a caustic scrubber containing a 5 percent NaOH solution. The oxygen content of the off-gas is monitored continuously with an oxygen meter, and a dry gas meter

Table 1. Selected Operating Parameters of the Wet Air Oxidation Process

Operating Parameter	Steady-State Conditions
Reactor inlet temperature	430–470°F
Reactor outlet temperature	440–480°F
Reactor pressure	1700 psig
Waste feed rate	2.5–3.0 gal/h
Dilution water feed rate	2.5–3.0 gal/h
Dilute nitric acid feed rate	0.5–0.6 gal/h
High-pressure air injection rate	60–80 scfh
Residual oxygen content of the off-gas	16–20%

measures the off-gas flow rate prior to release to the atmosphere. Off-gas samples collected before and after the scrubber are analyzed for methane and total hydrocarbons by a gas chromatograph.

Treatment of Oxidized Liquor

In order to discharge the oxidized liquor to the POTW, the wastewater required neutralization and metals precipitation. This was accomplished by the following process by use of nitric acid and sodium sulfide:

1. The oxidized liquor was pumped into a 100-gallon holding tank.
2. Nitric acid was added for pH adjustment. Two types of acid were used: (a) spent acid from the reactor acid wash or (b) concentrated, 38 degree Baumé acid.
3. Sufficient acid was added to lower the pH to 8.0 to 8.5.
   NOTE: Care must be taken when the acid is added because a large quantity of carbon dioxide is liberated and the reaction is somewhat violent.
4. Sulfide in the form of sodium hydrosulfide (NaHS) or sodium sulfide (Na_2S) was added to the tank and mixed.
5. The sulfide precipitates are very difficult to filter; therefore, diatomaceous earth was added as a filter aid to the slurry to improve filterability.
6. A sample of the slurry from the tank was filtered and a chip of sulfide was added to the filtrate. If the filtrate remained clear, the metal ion precipitation was complete. If a brown precipitate appeared, more sulfide was required. Additional sulfide was added to the holding tank until the filtrate remained clear when a sulfide chip was added.
7. A small plate and frame filter press was used for the filtering. The press cloths were precoated with diatomaceous earth prior to filtering the sulfide solids to improve filterability and prevent the sulfide solids from blinding the filter cloth. Filter cake generated from the treatment of the oxidized liquor was disposed of at a hazardous waste landfill.

Treatment Operational Problems

The most significant operational problems were plugging and/or scaling within the oil preheater and reactor. These problems were anticipated prior to the initial run because of high suspended and dissolved solids concentrations in the raw waste that did, in fact, result in the termination of the initial run. There were other operation problems (valve and gauge failure); however, they were more than likely due to the fact that the pilot systems had not been operational for over a year. After the initial run termination, modifications to the reactor (removal of internal mixing baffles), the oil preheater (repiping around the preheater), and further dilution of the raw waste allowed for the successful completion of the second run. The sampling program required 24 hours of steady-state operation. An engineering representative of Zimpro/Passavant estimated that the system could be operated for 7 to 10 days before requiring a complete system acid purge.

Following the completion of the sampling period, the treatment system was allowed to cool, and then the reactor was broken down to visually check the plugging/scaling effects of the second run. A brown, sand-like deposit was found at the bottom of the reactor. This material did not adhere to the reactor walls, but was deposited loosely at the reactor bottom. This would indicate that it was held in suspension in the reactor during operation and given sufficient operating time, would probably discharge from the reactor with the oxidized liquor [3].

A second type of material found in the reactor was a scale deposit on the reactor walls. This material adhered to the reactor walls and would not discharge from the reactor during operation. Removal of this scale would be required by the previously mentioned acid purge after a yet undetermined operating period. Neither the loose material in the bottom of the reactor nor the reactor scale should prevent effective treatment of F007 waste by the wet air oxidation technique. However, it could lower the heat transfer coefficient.

RESULTS AND CONCLUSIONS

As previously mentioned, in order to evaluate the effectiveness of wet air oxidation in this presentation, a "typical" F007 waste constituent characterization had to be generated for the feed stream. Incorporating this "typical" waste data in the calculation of percent removal for cyanide, 99.9 percent removal is observed for total and amenable cyanide as seen in Table 2. The destruction of cyanide in the wet oxidation process increased ammonia concentrations from minimal in the raw waste to an average of 5400 mg/L in the oxidized liquor. Cyanide was not detected in the off-gas scrubber water above its practical quantitation limit of 0.25 mg/L.

As seen in Table 2, a considerable amount of the copper and zinc is apparently removed by the wet ox-

Table 2. Percent Removals Across Oxidation/Precipitation Process

	F007* Raw Feed mg/L	Oxidized Liquor mg/L	Percent Removal	Filter Press Filtrate mg/L	Percent Removal
CN, total	33,000	2.51	99.9	1.65	99.9
CN, amenable	31,000	0.05	99.9	ND	100
Cu	5,000	802	84.0	2.45	99.9
Zn	10,000	6.47	99.9	2.45	99.9

*Rounded averages from other F007 wastes.
ND = not detected.

idation process. A mass balance for the process, including concentrations of the metals in the scale, bottom solids, and the acid wash following completion of the test run, reveals that the metals are concentrated in the scale and the bottom solids. In the mass balance calculations, metals concentrations from the actual raw waste were used. However, as previously pointed out, the raw data are not presented because of a claim of CBI by the generator.

Conclusions from this study are as follows:

1. Wet air oxidation is an effective treatment method for the destruction of cyanides in F007 wastes, including complexed cyanides.
2. Wet air oxidation, when followed by sulfide precipitation of metals, is an effective treatment system for the complete treatment of F007 wastes.
3. Engineering judgment and years of experience predict that other cyanide wastes containing metals, both liquids and sludges, could effectively be treated, after appropriate concentration or dilution, by the wet air oxidation/precipitation technology.

REFERENCES

1. USEPA, Office of Solid Waste, *Proposed Best Demonstrated Available Technology (BDAT) Background Document for Cyanide Wastes* (December 1988).
2. USEPA, 1988, Office of Solid Waste, *On-Site Engineering Report of Treatment Technology Performance and Operation for Wet Air Oxidation of F007 at Zimpro/Passavant*, Incorporated in Rothschild, Wisconsin, Washington, DC.
3. Zimpro/Passavant, *Final Report for the Pilot Plant Demonstration Study on Wet Air Oxidation of F007 Electroplating Cyanide Wastes*, Rothschild, Wisconsin (June 1988).

2.20

In Place Treatment of Contaminated Soil at Superfund Sites: A Review*

M. Roulier,[1] J. Ryan,[1] J. Houthoofd,[1] H. Pahren,[1] F. Custer[1]

ABSTRACT

The United States Environmental Protection Agency's (USEPA) Superfund research program is developing methods for in place (in situ) removal of contaminants from soils and for in place treatment of contaminated soils. This work is motivated by the high cost of managing large volumes of soil with low levels of contamination and because of the need to comply with provisions of the Superfund Amendments and Reauthorization Act (SARA) and the Resource Conservation and Recovery Act (RCRA). This paper summarizes available information sources and improvements in technology since 1984 when USEPA issued a summary report on in place soil treatment. There have been only a few instances of in place treatment based on aqueous solution chemistry, and these have involved primarily organic contaminants. Biodegradation has been successful for some organic compounds, and stabilization/solidification is increasingly successful for inorganics and some organics. In place vapor-phase removal processes such as vacuum extraction, steam stripping, and microwave heating appear most promising for low-solubility, low boiling point organic compounds. Improvements are needed in methods for delivering and mixing treatment materials in soil and in methods for recovering unreacted materials and reaction products.

INTRODUCTION

Work on in situ (in place) treatment of contaminated soils at Superfund sites is motivated by the need to comply with current regulations and by the high costs of treating large volumes of contaminated soil often encountered at sites. The provisions of the Comprehensive Environmental Response Compensation and Liability Act (CERCLA) as amended by SARA require the maximum possible treatment of wastes and contaminated soils at Superfund sites and restrict on-site containment and off-site disposal. The provisions of RCRA as amended by the Hazardous and Solid Waste Amendments (HSWA) establish treatment standards and disposal limitations that further restrict the disposition and encourage treatment of contaminated materials removed from Superfund sites.

A number of accepted technologies (e.g., incineration, stabilization/solidification, chemical treatment) are available or are being tested for the broad range of Superfund wastes and soils [1,2,3,4]. These technologies are cost-effective when applied to the concentrated wastes and highly contaminated soils that have been excavated. They have not been shown to be cost-effective for the large amounts of slightly contaminated soil that are often encountered after removal of wastes or near-surface soils. The volumes of soil remaining are usually large, and the cost of moving such soils for treatment is excessive. The alternatives for reducing these costs are to remove contaminants without moving the soils (in situ removal) or to treat the contaminants in place (in situ treatment).

The USEPA's first report [5] in 1984 on in place treatment described a large number of chemical and physical processes (e.g., oxidation, reduction, and precipitation) that could potentially be used in situ to immobilize or detoxify contaminants in soils. The authors noted that the majority of these processes were conceptual or had been tested only in the laboratory. Another USEPA report [6] published two years later evaluated in situ (in place) methods for stabilizing (treating) waste deposits. This report con-

---

*This paper appeared in *Remedial Action, Treatment and Disposal of Hazardous Waste Proceedings of the 15th Annual Research Symposium*, April 10–12, 1989, Cincinnati, OH.

[1]Risk Reduction Engineering Laboratory, U.S. Environmental Protection Agency, Cincinnati, OH 45268

sidered treatment materials (reactants), methods for delivering the treatment material, and methods for recovering the products of the reaction. The report noted that ". . . the combination of injection, reaction and recovery as a system for in situ remediation has scarcely been practiced and is in its infancy as an integrated technology" [6].

Most of these proposed treatments involved aqueous solution chemistry and it was hoped that workable in place treatment technologies would be developed along these lines. Instead, the major in place developments since that time have been in biodegradation, stabilization/solidification, and removal of contaminants. This trend is likely to continue with contaminants being increasingly recovered for aboveground treatment rather than being treated and left in place.

RECENT WORK

Biological and Physical Treatment

Biodegradation [1,2,3,7,8,9,10,11,12,13,14] and stabilization/solidification [1,3,4,10,15,16,17,18,19,20,21] are rapidly developing technologies that are distinquished, within the Superfund research program, as work areas separate from in place treatment. Only a few of the many recent references are listed. These technologies are being applied both in place and in aboveground reactors and batch plants. Biodegradation is effective only for organic contaminants; stabilization/solidification (S/S) processes are most effective for inorganic materials but are being developed to stabilize wastes containing organic contaminants. Experience with biodegradation and S/S suggests that these technologies will achieve expanded coverage of waste types and improved performance.

Both these technologies are limited by the problem of delivering materials to subsurface soils and achieving uniform mixing. Two commercial in place mixing technologies are available [15,18,22,23]; the cost and performance of these are being evaluated [3,4] in the Superfund Innovative Technology Evaluation (SITE) program. In place mixing technologies will also be appropriate for in place vapor-phase removal of organic compounds.

Chemical Treatment

There have been only a few examples of in place treatment of contaminated soil by chemical processes. Successful projects include treatment of polymer waste in an impoundment [24] and the treatment of a spill of acrylonitrile [25], various pesticides [26,27,28], and arsenic compounds in groundwater [29]. The limited amount of in place treatment appears to be due to difficulty in applying treatment materials uniformly and recovering unreacted materials and reaction products. There is also a reluctance to apply treatment materials that are dangerous substances (e.g., hydroxides and hypochlorites). The contaminants treated have generally been low-solubility organic chemicals. Organic chemicals with high aqueous solubility or high vapor pressures are transported out of the soil by natural processes or are easily removed in water or in the vapor phase for treatment above ground. Most inorganic contaminants are not treatable (cannot be truly degraded). Stabilization/solidification has been used for most of these contaminants but there is a controversy whether S/S processes are chemical treatments that form new low-solubility compounds or merely occlude the contaminants through physical processes and retard their release.

Delivery and Recovery

The limited success of in place treatment has led to increased work on processes for adding treatment materials and distributing them in soil and on processes for removing contaminants for treatment above ground without disturbing the soil (in place removal). The Superfund research program describes these collectively as innovative delivery and recovery processes. A recently completed review [30] identified seventeen processes, proposed or being used in other industries, that could be used for delivery or recovery in remediation of Superfund sites. The authors noted that several of the technologies were commercially available in industries such as petroleum production, but none were fully proven for use in waste site remediation. Vapor (vacuum) extraction is being used to decontaminate soils affected by leaking underground storage tanks and has successfully removed a variety of volatile organic compounds from soil.

Liquid-Phase Removal

Withdrawal of contaminated groundwater for treatment (pump and treat) was the first in place removal process used at Superfund sites. The Superfund research program is attempting to improve the pump-and-treat process through the use of intermittent (pulse) pumping for minimizing the amount of water removed per unit of contaminant and for avoiding hydrogeologic "dead zones" observed in many pump-and-treat operations. The engineering component of this research is being conducted by the USEPA's Risk

Reduction Engineering Laboratory (RREL) in Cincinnati, Ohio [31] and the evaluation/interpretation component by the USEPA's Robert S. Kerr Environmental Research Laboratory in Ada, Oklahoma [32]. These efforts are likely to improve removal of some inorganics and soluble organic compounds but not most inorganic cations or low-solubility organic compounds that are highly partitioned onto the solid phase in soil-water systems.

The aqueous liquid phase with chemicals added has also been used as a displacing solution to extract contaminants from soils and is called *soil flushing*. A pilot test for removal of organics [33] was less successful than full-scale operations for removal of cadmium from contaminated soils [34] and mixed metals and volatile fatty acids from a sludge lagoon [35]. Except in confined or well-controlled hydrogeologic settings, soil flushing will be limited for the same reasons as chemical treatment: difficulty in applying and recovering materials and reluctance to bear responsibility for adding dangerous chemicals to soil. Several commercial processes are available and are being tested under the SITE program [3,4].

Liquid Control

Two control procedures are being tested for use with liquid-phase removal. The movement of cations and soil water in response to an applied direct current gradient, called electrokinetics or electro-osmosis, has been adapted from applications in the mining industry and is being considered as a means for directing the movement of water and contaminants during removal/treatment in low-permeability soils [2,36]. The RREL sponsored one field test [37] and the SITE Emerging Technologies Program is testing the process when used in conjunction with an acoustic field [4]. In place freezing techniques used in the construction industry are being considered for temporarily making soil impermeable or for concentrating contaminants ahead of a slowly moving freezing front [38,39]. Both of these control procedures are in the early stages of testing to determine their usefulness at Superfund sites.

Gas-Phase Removal

For many low-solubility organic compounds, equilibrium thermodynamics indicates that the aqueous liquid phase will contain greater amounts of organic contaminants than an equal volume of the gas phase. However, the rates of transfer from the solid phase to the liquid phase are often slow relative to rates of transfer into the gas phase. Additionally, it is easier, because of relative densities and viscosities, to move large volumes of gas through the subsurface than it is to move equal volumes of liquid. This combination of greater convective flow and more rapid phase transfer makes gas-phase removal an efficient process for many organic compounds, particularly in vadose (partially saturated) zones where it is difficult to remove water.

Gas-phase removal, called *in situ vacuum extraction*, *air stripping*, *volatilization*, *vapor extraction*, or *forced-air venting*, is a commercially available process that has been used most extensively in decontaminating soils affected by leaking underground storage tanks [21,40,41,42,43,44] that often contain fuels and other volatile substances. A combination of air injection and vacuum venting wells are used to induce air flow through the contaminated soil. Performance data for this technology have been recently summarized [45] and detailed cost analyses were conducted for several typical systems [46]. Several gas-phase removal processes are being tested under the SITE program [3,4,22,23]. Although gas-phase removal is used almost exclusively in vadose zone soils, there is at least one site in Europe [10] and one in the United States [47] where this process has been used to simultaneously remove organics from groundwater and soil by injecting air into the saturated zone and capturing the stripped organics by using a vacuum system.

Soil Heating

Simple gas-phase removal is most effective for organic compounds with relatively low boiling points. Several methods are being tested for heating contaminated soil to increase vapor pressure and allow gas-phase removal of higher boiling point compounds. Radio frequency (microwave) heating has been used successfully in a U.S. Air Force pilot-scale field test for removal of fuel oil [48,49] and a full-scale test is being planned. The RREL is sponsoring a laboratory test of radio frequency heating for removing the higher boiling compounds, creosote and pentachlorophenol, from soils contaminated with wood-treating fluids. Steam injection, developed for petroleum recovery [30], is being adapted for use with vacuum extraction at hazardous waste sites [50]. It has been tested at a small scale in the field for solvent removal [51] and will be examined at a larger scale in a RREL field study that is being initiated. A combination of steam and hot air are used to heat soil for several of the gas-phase removal technologies being tested under the SITE program [3,4,22,23]. Another soil heating technology melts (vitrifies) the soil [52,53] to destroy most of the organic compounds and captures the remainder with a vacuum hood on the surface. This technology had been proposed for use at one Superfund site and is being tested under the SITE program [3,4] for treatment efficiency and for its ability to control the loss of volatiles to surrounding soils.

SUMMARY

Since EPA's first report on in place treatment, the technologies for in place removal, particularly in the gas phase, have developed much more rapidly than those for in place chemical treatment. Contaminants will be increasingly recovered for treatment above ground rather than being left in place after treatment. Biodegradation and stabilization/solidification can successfully treat some organic and inorganic contaminants in place, and these two technologies will continue to improve their ability in this regard. Further work on the development of methods for delivering and mixing treatment materials in soil and for recovering unreacted materials and reaction products would expedite improvements in all in place treatment and removal processes.

REFERENCES

1. U.S. Environmental Protection Agency. *Handbook: Remedial Action at Waste Disposal Sites (Revised), EPA/625/6-85/006.* USEPA Hazardous Waste Engineering Research Laboratory, Cincinnati, OH and USEPA Office of Emergency and Remedial Response, Washington, DC (1985).
2. Sanning, D. E. and R. F. Lewis. *Technologies for In Situ Treatment of Hazardous Wastes, EPA/600/D-87/014, NTIS-PB87-146007.* USEPA Hazardous Waste Engineering Research Laboratory, Cincinnati, OH (January 1987).
3. U.S. Environmental Protection Agency. *Technology Screening Guide for Treatment of CERCLA Soils and Sludges, EPA/540/2-88/004.* USEPA Office of Emergency and Remedial Response, Washington, DC (1988).
4. U.S. Environmental Protection Agency. *The Superfund Innovative Technology Evaluation Program: Technology Profiles, EPA/540/5-88/003.* USEPA Risk Reduction Engineering Laboratory, Cincinnati, OH (1988).
5. U.S. Environmental Protection Agency. *Review of In place Treatment Techniques for Contaminated Surface Soils (Volume 1: Technical Evaluation), EPA/540/2-84/003a.* USEPA Municipal Environmental Research Laboratory, Cincinnati, OH and USEPA Office of Solid Waste and Emergency Response, Washington, DC (1984).
6. U.S. Environmental Protection Agency. *Systems to Accelerate In Situ Stabilization of Waste Deposits, EPA/540/2-86/002.* USEPA Hazardous Waste Engineering Research Laboratory, Cincinnati, OH (1986).
7. Lee, M. D., J. M. Thomas, R. C. Borden, P. B. Bedient, J. T. Wilson and C. H. Ward. *Biorestoration of Aquifers Contaminated with Organic Compounds. CRC Critical Reviews in Environmental Control,* 18(1):29–89 (1988).
8. Soczo, E. R. and J. J. M. Staps. "Review of Biological Soil Treatment Techniques in The Netherlands," in *Contaminated Soil '88 Volume 1, Proceedings of the Second International TNO/BMFT Conference on Contaminated Soil, 11–15 April 1988, Hamburg, Federal Republic of Germany.* K. Wolf, J. van den Brink, and F. J. Colon, eds. The Netherlands:Kluwer Academic Publishers, pp. 663–670 (1988).
9. Mischgofsky, F. H. and R. Kabos. "1988 General Survey of Site Clearance Techniques: Trend Towards In Situ Treatment," in *Contaminated Soil '88 Volume 1, Proceedings of the Second International TNO/BMFT Conference on Contaminated Soil, April 11–15, 1988, Hamburg, Federal Republic of Germany.* K. Wolf, J. van den Brink, and F. J. Colon, eds. The Netherlands:Kluwer Academic Publishers, pp. 523–533 (1988).
10. U.S. Environmental Protection Agency. *Assessment of International Technologies for Superfund Applications: Technology Review and Trip Report Results, EPA/540/2-88/003.* USEPA Office of Solid Waste and Emergency Response, Washington, DC (1988).
11. Wetzel, R. S., D. J. Sarno, C. M. Durst, B. C. Vickers, P. A. Spooner, J. R. Payne, W. D. Ellis, M. S. Floyd and Z. A. Saleem. *In Situ Biological Degradation Test at Kelly Air Force Base, Volume I: Site Characterization, Laboratory Studies and Treatment System Design and Installation (ESL-TR-85-52-I).* U.S. Air Force Engineering & Services Center, Tyndall Air Force Base, FL 32403 (1986).
12. Wetzel, R. S., C. M. Durst, D. H. Davidson and D. J. Sarno. *In Situ Biological Treatment Test at Kelly Air Force Base, Volume II: Field Test Results and Cost Model (ESL-TR-85-52-II).* U.S. Air Force Engineering & Services Center, Tyndall Air Force Base, FL 32403 (1987).
13. Wetzel, R. S., C. M. Durst, D. H. Davidson and D. J. Sarno. *In Situ Biological Treatment Test at Kelly Air Force Base, Volume III: Appendices (ESL-TR-85-52-III).* U.S. Air Force Engineering & Services Center, Tyndall Air Force Base, FL 32403 (1987).
14. Olfenbuttel, R. F., E. Heyse, D. C. Downey and T. L. Stoddart. "Lessons Learned: A Basis for Future Success," in *Contaminated Soil '88 Volume 1, Proceedings of the Second International TNO/BMFT Conference on Contaminated Soil, April 11–15, 1988, Hamburg, Federal Republic of Germany.* K. Wolf, J. van den Brink and F. J. Colon, eds. The Netherlands:Kluwer Academic Publishers, pp. 471–479 (1988).
15. Ghassemi, M. "Innovative In Situ Treatment Technologies for Cleanup of Contaminated Sites," *Journal of Hazardous Materials,* 17:189–206 (1988).
16. Puglionesi, P. S., J. Kesari, M. H. Corbin and E. B. Hangeland. "Heavy Metals-Contaminated Soils Treatment," in *Superfund '87, Proceedings of the 8th National Conference.* The Hazardous Materials Control Research Institute, Silver Spring, MD, pp. 380–384 (1987).

17. Smith, R. L., D. T. Musser and T. J. DeGrood. "In Situ Solidification/Fixation of Industrial Wastes," in *Management of Uncontrolled Hazardous Wastes Sites*. The Hazardous Materials Control Research Institute, Silver Spring, MD, pp. 231–233 (1985).
18. Stinson, M. K. and S. Sawyer. "In Situ Treatment of PCB-Contaminated Soil," in *Superfund '88, Proceedings of the 9th National Conference*. The Hazardous Materials Control Research Institute, Silver Spring, MD, pp. 504–507 (1988).
19. Wiles, C. C. and H. K. Howard. "U.S. EPA Research in Solidification/Stabilization of Waste Materials," in *Land Disposal, Remedial Action, Incineration and Treatment of Hazardous Waste—Proceedings of the Fourteenth Annual Research Symposium, EPA/600/9-88/021*. USEPA Risk Reduction Engineering Laboratory, Cincinnati, OH, pp. 126–135 (1988).
20. U.S. Environmental Protection Agency. *Handbook for Stabilization/Solidification of Hazardous Wastes, EPA/540/2-86/001*. USEPA Hazardous Waste Engineering Research Laboratory, Cincinnati, OH (1986).
21. White, D. C., J. R. Dunckel and T. D. Van Epp. "Applying Alternative Technologies at Superfund Sites," in *Management of Uncontrolled Hazardous Wastes Sites*. The Hazardous Materials Control Research Institute, Silver Spring, MD, pp. 361–364 (1986).
22. La Mori, P. N. "In Situ Treatment Process for Removal of Volatile Hydrocarbons from Soils: Results of Prototype Test," in *Proceedings, Second International Conference on New Frontiers for Hazardous Waste Management*. USEPA Hazardous Waste Engineering Research Laboratory, Cincinnati, OH, pp. 503–509 (1987).
23. Van Tassel, R. and P. N. LaMori. "Innovative In Situ Soil Decontamination System," in *Superfund '87, Proceedings of the 8th National Conference*. The Hazardous Materials Control Research Institute, Silver Spring, MD, pp. 396–402 (1987).
24. Bort, M. R. "In Situ Stabilization of Viscoelastic Polymer Waste," in *Management of Uncontrolled Hazardous Wastes Sites*. The Hazardous Materials Control Research Institute, Silver Spring, MD, pp. 152–156 (1985).
25. Harsh, K. M. "In Situ Neutralization of an Acrylonitrile Spill," in *Proceeding of 1978 Conf. on Control of Haz. Materials Spills, April 11–13, Miami, FL*. Information Transfer Inc., Rockville, Maryland, pp. 187–189 (1978).
26. King, J., T. Tinto and M. Ridosh. "In Situ Treatment of Pesticide Contaminated Soils," in *Management of Uncontrolled Hazardous Wastes Sites*. The Hazardous Materials Control Institute, Silver Spring, MD, pp. 243–248 (1985).
27. Paulson, D. L., R. Honeycutt, Jr., J. Lebaron and V. Seim. "Degradation of High Concentration of Diazinon in Soil by Parathion Hydrolase," in *1984 Hazardous Material Spills Conference Proc.-Prevention, Behavior, Control, and Cleanup of Spills and Waste Sites*. J. Ludwigson, ed. Government Institutes, Inc., Rockville, MD, pp. 92–97 (1984).
28. Ryckman, M. D. "Detoxification of Soils, Water and Burn Residues from a Major Agricultural Chemical Warehouse Fire," in *Proceedings of the 5th National Conference on Management of Uncontrolled Hazardous Waste Sites*. Hazardous Materials Control Research Institute, Silver Spring, MD, pp. 420–426 (1984).
29. Stief, K. "Remedial Action for Groundwater Protection Case Studies within the Federal Republic of Germany," in *Proceedings of the 5th National Conference on Management of Uncontrolled Hazardous Waste Sites*. Hazardous Materials Control Research Institute, Silver Spring, MD, pp. 565–568 (1984).
30. Murdoch, L., B. Patterson, G. Losonsky and W. Harrar. *A Review of Innovative Technologies of Delivery or Recovery for the Remediation of Hazardous Waste Sites*. Final report to the USEPA Risk Reduction Engineering Laboratory for work assignment 1-11, Contract 68-03-3379. Center Hill Research Facility, Cincinnati, OH 45221.
31. Beljin, M. S. and J. F. Keeley. *In Situ Delivery and Recovery: Pulse (Cyclic) Pumping, Phase I*. Contract 68-03-4038 with the USEPA Risk Reduction Engineering Laboratory, Cincinnati, OH, Work Assignment 2-WM32.0, University of Cincinnati (1989).
32. Keeley, J. F. *Remedial Performance Evaluation for Pump and Treat*. Cooperative Agreement CR812808 between the USEPA Robert S. Kerr Environmental Research Laboratory, Ada, OK and the National Center for Groundwater Research, Rice University (1988).
33. Nash, J. H. Project Summary: *Field Studies of In Situ Soil Washing, EPA/600/S2-87/110*, February 1988. USEPA Hazardous Waste Engineering Research Laboratory, Cincinnati, OH (1988).
34. Urlings, L. G. C. M., V. P. Ackerman, J. C. V. Woundenberg, P. P. v.d. Pijl and J. J. Gaastra. "In Situ Cadmium Removal—Full-Scale Remedial Action of Contaminated Soil," in *Contaminated Soil '88 Volume 1, Proceedings of the Second International TNO/BMFT Conference on Contaminated Soil, 11–15 April 1988, Hamburg, Federal Republic of Germany*. K. Wolf, J. van den Brink, F. J. Colon, eds. The Netherlands: Kluser Academic Publishers, pp. 911–920 (1988).
35. Legiec, I. A. and D. S. Kosson. "In Situ Extraction of Industrial Sludges," *Environmental Progress*, 7(4): 270–278 (1988).
36. Herrmann, J. G. *Status Report on Electrokinetics Technology*. January 23, 1989 memorandum from J. G. Herrmann, USEPA Risk Reduction Engineering Laboratory, Cincinnati, Ohio to J. Kingscott, USEPA Office of Solid Waste and Emergency Response, Washington, DC. (1989).
37. Horng, J. and S. Banerjee. "Evaluating Electrokinetics as a Remedial Action Technique," in *Proceedings, Second International Conference on New Frontiers for Hazardous Waste Management, EPA/600/9-87/018F*. USEPA Hazardous Waste Engineering Research Laboratory, Cincinnati, OH, pp. 65–77 (1987).
38. Iskandar, I. K. and J. M. Houthoofd. "Effect of Freezing on the Level of Contaminants in Uncontrolled Hazardous Waste Sites—Part 1: Literature Review and Concepts," in *Proceedings, Eleventh Annual Research Symposium, EPA/600/9-85/013, Cincinnati, OH, April 29–May 1, 1985*, pp. 122–129 (1985).
39. Iskandar, I. K. and T. F. Jenkins. "Potential Use of Artificial Ground Freezing for Contaminant Immobilization," in *Proceedings, International Conference on New Frontiers for Hazardous Wastes Management, EPA/600/9-85/025*. USEPA Hazardous Wastes Engineering Research Laboratory, Cincinnati, OH, pp. 128–137 (1985).

40. Anastos, G. J., P. J. Marks, M. H. Corbin and M. F. Coia. "In Situ Air Stripping of Soils Pilot Study," U.S. Army Toxic and Hazardous Materials Agency, Report (AMXTH-TE-TR-85026). Aberdeen Proving Ground (Edgewood Area), MD (1985).

41. Electric Power Research Institute. *Remedial Technologies for Leaking Underground Storage Tanks (EPRI CS-5261)*. Electric Power Research Institute, Palo Alto, CA (1987).

42. U.S. Environmental Protection Agency. *Underground Storage Tank Corrective Action Technologies, EPA/625/6-87/015*. USEPA Office of Solid Waste and Emergency Response, Washington, DC (1987).

43. Glynn, W. and C. Fan. "Underground Storage Tank Corrective Action: Application and Field Evaluation of Vacuum Extraction Technology," in *Land Disposal, Remedial Action, Incineration and Treatment of Hazardous Waste—Proceedings of the Fourteenth Annual Research Symposium, EPA/600/9-88/021*. USEPA Risk Reduction Engineering Laboratory, Cincinnati, OH, pp. 12–24 (1988).

44. U.S. Environmental Protection Agency. *Cleanup of Releases from Petroleum USTs: Selected Technologies, EPA/530/UST-88/001*, USEPA Office of Underground Storage Tanks, Washington, DC (1988).

45. Hutzler, N. J., B. E. Murphy and J. S. Gierke. *State of Technology Review: Soil Vapor Extraction Systems*. Interim Final Report to the USEPA Risk Reduction Engineering Laboratory for Cooperative Agreement CR-814319, Michigan Technological University, Houghton, Michigan 49931 (1989).

46. Metzer, N., M. Corbin and S. Cullinan. *In Situ Volatilization (ISV) Remedial System Cost Analysis*. (AMXTH-TE-CR-87123) U.S. Army Toxic and Hazardous Materials Agency, Edgewood Area, Aberdeen Proving Ground, MD 21010 (1987).

47. Webster, K. T. Aquadetox/VES Integrated System. AWD Technologies, Fairfield, NJ (1988) (personal communication).

48. Dev, H., P. Condorelli, J. Bridges, C. Rogers and D. Downey. "In Situ Radio Frequency Heating Process for Decontamination of Soil," in *Solving Hazardous Waste Problems Learning from Dioxins*. Division of Environmental Chemistry at the 191 Meeting of the American Chemical Society, New York, NY, pp. 329–339 (April 13–18, 1986).

49. Dev, H., G. C. Sresty, J. E. Bridges and D. Downey. "Field Test of the Radio Frequency In Situ Soil Decontamination Process," in *Superfund '88, Proceedings of the 9th National Conference*. The Hazardous Materials Control Research Institute, Silver Spring, MD, pp. 498–502 (1988).

50. Lord, A. E., Jr., R. M. Koerner and V. P. Murphy. "Laboratory Studies of Vacuum-Assisted Steam Stripping of Organic Contaminants from Soil," in *Land Disposal, Remedial Action, Incineration and Treatment of Hazardous Waste, Proceedings of the Fourteenth Annual Research Symposium, EPA/600/9-88/021*. USEPA Risk Reduction Engineering Laboratory, Cincinnati, OH, pp. 65–92 (1988).

51. Udell, K. S. Results of a Pilot Study of Vacuum Extraction and Steam Injection at Solvent Service, Inc., San Jose, California. University of California, Berkeley, CA (1988) (personal communication).

52. FitzPatrick, V. F., C. L. Timmerman and J. L. Buelt. "In Situ Vitrification—An Innovative Thermal Treatment Technology," in *Proceedings, Second International Conference on New Frontiers for Hazardous Waste Engineering Research Laboratory, Cincinnati, OH*, pp. 305–322 (1987).

53. FitzPatrick, V. F. "In Situ Vitrification and Innovative Melting Technology for the Remediation of Contaminated Soil," in *Contaminated Soil '88 Volume 1, Proceedings of the Second International TNO/BMFT Conference on Contaminated Soil, April 11–15, 1988, Hamburg, Federal Republic of Germany*. K. Wolf, J. van den Brink, and F. J. Colon, eds. The Netherlands:Kluwer Academic Publishers, pp. 857–859 (1988).

Control of Air Emissions from Soil Venting Systems

F. A. M. Buck,[1] Craig A. Smith[2]

INTRODUCTION

The cleanup or remediation of soils and ground water contaminated by VOCs (volatile organic compounds, e.g., gasoline) has become a great concern of state and local regulatory agencies. The cleanup requirements set by the agencies are usually determined after evaluation of the VOC underground plume. Important considerations are contaminant type, contamination extent (vertical and horizontal), proximity to ground water, environmental threats, potential damage to public health and safety, and the available technology for remediation.

EXTENT OF THE PROBLEM

The number of VOC-contaminated sites in the United States is not known. In the four southern California counties comprising the South Coast Air Quality Management District (AQMD), however, it is estimated that over 1800 sites are monitored per year [1]. About 30 percent of these sites, or 540 per year, have VOC contamination at a level that requires remedial action under South Coast regulations. Without remedial action, it is estimated that there would be 10 tons per day of VOC emissions. Under South Coast AQMD regulations, the emission rate should be reduced by a minimum of 90 percent [2].

CONTROL TECHNOLOGIES

A number of methods are available for removing VOC contamination from soil. They include the following:

1. Excavation and transport
2. Excavation and treatment of the excavated soil at the site
3. On-site chemical treatment
4. On-site biodegradation
5. Vapor extraction and treatment of the recovered VOC

There is often an overlap in methods 4 and 5; that is, the vapor extraction process induces a flow of fresh air into the soil and this seems invariably to stimulate aerobic biodegradation by naturally occurring soil organisms.

The selection of a remedial method depends on practical on-site conditions and restrictions, on cost-effectiveness, and on the ability to meet the regulations of agencies having jurisdiction, such as the regional fire department, department of health, water quality control board, air pollution control district, and others.

VAPOR EXTRACTION

As a recently developed technology, vapor extraction (also referred to as soil venting) is receiving widespread acceptance for the treatment of VOC-contaminated soils. Simply described, a vacuum pump or fan is connected to one or more properly designed extraction wells that are typically installed to penetrate the contaminant plume near the zone of highest VOC concentration. When suction is applied to the well(s), it induces a subsurface air flow radially toward perforations in the well casing. Ventilation wells may be placed at selected locations to help direct the flow of induced air toward the extrac-

---

[1] King, Buck & Associates, Inc., 2384 San Diego Avenue, Suite 2, San Diego, CA 92110
[2] Nachant Environmental, Inc.

tion wells (see Figure 1). The design of a collection system that cleans up the contaminated site in minimum time and at minimum equipment cost is dependent upon the skills of experienced hydrogeologists.

Vapor extraction equipment has been developed to handle a wide range of vapor extraction rates, to greater than 28 nm³/min (1000 scfm). Early equipment usually had low capacities of about 0.3 nm³/min (10 scfm), but lengthy cleanup times inspired the development of equipment capable of higher extraction rates. However, with the higher extraction rates come potentially higher emission rates of VOCs to the atmosphere. This led to more concern over air quality effects.

Many agencies responsible for controlling air pollution have mandated treatment of the recovered vapors. Air pollution control districts in southern California have been in the forefront with strict regulations controlling the process of soil venting. The South Coast Air Quality Management District issued Rule 1166, "Volatile Organic Compound Emissions from Decontamination of Soil," in August 1988. The applicability is to "limit the emissions of VOC" from soil contaminated with VOCs, and the regulation contains requirements that control VOC emissions during any type of mitigation measure, including vapor extraction [2].

EMISSION CONTROL SYSTEMS

Several potential treatment processes are available to meet pollution control standards using vapor extraction. These methods include the following:

1. Refrigerated condensation to recover VOCs as liquids
2. Adsorption of VOC on activated charcoal
3. Burning (oxidation) of VOCs by thermal oxidation, catalytic oxidation, or by a two-staged thermal and catalytic oxidation. The two-stage oxidation involves the induction of the extracted vapor into the intake of an internal combustion engine having a catalytic converter on the engine exhaust and using thermal oxidation when the VOC content of extracted vapor is high and a catalytic oxidizer when the VOC content is moderate to low.

If the VOC is a hydrocarbon, or if its combustion products do not contain appreciable amounts of corrosive or noxious compounds such as HCl, burning of the extracted vapor (by thermal or catalytic oxidation) is usually the process preferred economically.

A few process characteristics are offered supporting the general conclusion that oxidation processes are the preferred choice. The product recovered by refrigerated condensation would be gummy, partially oxidized VOCs that present a disposal problem. It could be burned; therefore, it would be better to burn it in the vapor form without using a costly refrigeration process as an intermediate step.

As for adsorption on activated charcoal, the process has competitive economics only at very low VOC concentrations in the extracted vapor. (For an operating cost comparison see Figure 5).

Burning the VOC in an internal combustion engine (ICE) may seem to be attractive, but problems arise from some basic characteristics of ICEs. First, the modern ICE is a complex, finely tuned machine that requires highly refined fuels and lubricants that meet stringent specifications. Purity specifications rule out materials that would form gums or lacquers in the carburetor, on valves, etc., but the soil vent vapor is just such a reactive, gummy fuel. Moreover, the combustion process must be completed efficiently in a fraction of a second. Experience has shown that extracted vapor has large variations in combustion properties. Not only does the VOC content vary but, as shown by analyses, carbon dioxide content may range to 6 percent or more, and oxygen may be much lower than the twenty-one percent by volume expected in fresh air. Extracted vapor with such composition has a significantly different flame propagation speed than normal fuel air mixtures. Operating conditions of thermal oxidizers have to be modified to burn process gas efficiently if it has high CO_2 and low O_2 concentrations. The problems are magnified in an ICE, especially if power output is a consideration.

EMISSION CONTROL BY OXIDATION

VOCs in the extracted air can be oxidized efficiently in properly designed burners if the concentration of combustibles is above their lower flammability limit (LFL) in air. At VOC concentrations

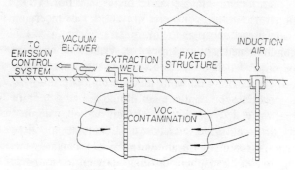

FIGURE 1. In situ vapor extraction.

below the LFL, enrichment fuel (e.g. natural gas or propane) must be added to maintain efficient combustion. There is no lower limit on VOC concentration below which thermal oxidation will not effectively destroy the VOC; the requirement is merely to add enough enrichment fuel to maintain a stable, efficiently burning flame. Experience has shown that a premix of air and propane or natural gas will burn efficiently at a fuel content 20 to 50 percent above the LFL. This permits an estimation of the maximum fuel consumption for a premix-type thermal oxidizer in an idealized case. For example, the fuel required for an efficient flame with 100 scfm air (no VOC) would be 2.8 percent by volume of propane ($=1.33 \times$ LFL). Propane consumed therefore would be 2.8 scfm, or 4.6 gal/h. VOCs in the extracted vapor will reduce the fuel requirement below the "fresh air" case but, as mentioned earlier, other factors such as the concentrations of oxygen, carbon dioxide, and water vapor will affect the combustion process. It is not always accurate to estimate the supplementary fuel consumption by considering only the VOC content of the extracted vapor.

At VOC levels at or below about 30 percent of the LFL, oxidation over catalysts can be used for efficient VOC destruction. There are two boundary limits to the catalytic process: (1) if the VOC concentration is too high, the heat released during oxidation will cause high-temperature destruction of the catalytic oxidizer, or (2) if the VOC concentration is too low, not enough heat is released to maintain the catalyst at a temperature needed for efficient oxidation.

In the first situation, where the exotherm (heat released during oxidation of the VOC) is too large for the system, it is obvious that dilution air can be admitted to the suction of the vacuum/compressor unit to dilute the VOC content and lower the exotherm. There is an unwanted result of this dilution, however. The system capacity is set by one or more limitations on the system, e.g., by the capacity of the vacuum pump, or by the design space velocity of the process gas across the catalyst. At the maximum flow rate of the process gas, dilution of the extracted vapor with fresh air (to contain the exotherm) means that the rate of extraction of vapor from the soil must be reduced. Therefore, this reduction will add to the time and cost required for site remediation. To avoid reducing the extraction rate of vent gas, it is desirable to oxidize high-VOC-content vapor in a thermal oxidizer and to use the catalytic oxidizer with vapor of lower VOC content that has an exotherm compatible with the catalytic system.

The second operating limit on VOC concentration leads naturally to some type of preheater for the process gas before it enters the catalyst bed(s). Usually, a preheat system is controlled by a process controller reading the temperature of one or more thermocouples in the catalytic reactor. For efficient oxidation of most VOCs, the minimum temperature at catalyst entry will be set between 315 and 370°C (600 to 700°F). Energy conservation in a catalytic oxidizer is standard. Usually a product-to-feed gas heat exchanger recaptures 50 percent or more of the sensible heat of the effluent gas from the catalytic reactor before the effluent is discharged to the atmosphere.

An isometric block diagram of a two-stage thermal/catalytic oxidation system is shown by Figure 2.

OPERATING DATA

Efficiency of the emission control system is calculated by measuring total hydrocarbons (THCs) in the effluent from the treatment system and comparing the result to THCs in the extracted vapor. From the point of view of the emission control system, the latter stream is called the influent. THCs can be determined by laboratory analyses of gas samples taken at the operating site, or by continuous, on-line analysis of the influent and effluent. A variety of flame ionization detectors (F.I.D.) are available commercially that have proven to be satisfactory for measuring THCs.

Typically, when an emission control system has been permitted by a local air quality control agency, a condition of the permit requires that a formal performance test be made, often by a third party testing laboratory. The performance test must demonstrate that the equipment is controlling emissions below the limits specified in the agency's permit.

The following operating and analytical data were obtained during agency-approved performance tests. The test reported in Table 1 was run in 1988, that of Table 2 in 1987; both were in the South Coast AQMD.

Vapor compositions from two vapor extraction sites are shown in Figure 3. The data illustrate the broad range in THC concentrations typically encountered. The top curve shows data from an operation during which the extracted vapor was in the flammable range for more than 20 days (the LFL of the higher boiling gasoline vapors in air is about 12,000 ppmv, or 1.2 percent by volume). The lower curve shows data obtained during operation at a different site. In this case the THCs in the extracted vapor fell steadily; after 240 days it had declined from its initial value above 10,000 ppmv to 500 ppmv.

To estimate the amount of THC potentially emitted to the atmosphere during vapor extraction, refer to Figure 4. In this chart, the conversion from concen-

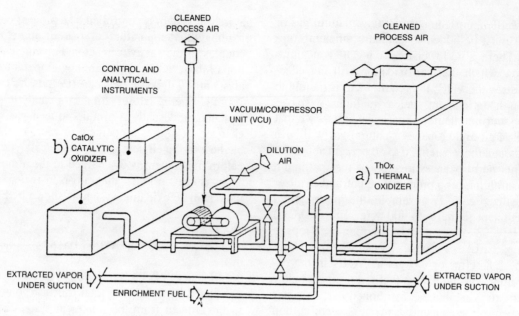

FIGURE 2. Two-stage thermal/catalytic oxidation process.

tration of gasoline vapor to emissions in pounds per day is based on computed physical properties of a surrogate gasoline. The surrogate gasoline was taken to be one-third each of isooctane, methylcyclohexane, and toluene (an alkane, a cycloalkane, and an aromatic). It is assumed to be representative of the weathered gasolines typically encountered in vapor extraction projects. The surrogate gasoline has a vapor density, under standard conditions, of 0.25 lb/ft^3.

From the lower curve of Figure 3, and the correlation of Figure 4, it is calculated that vapor extraction at 100 scfm would have emitted 15,800 lb (2400 gal) of gasoline vapors over the 240-day period. Emission control, at a minimum 90 percent destruction efficiency of THCs, therefore reduced atmospheric emissions by 14,000 lb or 2170 gal.

OPERATING COSTS

The base comparison case for soil extraction processes is remediation by soil excavation and transport. There are at least four intrinsic disadvantages

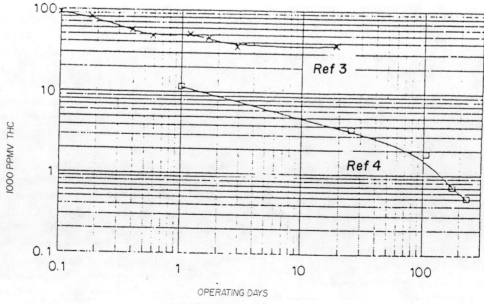

FIGURE 3. VOC decline curves.

Table 1. Performance of a Thermal Oxidizer, Vapor Extraction of a Gasoline-Contaminated Site

ThermOx Inlet:	
Flow rate, contaminated air	95 cfm
Flow rate, natural gas (enrichment fuel)	6 cfm
Total hydrocarbons (as propane)	2.56%
benzene, mg/m³	7200
ThermOx Effluent:	
Volumetric flow rate	2508 dscfm
Total hydrocarbons (as propane), ppmv	11.0
benzene, mg/m³	ND*
oxides of nitrogen, ppmv	2.5
carbon monoxide, ppmv	22.5
Calculated destruction efficiency of THCs:	98.8%

*Detection limit 5.35 mg/m³.

Table 2. Performance of a Catalytic Oxidizer, Vapor Extraction of a Gasoline-Contaminated Site

	CatOx Inlet	CatOx Outlet
Flow rate, dscfm	72	—
NOx, ppm	<1	<1
CO, ppm	<15	<10
Hydrocarbons:		
CH_4, ppm	—	30
Nonmethane, ppm as C_1	14,400	573
Hydrocarbon destruction efficiency, %		95.8

with this type of mitigation: (1) weathering of the VOC contaminant is uncontrolled during both digging and trucking of the contaminated soil; (2) in the long run, contaminants have merely been taken somewhere else; (3) in most cases, the waste generator retains liability for material admitted to a landfill; and (4) many states require that annual fees be paid indefinitely for maintenance of the landfill. A contractor in southern California, who practices both vapor extraction and excavation/transport, reports that costs in excess of $300 per cubic yard are common for the process of removal, transportation, and landfill acceptance [5]. Excavation of contaminated soil deep beneath the surface or beneath fixed surface improvements (e.g., roads or buildings) would be much more costly if not impossible. At the lower figure of $300 per cubic yard, it is not unusual for modest-sized remediation projects to cost $100,000 or more.

Total daily operating costs for a two-stage thermal/catalytic oxidizer of 100 scfm capacity and comparative costs for emission control by adsorption of THCs on activated charcoal at 100 scfm are given in Figure 5. Total costs include utilities and materials, operating labor, capital payback, and an illustrative cost for obtaining operating permits. Project management and overhead costs are not included. It is seen that as a broad average the *total* operating costs (capital payback included) for the oxidation process run between $200 and $300 per day.

The required rate of capital payback is a subjective figure for which each project manager will set his/

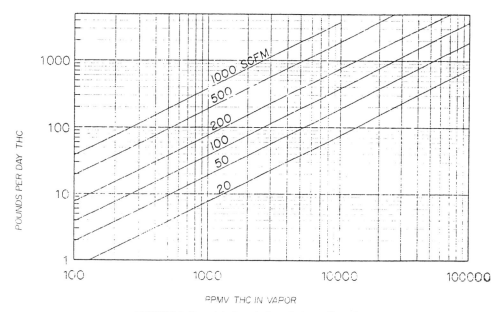

FIGURE 4. Potential air emissions during soil venting.

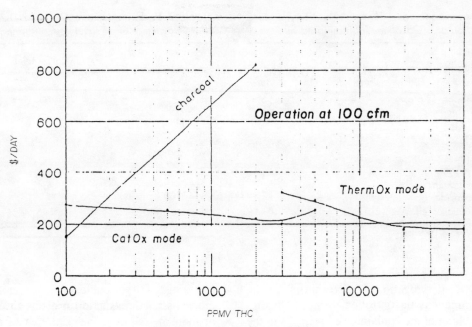

FIGURE 5. Total daily operating costs.

her own target. The capital payback rate included in the total costs of Figure 5 was obtained from the quotes of a commercial leasing company for a two-year equipment lease to a customer of sound credit rating. Annual costs were divided by 330 operating days per year to derive daily rates.

Since direct operating costs depend on the combustible content of the extracted vapor, the costs are plotted as a function of the THC content of the vapor. Minimum daily costs are experienced with the thermal oxidizer when the vapor is flammable, and costs increase as the vapor gets leaner because more supplementary fuel is required. Minimum daily costs are experienced with the catalytic oxidizer at about 3,000 ppmv THC in the extracted vapor, and operating costs increase as the vapor gets leaner and more preheat is required to keep the catalyst at an efficient operating temperature.

Comparable cost figures for charcoal adsorption show that this process, despite its lower capital cost, has daily operating costs much higher than the oxidation process. The principal operating cost with activated carbon is the replenishment of the charcoal when it approaches saturation with one or more hydrocarbon components and breakthrough is imminent. Charcoal adsorption is a lower cost process only if the THC content of the vent gas is less than 200 ppmv.

The cost of operating a vapor extraction system and of controlling the potential emissions by oxidation of the VOC can be estimated roughly from Figure 5. Daily costs would be based on the average of $275/day (including capital recovery). Then, for example, if the on-site cleanup time were projected to be 6 months, the estimated cost would be about $50,000 for a process that reduced emissions to the limits set by southern California air pollution control agencies.

NOMENCLATURE AND SI CONVERSION FACTORS

scfm = standard cubic feet per minute
dscfm = dry standard cubic feet per minute
nm = normal meters
100 scfm = 2.83 nm³/min
1.0 lb/d = 0.454 kg/d
1.0 gal/h = 3.785 L/h

REFERENCES

1. Basilio, L. Rule Development Division, South Coast Air Quality Management District, Supplemental Staff Report on Proposed Rule 1166, May 16, 1988.
2. *Rule 1166*, South Coast Air Quality Management District, El Monte, CA 91731, August 1988.
3. Fall, E. W. and W. E. Pickens. "In situ Hydrocarbon Extraction," Focus Conference on Southwest Groundwater Issues, Albuquerque, March 1988.
4. Irwin Environmental Group, Long Beach, CA, private communication, 1988.
5. Nachant Environmental Inc., La Mesa, Ca, private communication, 1988.

Landfill Leachate Control Treatment

Trevor P. Castor[1]

INTRODUCTION

A great concern exists today because hazardous wastes buried in landfills across the country are escaping into the environment. There are hundreds of hazardous waste landfills in the United States with documented leakage or concern about leakage. Known remedial measures to prevent the escape of hazardous waste from landfills are costly to implement and of uncertain effectiveness. There is a critical need for affordable and effective techniques to seal leaking landfills and curtail continued environmental contamination.

Currently, efforts to control the waste from leaking landfills include removing the waste, solidifying the waste, detoxifying the waste, and containing the waste. The particular technique chosen varies with the character of the individual site. For example, thousands of pounds of material including chloroform, benzene, carbon tetrachloride, chlorinated phenols, and dioxin are present at the S-Area landfill in Niagara Falls, New York. Because the cost and logistics involved in removing these materials to another location are prohibitive, an extensive containment and treatment plan has been developed for this site. The containment and treatment plan calls for the installation of containment walls from the surface down to the bedrock around the perimeter of the site and for the placement of an impermeable cap over the top of the site. The perimeter walls will stop the flow of materials out and the flow of water into the site at its boundaries, while the cap will stop surface water and precipitation from percolating through the site and leaching materials out. The bedrock or clay base must be relied upon to arrest the flow of waste materials out of the site into the groundwater-containing layer.

Current containment technologies available for forming vertical barriers around leaky landfill sites include slurry walls, grout curtains, and pilings. To be effective, the vertical barriers must be attached to a low-permeability layer at the bottom. Typically, this layer is the existing bedrock or clay. Sites where the bedrock or clay is fractured or too far below the surface cannot be treated in this manner. Even where the vertical barriers are attached to a low-permeability layer, concern exists regarding the effectiveness of conventional containment techniques. Further expensive control measures are taken such as dewatering the site with collection wells and providing pumps around the perimeter walls to induce the flow of groundwater up through the underlying strata of the site to stop any flow of contaminated water out of the landfill site. Furthermore, comprehensive monitoring is typically conducted to detect failure of the system. To date, no adequate technique for sealing the entire base of a landfill site has been developed.

This chapter reviews the development of an effective method for sealing the entire base of a landfill and thus preventing the leaching of toxic wastes into underground supplies of drinking water. EngViro Services, Inc. of Arlington, Massachusetts is currently developing a landfill leachate control treatment (LLCT) process [1] to seal the bottom of existing landfills, and thus to prevent the vertical leaching of hazardous materials into groundwater supplies. The LLCT process also has applicability to leaking underground storage tanks [2], spills and dumps of hazardous waste, lagoons that store waste chemicals and other toxic materials, and mill tailings that con-

[1] EngViro Services, Inc., Arlington, MA 02174

stitute either a heavy metal contamination problem or a diffuse source of low-level radioactivity.

PROCESS DESCRIPTION

The landfill leachate control treatment (LLCT) process involves the selective emplacement of an impermeable polymeric barrier in the semi-permeable substructure, and faulted or fractured geologic barriers that underlie leaking waste disposal sites. The LLCT barrier, designed to be resistant to the toxic waste leachate, is emplaced to prevent the downward vertical migration and subsequent transportation of the waste away from the landfill or waste disposal site into the surrounding environment.

The impermeable polymeric barrier can be created in situ within the subterranean formation by the *delayed* or *controlled* cross-linking of dilute aqueous solutions of synthetic or biologically produced polymers. The low-viscosity polymer solutions can be readily injected into the formation, preferentially seeking out the higher permeability zones in sands with severe matrix heterogeneities, faults, or other thief zones. The emplaced polymeric solutions will adsorb onto the rock surface and, with time, cross-link into gels that form a network across pore openings. Cross-linking control is accomplished either through the catalytic release of the cross-linking agent or through a multistage injection process involving repeated sequences of polymer and cross-linking agent. The gelation time can be varied from a few minutes to several days. The treatment radius can be thus varied from a few feet to as much as one thousand feet.

The LLCT process can be used in consolidated and unconsolidated sandstones, dolomites, and fractured limestones below a leaking landfill site. The dilute aqueous polymer solutions can be emplaced by a variety of techniques depending on the geologic and lithologic characteristics of the site. Preferred techniques include direct injection into the semi-permeable matrix or by an indirect technique called "hydraulic fracturing" through which a pancake-like horizontal fracture can be created as shown in Figure 1. Hydraulic fracturing, a commercially available and widely used technology in the petroleum industry, can be described as the process of creating a fracture or fracture system in a porous medium by injecting fluid under pressure through a wellbore in order to overcome native stresses and to cause material failure of the porous medium. The injection pressures must be high enough to overcome the earth's stresses, between 0.3–1.0 psi per ft of depth (6.9×10^3–2.3×10^4 N/m² per m) at the depth of the intended fracture, and any frictional losses in the pumping system and the injection string.

The planar orientation of the fracture depends on the depth of the fracture zone because the "frac" is created perpendicular to the plane of maximum principal stress. A barrier beneath a waste disposal site would typically be placed at a depth of less than 1,000 ft (300 m). At such shallow depths, the created fracture will most probably be horizontal since the stress in this plane is less than that in the vertical plane. In a shallow fracture, the earth's overburden is actually lifted. The propogation of the fracture is usually symmetrical; the extent of the fracture depends in part on the rate and volume of fluid injected and the fluid loss properties of the fractured zone, i.e., the extent of the fracture depends in part on the *net* rate and volume of fluid injected into the created fracture.

On release of injection pressure, fracture conductivity can be maintained by the simultaneous injection of sufficient quantities of a proppant (propping agent) within the fracture fluid. Usually, the proppant is a solid material whose presence maintains the fracture by preventing the fracture faces from coming together or "healing" when the pressure gradient is released. The proppant usually is suspended in the fracture fluid prior to injecting the fluid to create the fracture. The propped zone is typically between about 0.1 to 0.25 inches thick (0.3–0.6 cm) thick after release of the fracture pressure gradient.

The polymeric solutions and cross-linking or gelling agents will be pumped under moderate pressures (100–500 psig) into the subterranean formation at a selected depth and will thus be used as the hydraulic fracturing fluid to part and lift the formation. The hydraulic fracturing fluid will be of sufficient viscosity to support a proppant that consists of well-sorted, clean sand grains or glass beads and that could include ion-exchange materials with a selective capacity to detoxify wastes and immobilize toxic species such as heavy metal ions. The polymeric solution will be low enough in viscosity to allow significant penetration into the faces of the fracture, as illustrated in Figure 1, for the purpose of filling void spaces, vugs, fissures, and any offsetting fractures.

Upon release of surface pressure, the formation will self-heal onto the mixture of polymeric solutions, cross-linkers, and proppants, leaving an impermeable barrier upon gelation of the mixture. The cased injection well(s) will subsequently be plugged with cement above the fractured zone. This enhanced technique will be utilized to ensure uniformity of coverage and thickness of the emplaced LLCT barrier formulations and will also minimize the competition for cross-linking sites by other chemicals in

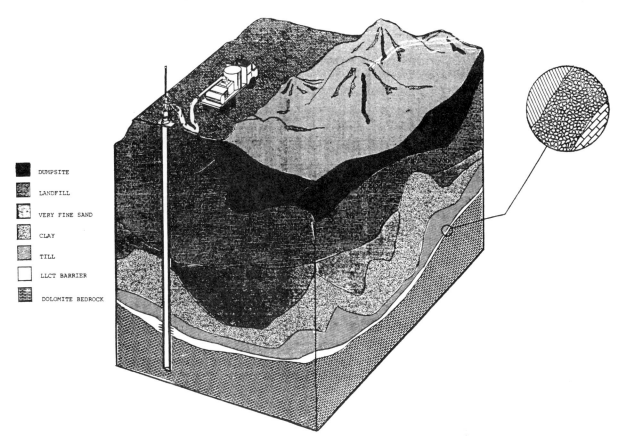

FIGURE 1. Proposed landfill leachate control treatment (LLCT) emplacement.

soils saturated with toxic waste leachates, thus maximizing the integrity of the emplaced LLCT barrier. Figure 1 is a conceptualization of the LLCT process emplaced below the S-area waste disposal site in Niagara Falls, New York. The LLCT barrier material is emplaced in a horizontal fracture located in the till zone just above the dolomitic bedrock in which the groundwater flow transports toxic waste leachates into a nearby river.

The selection of a suitable barrier solution and gelling agent will depend largely on the soil characteristics of the landfill base and the characteristics of the wastes in the landfill. The gelled solution must be impermeable to and resistant to degradation by the toxic waste leachates. The major degradation mechanisms are ionic degradation, microbial decomposition, dissolution effects, nucleophilic attacks, thermal alteration of reaction kinetics, mechanical shear stressing, and hydration or aqueous-phase swelling. In addition to degradation of the gel, the in situ gelling or cross-linking of the barrier material may be impaired in the presence of toxic waste leachates. This process weakness can, however, be prevented because the preferred LLCT process uses a barrier solution that incorporates a clean, well-sorted proppant.

Polymers that are applicable for use in the LLCT process include polysaccharides, polyacrylamides, polyacrylonitrile, polyacrylic acid, carboxymethyl cellulose, polyvinyl alcohol, polystyrene sulfonates, polyalkenes, hydroxyalkyl locust bean gum, and guar gum. Other polymers that may be particularly suitable for use in the LLCT process are polyvinylpyrrolidone (PVP), sulfonated/aliphatic copolymers, phenol-formaldehyde and their derivatives (phenolics) and polysiloxanes. PVP has resistance to salts, acids, and thermal degradation. Sulfonated aromatic aliphatic copolymers are inert toward acids and bases and exhibit good thermal stability. Phenolics exhibit excellent stability and resistance to hydrolytic degradation. Polysiloxanes exhibit good thermal stability and resistance to chemical and bacterial degradation and fungal growth.

Natural and biologically produced polymers such as polysaccharides are preferred over synthetic ones such as polyacrylamides in terms of environmental compatibility. The polymer or mix of polymers used will be selected primarily on the basis of their toxicity levels and the toxicity levels of their breakdown and/or reaction products; secondly, on their abilities to form an impermeable barrier in the leaky landfill site—such abilities will largely depend on the soil characteristics of the landfill base and the characteristics of the wastes in the landfill; and finally, on their long-term resistance to the toxic waste leachates in the site being bottom sealed. The concentration of the polymer in the injections will be between 1,000 to 40,000 ppm, more preferably from about 5,000 to 30,000 ppm.

There are two major generic categories of polymers: biologically produced polymers such as polysaccharides and synthetically produced polymers such as polyacrylamides, both of which are high potential candidates for the LLCT process. Polysaccharide gel systems are relatively stable in the presence of strong nucleophiles and have excellent compatibility with organic solvents and oil wastes. However, they have poor stability in the presence of hydrating agents, inorganic solvents, strong bases, and strong acids, except for hydrochloric acid which appears to strengthen some polysaccharide gel systems. Polyacrylamide gel systems are relatively stable in the presence of strong nucleophiles, organic solvents, hydrating agents, and inorganic salts. Polyacrylamides have excellent compatibility with oily wastes and appear to have poor stability in the presence of strong bases, strong acids, inorganic solvents, and dehydrating agents.

The gelling agent selected will depend in part upon the polymer selected for the barrier solution, the soil conditions, and the amount of delay between injection and gelling that is desired. Suitable polymer cross-linking agents include multivalent metal cations such as Fe^{2+}, Fe^{3+}, Al^{3+}, Ti^{4+}, Sn^{4+}, Ca^{2+}, Mg^{2+}, and Cr^{3+}. Gelling control may be accomplished through the catalytic release of the cross-linking agent. For example, redox reactions controlling the release of the cross-linking agent have been used successfully in the oil industry. The cross-linking agent makes up about 50 to 200% and preferably 100 to about 150% of the stoichiometric amount required to completely cross-link the amount of polymer used. Any excess heavy metal ion cross-linking agent may itself be a waste problem and should be avoided since unreacted toxic ions may leach out of the gel.

The viscosity of the barrier solution before the proppant is added should be low enough so that the solution will penetrate the areas that have not been fractured and through the face of the fracture so that the treated area will become 0.5–2.0 feet (15–61 cm) thick. Such a viscosity also allows the filing of void spaces, vugs, fissures, and any offsetting fractures from the fracture face. The barrier solution should be of sufficient viscosity to support the proppant. Preferably, the viscosity of the solution should be between 1 and 5 centipoise before adding the proppant. The proppant may be any conventional proppant or mixture thereof that is sized between 10 and 40 mesh. The amount of proppant required will vary with the particular site conditions and the particular barrier material selected. For example, the depth of the intended fracture, the viscosity of the barrier solution, and the cross-linking agent selected may all influence the amount of proppant required. Ideally, the amount of proppant should be sufficient to maintain the fracture space permeable to fluids. The preferred proppant is sand, glass beads, or a waste active material. For example, the proppant may be an ion-exchange compound such as polystyrene and, when such a compound is used, the gel will have the ability to immobilize certain toxic leachates that contact or penetrate the barrier gel. Other waste-active materials include adsorbents, molecular sieves, and molecular exclusion or entrapment materials.

The application of in situ polymeric gels has the potential to dramatically reduce material migration out of landfills by controlling hazardous material losses from semipermeable and fault or thief zones. Additional treatment can also be designed to reduce the bulk porosity of permeable formations. Together, these techniques could substantially reduce hazardous waste losses from leaking landfills. These solutions can be injected uniformly over a larger drainage area than those found in existing toxic waste control processes due to their low initial viscosities. The LLCT process has potential for use as a temporary barrier as well as a permanent one because retreatment or layering can be readily effected.

There is no current technology on the market for sealing the bottoms of hazardous waste or landfill sites. There are containment technologies available that form vertical barriers around such sites. Slurry walls, grout curtains, and pilings all employ a semipermeable barrier around a waste site. They must, however, be attached to a low-permeability layer at the bottom, typically taken to be the existing bedrock or clay. Sites where the bedrock or clay is fractured or too far below the surface cannot be treated in this manner. By using LLCT technology, a low-permeability barrier for attachment can be created in situ and at costs that are comparable to those of the most inexpensive groundwater barriers used. The estimated cost of bottom sealing the 7.4 acre (3 hm²) S-area site in Niagara Falls, New York by the LLCT process ranges from $50,000 to $120,000 per acre; this cost is approximately 20 to 50% of the cost to construct a slurry wall [3]. Other technologies for containing a landfill site include groundwater pumping, subsurface drains, runon/runoff controls, and surface seals/caps. These technologies can be combined with the LLCT process to completely seal a site and condition its environment. The hazardous wastes can then be degraded by the injection of selective microbial species that have been first isolated and then manipulated or genetically engineered to maximize degradation of the toxic wastes, now contained in a sealed landfill site or "bioreactor."

STATE OF DEVELOPMENT

Two polymeric materials, polysaccharides manufactured by Pfizer Chemical Company, Groton, Connecticut and polyacrylamides manufactured by American Cyanamid Company, Wayne, New Jersey, have been evaluated for use as LLCT barrier materials.

Polysaccharides are biopolymers formed by the fermentation of a carbohydrate substrate. One of the most commonly used is xanthan gum which is a high molecular weight polysaccharide containing glucose, mannose, glucuronic acid, pyruvate, and acetate moieties. Xanthan gum, a heteropolysaccharide, is produced by the *Xanthomonas campestris* fermentation of a carbohydrate substrate with a protein supplement and an inorganic nitrogen source. The biopolymer is produced as an extracellular slime that forms on the surface of *Xanthomonas campestris* cells. The backbone of the xanthan molecule consists of glucose linkages like those of cellulose with mannose and glucuronic acid units in the side chains. Some of the mannose units are modified with acetyl or pyruvic ketal groups. In its natural form, xanthan exists as a multistranded helix. Although the xanthan biopolymers are anionic in nature, the charge density over the molecule is much less than that found in polyacrylamide copolymers. These biopolymers, with molecular weights of approximately 1 million, are currently used as thickeners for oil field drilling and enhanced oil recovery and have been approved for use as food additives, e.g., thickeners in salad dressings, ice cream, etc. Xanthan gum has an oral lethal dose (LD_{50}) of 5,000 mg/kg for rats, 20,000 mg/kg for dogs, and 1,000 mg/kg for mice.

Polysaccharides can be gelled when a sufficient helix is formed to provide the cross-links for a continuous network. Solutions of xanthan gum have been gelled by adding 5 to 25 wt% of finely divided zinc metal that displaces hydrogen atoms and 50 to 200 wt% of a multivalent salt such as aluminum sulfate, ferrous sulfate, or chromium chloride.

Polyacrylamides are synthetically manufactured by the copolymerization of the acrylamide molecule via a free-radical mechanism that is catalyzed by strong basic or aprotic solvents. The acrylamide molecule, $H_2C=CHCONH_2$, contains two reactive centers and is readily copolymerized because of its electron-deficient double bond. Polyacrylamide copolymers are usually made up of polyacrylamide and its salts. The composition of the polyacrylamide copolymers determines the degree of hydrolysis and the ionic character of a polymer chain. The ionic character can be varied from 100% cationic by use of quaternary ammonium sulfate (methylated) salts of acrylamide through 100% nonionic by use of polyacrylic acid. The nonionic polyacrylamides, which usually have less than 10% activity, are compatible with a wide variety of brines, whereas the anionic polymers are compatible only with fresh water or soft water brines [4]. The molecular weights of the polyacrylamide copolymers are 5 to 20 million. Polyacrylamide, with an acute oral (rat) LD_{50} of 32,000 mg/kg is considered to be relatively nontoxic; the acrylamide monomer is, however, a severe neurotoxin and a cumulative poison in man and therefore represents the main toxicological concern regarding polyacrylamides.

Cross-linking agents include multivalent salts such as magnesium chloride, aluminum citrate, aluminum sulfate, and aluminum nitrate. Cross-linking agents work by acting like catalysts that compound or complex the polymer by attracting reactive sites of the polymer molecule.

Polysaccharide and polyacrylamide in situ polymer cross-linking processes were evaluated for gel strength and stability in the presence of twenty-eight representative toxic chemical leachates that might commonly be found in leaky landfill sites. The gel strength and stability studies were performed in 500-cc sample jars that contained the gelled polymer in contact with aqueous and/or organic phases containing various concentrations of different classes of toxic substances. The sample jars were filled with the cross-linked polymer (100–250 cc) and the representative toxic chemical leachates and then sealed with a screw-on cap in order to minimize oxygen intrusion. Relative gel strength was quantified as a function of time with a probe penetrometer, and gel stability was qualitatively evaluated as a function of time. The penetrometer readings were performed with a Precision 73510 probe penetrometer with a 6.2-g cone, having a total rod assembly weight of 53.7 g. The tip of the penetrometer cone was placed above the surface of the gel, just touching it. The rod assembly was then released manually for approximately one second and the depth of penetration was read off the dial in intervals of 0.1 mm. Penetrometer readings were performed at regular intervals. Multiple readings were performed on each sample with each reading or surface penetration made on a virgin portion of the gel/leachate interface. Penetrometer readings are listed in Table 1 for compatibility tests

Table 1. Penetrometer Readings for Polysaccharide Gels in Bulk Phase Compatibility Tests

Chemical (conc. in wt%)	Time in Days						
	7	19	21	27	32	40	54
control	215	240	220	225	210	250	max
water (100%)	240	385	415	max	—	—	—
Strong Acids							
formic acid (10%)	—	max	—	—	—	—	—
sulfuric acid (50%)	max	—	—	—	—	—	—
hydrochloric acid (10%)	185	180	170	180	170	70	60
Strong Bases							
sodium hydroxide (4%)	—	—	—	—	—	—	—
sodium carbonate (1%)	—	—	—	—	—	—	—
Strong Nucleophiles							
sodium cyanide (1%)	—	—	—	—	—	—	—
sodium iodide (1%)	170	170	150	180	170	100	160
Solvents							
toluene (100%)	250	220	230	230	210	260	230
trichloroethylene (100%)	240	190	175	220	190	200	320
methyl ethyl ketone (100%)	180	180	200	225	250	230	230
acetic acid (10%)	—	—	—	—	—	—	—
ethanol (100%)	180	195	200	210	190	120	190
propionic acid (10%)	205	270	380	370	max	—	—
methanol (100%)	205	200	170	215	190	220	190
formamide (10%)	255	255	260	270	180	280	320
Dehydrating Agents							
formaldehyde (10%)	200	210	190	190	170	140	160
ethylene glycol (100%)	185	170	180	190	200	200	170
Hydrating Agents							
biosoft (1%)	—	—	—	—	—	—	—
Inorganic Salt							
copper sulfate (1%)	max	—	—	—	—	—	—
Oily Waste							
kerosene (100%)	230	275	190	180	230	300	360

of polysaccharides gelled with an inorganic salt. The readings listed in Table 1 are graphed in Figures 2 and 3. A softening in the gel is indicated by a positive penetration rate (an increase in penetrometer readings); an increase in gel strength can be inferred from a decrease in penetrometer readings. The penetrometer readings typically had a variance of ±25% because of the influence of uncontrollable factors such as the presence of other surface abrasions and edge effects. The pH of the solutions in contact with the gels were measured at the start, in the middle, and at the end of the testing period.

The LLCT process used in our experiments is based on injecting alternate slugs of the polyacrylamide, a cross-linker solution (aluminum citrate), and another slug of polyacrylamide into the porous medium. This process was modified so that the polymer and cross-linker were mixed in bulk and then allowed to gel in situ. This modification was made in order to conduct the bulk-phase compatibility tests, and to tailor the process for sealing large magnitude voids. The solution polymer (Cyanagel 100) was mixed in water in concentrations ranging from 2.8 to 3.2 wt% and stirred with a Teflon rod until dissolved. The cross-linker (Cyanaperm XL) was then added in concentrations ranging from 400 to 2,000 ppm. In later formulations, the concentrations of both the solution polymer and the cross-linker were increased, and the pH of the resulting mixture was adjusted. For making up the emulsion polymer (Cyanagel 150 and 240) solutions, a Waring blender was used. The water was measured and poured into the blender's container, and the appropriate amount of the emulsion polymer was injected with a syringe after a vortex could be seen in the water in the container. The solution was mixed slowly for a few seconds, in order to avoid mechanical shearing of the polymer. The cross-linker was then added to the polymer emulsion until the desired concentration ratios (3,000–5,000 ppm of Cyanaperm 150 or 240

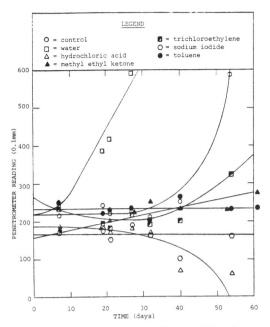

FIGURE 2. Penetrometer readings as a function of time for polysaccharide gels in contact with representative toxic waste leachates (batch #1).

and 400–2,000 ppm of Cyanaperm XL) were established. Several samples of solution and emulsion polymers with cross-linkers, each at different pHs and electrolyte concentrations, were made up and allowed to set over extended periods of time in order to produce a satisfactory cross-linked polyacrylamide gel system. All experiments were performed at room temperature. A firm gel of 6 wt% Cyanagel 100 and 2,700 ppm Cyanaperm XL was formed within twenty-four hours at a room temperature of 75°F, whereas it took four days in the laboratory to obtain a similar gel at laboratory temperatures between 45 and 65°F. The formulations will thus have to be customized in order to obtain appropriate gelation times at the temperatures found in landfill sites. Gel stability tests were performed using the solution polymer system consisting of 6 wt% Cyanagel 100 and 2,700 ppm Cyanaperm XL. Sixteen 500-cc sample jars were used with 250 cc of the polymer solution in each jar. The samples were allowed to gel for five days in the laboratory at temperatures ranging from 45 to 65°F.

Flow experiments were performed in a Hele-Shaw type flow cell (Figure 4) to evaluate the integrity and impermeability of the polymer barriers. The Hele-Shaw type flow cell was made up of 8 by 12 inch (20 by 30 cm) polycarbonate sheets separated by 0.5 inches (1.2 cm) with a polycarbonate frame. It had two ports in its detachable top, and two ports, one near the bottom and one in the middle, in each of the vertical frame members. All ports were connected to 0.25-inch (.6 cm) lines that were attached to a pump. All lines contained valves so that flow from the pump could be directed to each particular inlet or combination of inlets. In these experiments, the cell was packed with sand to simulate in situ conditions. The polymer gel was placed in the middle region of the cell. Groundwater flow was established in the bottom region of the cell. The chemicals representative of toxic wastes were added in the top region of the cell and the leaching of the chemicals was observed over a period of time. The results of the flow tests were then compared with the bulk-phase experimental results in order to evaluate differences in gel integrity and impermeability in a porous medium and in the bulk phase.

The performance characteristics of the selected LLCT barrier materials (polysaccharides and poly-

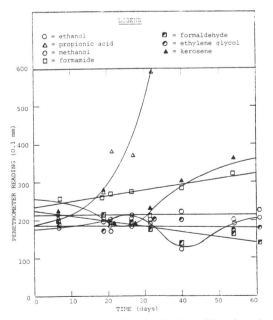

FIGURE 3. Penetrometer readings as a function of time for polysaccharide gels in contact with representative toxic waste leachates (batch #1).

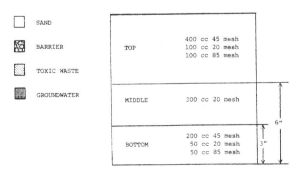

FIGURE 4. Typical cross-section of Hele-Shaw flow cell with typical Ottawa sand packing arrangements.

Table 2. Compatibility Rating of Potential LLCT Barrier Formulations with Representative Toxic Chemical Groups (Based on Bulk-Phase Testing)

#	Representative Toxic Waste	Polysaccharide Gel Systems	Polyacrylamides Chromium-Redox	Polyacrylamides Aluminum-Citrate
1	Strong acids			
	organic	poor*	poor*	poor*
	inorganic	poor**	poor*	poor*
2	Strong bases	poor*	poor*	poor*
3	Strong nucleophiles	good*	mediocre	good
4	Solvents			
	organic	excellent	good	good
	inorganic	poor	mediocre	poor
5	Dehydrating agents	excellent	good	poor
6	Hydrating agents	poor	good	good
7	Inorganic salts	mediocre	good	good*
8	Oily wastes	excellent	excellent	excellent

*Very pH dependent.
**Except in contact with hydrochloric acid which causes the gel to shrink and apparently strengthen.

acrylamides) are based on the results of testing these materials against eight groups of representative toxic chemicals in the bulk phase (jar tests), and in a porous medium (flow tests). The results of the bulk phase tests are summarized in Table 2 with compatibility ratings of poor, mediocre, good, and excellent for each chemical grouping. Both the polysaccharide and polyacrylamide formulations showed excellent compatibility with kerosene and a good compatibility with a variety of organic solvents. These formulations will thus be very useful in containing underground or aboveground tanks that store petroleum products, organic chemicals, or solvents.

The LLCT formulations tested showed significant incompatibilities with representative aqueous chemical leachates at high pHs (greater than 10) and at low pHs (less than 4). Optimum compatibility was obtained around a neutral pH of 7 with low to moderate degradation occurring over the pH range of 4 to 10. The formulations appeared more stable to hydrocarbon/organic liquids than to aqueous solutions. The polysaccharide solutions, in particular, exhibited significant swelling in the presence of water; in some cases, such as 10 wt% formamide solutions, swelling was suppressed in the presence of this strong nucleophile. The resistance of the formulations to ionic degradation and strong hydrating agents appears to be improved in gel mixtures with higher concentrations of cross-linking agents.

The performance of the barrier materials appears to have been an order of magnitude better in the simulated landfill site setting than that observed during the bulk-phase compatibility testing. This observation is made on the basis of evaluating the compatibility between representative toxic waste leachates and the cross-linked barrier materials in a flow cell, e.g., testing an emplaced polysaccharide gel mixture against 10 N (50 wt%) sulfuric acid as shown in Figure 5. This test indicated good compatibility over an eleven-day period under simulated underground porous medium conditions even though the 10 N sulfuric acid destroyed the same polysaccharide gel within one day in the bulk phase. Theoretical considerations indicate that the concentration of a toxic chemical leachate in the porous medium may be as little as 20% of the composition in the bulk phase due to equilibrium adsorptive and dispersive effects in the porous medium. This concentration reduction effect of the porous medium could, in part, be responsible for the improved compatibility observed. Other factors, such as reduced interfacial area and mass transfer rates, may have also contributed to the improved compatibility of the LLCT process in the simulated porous medium.

The flow test experiments indicate that the LLCT

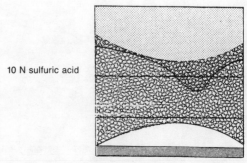

FIGURE 5. Profile of emplaced LLCT barrier formulation (cross-linked polysaccharide), with flowing groundwater and representative toxic waste leachate (10 N sulfuric acid) in flow test #4 after 11 days.

barrier material is best emplaced by reversing flow vis-à-vis a two point (or multiple point) injection system in which the two points (or multiple points) are opposite each other. The flow test experiments also indicate that mechanical shear stressing will impact the integrity of the emplaced gel system. This degradative impact may not, however, be representative of real world conditions because the simulated groundwater flow rates were orders of magnitude higher than nominal groundwater flow rates.

The major weakness of the tested LLCT barrier materials appears to be ionic degradation as the result of hydrolysis at low and high pHs. Utilizing an ionic degradation model, we predict that optimum compatibility (90% integrity) can be obtained over a pH range of 2 to 8 by increasing the molar concentration of the cross-linking agent (see Figure 6). The tested polysaccharide gel system was extremely sensitive to microbial decomposition, and exhibited swelling in contact with water and other hydrating agents. The microbial attack on the polysaccharide gel system was so pervasive that all test results after a twenty-day period could be attributed to microbial degradation effects. The susceptibility of polysaccharides to microbial degradation was predictable. This susceptibility is usually controlled by the addition of biocides such as formaldehyde. The tested polysaccharide gel systems did contain formaldehyde (around 200 ppm) that was either insufficient for microbial population control or depleted by vaporization during mixing and/or redistribution into the contacting toxic chemical phase. The polysaccharide gel system, cross-linked with sufficient biocide within its helix structure, should be much more resistant to microbial degradation. Xanthan gum did not exhibit good compatibility with hydrating agents primarily because it is a hydrophilic colloid.

The in situ cross-linking of the LLCT barrier material will most likely be impaired in the presence of toxic chemical leachates [5]. This major process weakness will be encountered in leachate-saturated soil conditions and can be averted by utilizing clean, well-sorted sand grains as a proppant during emplacement of the cross-linked polymeric barrier.

LLCT formulations, based on current experimental results, have shown excellent compatibility with kerosene and a good compatibility with a variety of organic solvents. These formulations will thus be very useful as secondary containment barriers for leaking underground and aboveground storage tanks that contain gasoline, organic chemicals, solvents, or other petroleum-related products. Laboratory testing indicates that the performance of the LLCT barrier materials is an order of magnitude better in a simulated landfill site setting than in the bulk-phase compatibility tests. Once fully developed, this technique will allow the in situ construction of a bottom seal to existing landfill sites and other hazardous waste storage sites without removing the stored materials. The process can be used to seal the entire bottom of the landfill sites without liners or just for sealing leaky areas in sites with broken liners. The process can be designed to be temporary or permanent; the barrier materials can be relayered, if necessary, to improve the effectiveness of the emplaced LLCT barrier. The LLCT process is intended to prevent and mitigate releases and resulting environmental impacts from high-hazard/high-risk hazardous waste storage situations. Criteria for usage of the tested barrier formulations are as follows:

1. Moderate values of pH (between 4 and 10)
2. High organic-phase concentrations
3. Low concentrations of inorganic salts and strong nucleophiles
4. Controlled microbial activity especially for polysaccharides
5. Minimum mechanical shear conditions for polyacrylamides
6. Low aqueous-phase saturation and rates of groundwater flow
7. Low concentrations of oxygen that can enhance microbial activity and act as a free-radical agent
8. Sufficient permeability to emplace the LLCT barrier materials
9. Low viscosities during emplacement to maximize penetration below the landfill site yet sufficient viscosity to support the proppant during hydraulic fracturing

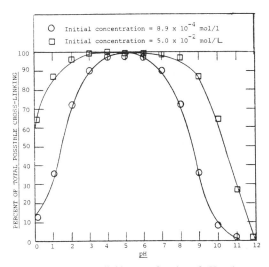

FIGURE 6. Percent cross-linking as a function of pH and concentration for cross-linked polyacrylamide gels.

10. Leachate-free conditions at the depth of emplacement
11. Adequate gel strength and stability at the temperature of emplacement
12. Adequate gel integrity in contact with specific concentration ranges of toxic waste leachates

The LLCT process has not yet been field tested at an actual landfill site or hazardous waste storage installation. Testing has been restricted to simulated conditions in the laboratory. In addition to its use in preventing and mitigating leaks of hazardous wastes into the environment, the LLCT process could be very useful in creating a nonporous "bioreactor" for the in situ biodegradation of toxic wastes. For further information about the LLCT process, please contact EngViro Services, Inc., 216 Sylvia Street, Arlington, MA 02174.

REFERENCES

1. Castor, T. P. U.S. Patent No. 4,790,688 (December 13, 1988).
2. Castor, T. P. et al. *Tappi Proceedings 1988 Environmental Conference, April 18–20, 1988, Charleston, South Carolina*, pp. 105–113 (1988).
3. OTA. *Superfund Strategy*, Washington, DC:U.S. Congress, Office of Technology Assessment, OTA-ITE-252, pp. 171–213 (1985).
4. Sparlin, D. D. "An Evaluation of Polyacrylamides for Reducing Water Production," *J. Pet. Tech.*, 29(8) (1976).
5. Spooner, P. A. et al. *Compatibility of Grouts with Hazardous Wastes*, EPA-600/S2-84-015 (1984).

2.23

Innovative Practices for Treating Waste Streams Containing Heavy Metals: A Waste Minimization Approach

Douglas W. Grosse[1]

INTRODUCTION

Innovative practices for treating waste streams containing heavy metals often involve technologies or systems that either reduce the amount of waste generated or recover reusable resources. With the land disposal of metal treatment residuals becoming less of an accepted waste management alternative, waste minimization practices have received increasing popularity in the primary metals industry.

When considering the percentage contribution of various industrial sectors to the total amount of hazardous waste generated, primary metals and fabricated metal products rank number 2 and 4 respectively for a combined percentage of 27.6 percent of the total [1]. Although the recycling of recoverable resources from metal and cyanide-bearing wastes accounts for as much as 48 percent of the total waste categories identified, much more can be done to reduce generated waste even further [2].

Many wastes generated by the primary metals industry result from the electroplating and metal treatment/fabrication industries which are profoundly affected by the land disposal restrictions program [3].

A list of these wastes along with the proposed ban schedule are shown in Table 1 [4].

Preferred management practices for reducing or eliminating generated wastes are source reduction (i.e., material substitution, recycle/reuse) and reduction of toxicity in order to meet "technology-based treatment standards." Either a specified allowable concentration level for constituents in waste residuals will be prescribed or specific treatment technologies may be mandated. In either case, the goal here is to eliminate environmental pollution. Major topics to be discussed in this section will focus on waste stream characterizations, process descriptions, and summarized results of demonstration projects and audits. Only state-of-the-art and innovative approaches will be considered for review. Particular emphasis will be placed on waste minimization approaches including source reduction, recovery techniques, and centralized treatment.

METAL-BEARING HAZARDOUS WASTE STREAMS

Any discussion of the various options that may be considered for managing hazardous metal wastestreams must be preceded by a description of the various representative waste streams, including waste generation, characteristics and sources.

Waste Generation

Although metals and cyanide are used in many different industrial applications, only a few generate waste significantly high in metal/cyanide concentrations to be affected by current land disposal restrictions. Those industries with the greatest number of generating facilities are plating and polishing operations, paints and allied products manufacturers, metal coating operations, motor vehicle parts and accessories manufacturers, and inorganic chemicals producers. Nevertheless, it is quite apparent that metal and cyanide wastes are generated by metal fabrication facilities that perform plating, polishing, and coating operations along with metal fabrication facilities that manufacture metallic parts and equipment. A tendency exists for these facilities to generate concentrated spent bath solutions which are likely to be

---

[1] Risk Reduction Engineering Laboratory, U.S. Environmental Protection Agency, Cincinnati, OH 45268

Table 1. Wastes Generated by the Electroplating and Metal-Finishing Industries [4]

Date of Final Regulation	Waste Code	Description
Degreasing Solvents		
Nov. 8, 1986	F001	Spent halogenated solvents used in degreasing
Nov. 8, 1986	F002	Spent halogenated solvents
Nov. 8, 1986	F003	Ignitable solvents (nonhalogenated)
Nov. 8, 1986	F004	Toxic solvents (nonhalogenated)
Nov. 8, 1986	F005	Ignitable and toxic solvents (nonhalogenated)
Other F-Wastes		
Aug. 8, 1988	F006	Wastewater treatment sludges from electroplating operations
Aug. 8, 1988	F007	Spent cyanide plating bath solutions from electroplating operations.
Aug. 8, 1988	F008	Plating bath residues from the bottom of plating baths from electroplating operations where cyanide is used in the process
Aug. 8, 1988	F009	Spent stripping and cleaning bath solutions from electroplating operations where cyanide is used in the process
June 8, 1988	F010	Quenching bath residues from oil baths from metal heat-treating operations
June 8, 1988	F011	Spent cyanide solutions from salt bath pot cleaning from metal heat-treating operations
June 8, 1988	F012	Quenching wastewater treatment sludge from metal heat-treating operations where cyanide is used in the process
Aug. 8, 1988	F019	Wastewater treatment sludge from the chemical conversion coating of aluminum
Ep Toxic Metal-Bearing Wastes		
May 8, 1990	D004	Arsenic
May 8, 1990	D004	Barium
May 8, 1990	D006	Cadmium
May 8, 1990	D007	Chromium
May 8, 1990	D008	Lead
May 8, 1990	D009	Mercury
May 8, 1990	D010	Selenium
May 8, 1990	D011	Silver

affected by tighter disposal limits. Although small-quantity generators comprise only a small fraction of generated hazardous wastes, they stand to suffer significant adverse impacts from higher land disposal restrictions since they are less capable of implementing alternative waste management techniques. Table 2 exhibits a partial listing of the number of small quantity generators relative to the waste quantity generated for each waste stream [5].

An estimated 7.9 billion gallons of metal-bearing wastes, 4.7 billion gallons of cyanide wastes, and 0.8 billion gallons of reactive wastes containing cyanide were generated in the United States in 1983. Likewise, wastes handled by treatment storage and disposal facilities (TSDF), on an annual basis, consisted of 5.8 billion gallons of metal-bearing wastes, 3.3 billion gallons of cyanide-bearing wastes and 0.76 billion gallons of reactive wastes containing cyanide.

Waste Characterization

For the purpose of consistency, a description of how metal and cyanide wastes are categorized will be detailed. Hazardous wastes can be segregated into five source designations, as a result of the Resource Conservation and Recovery Act (RCRA) legislation of 1976 [6]. The source categories are as follows:

D Wastes which are hazardous because they exhibit a particular hazardous characteristic, such as toxicity
F Wastes from nonspecific sources
K Wastes from specific sources
P Acutely hazardous constituents
U Toxic constituents

As illustrated in Table 3, the majority of metal/cyanide wastes generated in this country are charac-

Table 2. Number of Small-Quantity Generators and Waste Quantity Generated by Waste Stream [5]

Waste Stream (Description)	No. of Generators	Waste Quantity (Mt/yr)
Cyanide	1,972	2,146
Heavy Metal	238	568
Ink sludge (Cr or pb)	1,176	217
Photographic	26,236	18,431
Spent plating	5,382	5,786
Batteries (lead-acid)	197,627	369,097
Inks containing solvents/metals	4,360	1,622
Wood preservatives	196	719
Wastewater sludge w/heavy metals	1,684	2,404

teristic toxic metal wastes (D) and wastes that result from electroplating and metal treating processes (F).

Waste Sources

Although dominated by many small job shops and specialized manufacturing operators, electroplating and metal surface treatment processes generate significant quantities of metal/cyanide hazardous wastes from a variety of applications. These include electroplating, electroless depositions, conversion coating, anodizing, cleaning, milling, and etching. The primary sources of metal/cyanide hazardous wastes are found in the residuals generated from these processes. Some of these residuals include (1) hydroxide sludge from the pretreatment of rinse baths, (2) contaminated plating baths, (3) spent media and filter residues resulting from the regeneration process bath, and (4) spent degreasers, etching solutions, and complexing agents.

Metal ions typically used for electroplating and metal surface treatment processes include cadmium, zinc, lead, chromium, nickel, copper, vanadium, platinum, silver, rhodium, and titanium. These metals are put into plating solutions via anodes and metal-containing reagents for the purpose of depositing a layer of metal onto a basis surface with product use dictating the type and method of application. Electroless and immersion deposition processes contain many of the same constituents that are found in electroplating processes, including reducing agents (e.g., formaldehyde, sodium hydrosulfite, metabisulfite) and complexing agents (e.g., cyanide, ammonia, EDTA, tartrate, oxalate, gluconate, and amines). More often than not, basis metals will be "pickled" or cleaned prior to plating. Organics in the form of oils, greases and solvents (e.g., 1,1,1-trichloroethane, toluene, and xylene), resulting from pickling operations, can be generated.

Another significant source of metal/cyanide wastes result from printed circuit board (PCB) manufacturing. Tin, lead, and nickel solder plates are the most widely used resistant overplates. Tin is plated from both acid and alkaline baths. Chemical systems used for "etchback" consist of sulfuric acid, chromic acid, and chloroform-alcohol systems, to name a few. Rinses are contaminated from cyanide-containing baths and stripping solutions [8].

Other sources for metal wastes include the wood processing industry where an increase in popularity of chromated copper arsenate (CCA) wood treatment produces arsenic-containing wastes; inorganic pigment manufacturing producing pigments that contain chrome yellow and orange, molybdate chrome orange, chrome oxide green, cadmium sulfide, and white lead pigments; petroleum refining which generates conversion catalysts contaminated with nickel, vanadium, and chromium; and photographic operations producing film with high concentrations of silver and ferrocyanide. All of these generators produce a large quantity of wastewaters, residues, and sludges that can be categorized as hazardous wastes requiring extensive waste treatment.

Table 3. Generated Waste Distribution [7]

Hazardous Waste Group	Annual Volume (million gals)	Metals (% of Total Metals)	No. of Generators	Annual Volume (million gals)	Cyanides (% of all CN)	No. of Generators
D Wastes	3,685	46.9	3,860	Est. 750	—	Not avail.
F Wastes	3,920	49.9	2,091	3,920	83	2,091
K Wastes	219	2.8	402	572	12.1	10
P Wastes	28.3	.36	405	226	4.8	800
U Wastes	5.2	.07	365	3.92	.08	306
Total	7,858		6,586	4,723		2,781

Note: Total number of generators shown may not agree with sum of all categories due to double-counting in individual categories.

WASTE TREATMENT ALTERNATIVES

From a waste management perspective, treatment strategies for metal-bearing hazardous waste streams can be divided into two major classifications: waste treatment processes and waste minimization practices. These classifications are not rigid delineations, rather, they often overlap where a particular waste treatment process or recovery technique can be considered as an effective waste minimization (WM) practice. Although many emerging technologies and innovative ideas are continually being promoted with some warranting close scrutiny, this chapter will focus primarily on technologies or practices which aim to reduce waste generation at its source, recover valuable resources for reuse, and centralized treatment for ultimate disposal.

Source Reduction

It has long been recognized that reducing waste flows from industrial operations is a function of more efficient utilization of material inputs, substitution, automation, by-product utilization, and sophistication of production process control. Minimizing the generation of waste is not only beneficial to the environment, it can serve to reduce a firm's liability.

Basically, *source reduction* encompasses product substitution and source control (i.e., in-plant changes). *Product substitution* involves the replacement of a hazardous product with one that is nonhazardous or that produces a nonhazardous residue. *Source control* involves the alteration of input materials, process technology and/or production process. *Input material changes* can be further classified into purification, substitution, and dilution. *Technology changes* relate to production modification, equipment or layout changes, and conservation measures (e.g., energy, water use, etc.).

With specific application to the metal finishing industry, some of the cost-effective, in-plant measures used for reducing spent process wastes include: (1) concentration and reuse of waste contaminants (e.g., evaporators and precipitators), (2) conservation of water (e.g., rinsing techniques and solution segregation), (3) use of spent reagents for wastewater treatment, and (4) material substitution [9]. Each of these measures will be discussed in more detail.

Concentration and Reuse of Waste Contaminants

As a first step to recovering plating chemicals, many platers use evaporators. Traditionally, evaporators have been used to concentrate rinse water and recycle "drag-out" back to the plating bath. Two types of systems have been employed: (1) A closed-loop system, which totally eliminates a water discharge from the plating process and requires substantial make-up water to satisfy rinsing requirements, and (2) an open-loop system, which has lower operating costs as a result of less use for make-up water [10]. Although closed-loop systems tend to be more expensive, open-loop systems generate a wastewater discharge. Figure 1 illustrates a general schematic of a closed-loop system [11]. Typically, evaporators recover metals in solution form. Spent plating solutions and etchants are concentrated via vacuum and atmospheric evaporation for reuse in process solutions. Recovery is based upon the reduction of raw chemicals consumed in the process. Data collected from a metal finishing survey, investigating metal reduction and recovery techniques currently operating in Ontario, Canada, quantified recovery rates [12]. Table 4 presents the results of their findings. Recovery rates or "avoided losses" range from a low of 13 kg to a high of 275 kg of metals, daily. Annual cost savings were estimated from these values.

Conservation of Water

The most effective way of reducing drag-out losses, water use, and ultimately, waste treatment cost, is to conserve water. One way to accomplish this is to alter rinsing techniques. Variations can range from recycling the rinse water to installation of two or more rinse tanks in a counter-current configuration. Cost benefits can be obtained by employing up to three rinsing tanks in sequence. However, additional tanks offer little improvement on return of investment. Counter-current rinsing has the advantage of concentrating the rinse water. The percent reduction in flow and relative concentration of metal salts, using up to three tanks, at a rinsing ratio of 100:1, are shown as follows [13]:

Number of tanks	1	2	3
Percent reduction in flow	0	90	95
Relative conc. of salts	1	0.01	0.0001

A mathematical expression which can be used to calculate the required flow for series rinsing is [14]:

$$Q = [(C_p/C_n)^{1/n} + 1/n]\, \theta \qquad (1)$$

where:

Q = rinse tank flow rate
C_p = concentration of salts in the process solution
C_n = allowable concentration of salts in the rinse tanks

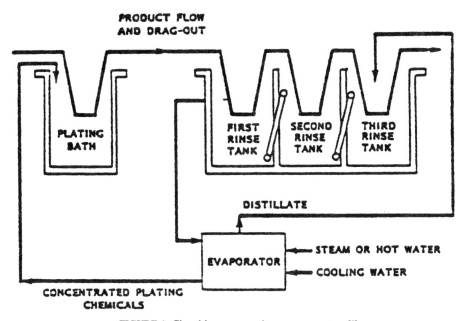

FIGURE 1. Closed-loop evaporative recovery system [1].

θ = drag-out rate
n = number of rinse tanks employed

Drag-out losses can also be minimized by reducing the amount of plating solution leaving the plating bath. The use of a drain board is the simplest method for achieving this aim. The board can capture drips of plating solution as racks and barrels leave the plating bath prior to rinsing.

Spray rinsing systems have shown to be very effective in reducing water usage and the ultimate generation of wastewater. One study evaluated the performance of a multiple spray rinse system as compared to a shuttle dip rinse system [15]. As much as a 93 percent reduction in total amount of water usage was achieved by the spray rinse system while yielding comparable pickle metal removal efficiencies (99.6 vs 99.8 percent).

Use of Spent Reagents

An example of how process modifications can be made to reduce wastewater and cost is associated with the use of spent reagents for wastewater treatment. Alkaline or acid cleaning solutions can be used for pH adjustment in waste treatment systems. Usually, these cleaning solutions are dumped when containment levels exceed acceptable concentrations. However, these spent solutions can be used to treat plant wastewater discharges; waste acid solutions are used to reduce chromate to the trivalent state while spent caustic soda solutions are used to adjust pH levels in precipitation tanks [16].

Table 4. Recovery Rates Using Evaporation Technology [12]

Rinse Solution	Estimated Recovery Rate (kg/d)	Metal* Value ($/kg)	Recovered Metal Value ($/yr)
Chromium Plating	13	3.52	15,100
	45**	3.52	52,272
	75	3.52	87,120
	85	3.52	98,756
Zinc Plating	100	1.37	45,210
Nickel Plating	34**	10.12	86,020
Chrome Acid Etch	275	3.52	319,440
	21	3.52	24,374
	95	3.52	110,352

*Chemical Marketing Reporter, April 6, 1987.
**Atmospheric evaporation was used as a recovery process; all others listed used vacuum evaporation.

Material Substitution

Material substitution is another alternative to be considered for in-plant modifications to reduce the amount of generated hazardous waste. Basically, potentially hazardous process chemicals are substituted with those that are nonhazardous. As a result, any generated waste will be treated and discharged to a publicly owned treatment works (POTW) instead of a TSDF. An example of this practice was recently demonstrated at a metal plating jobshop for a non-cyanide brass substitute [17]. This substitute has been used to plate plumbing fixtures, furniture, lamps, and other parts traditionally brass plated. Purportedly, this process utilizes a "nontoxic" resin coating to produce a brass-like electroplate without the need for a final spray lacquering procedure that has the potential for producing significant concentrates of volatile organic compounds (VOC) emissions. Some types of basis metals that have been successfully plated include bright nickel, aluminum steel, stainless steel, and tin.

Essentially, the brass substitute involves an electrocoating operation where a resin is "electrophoretically" applied with a colorant system. The process cycle begins with the immersion of the basis metal in a foam aqueous solution where organic resin particles are deposited by direct current. Next, the parts are transferred into a dye immersion solution followed by curing in a high-temperature oven (320°F). A finished product is produced having good corrosion and chemical resistance, hardness, and complete adhesion. Some of the disadvantages associated with this process are that finished surfaces are unable to pass the ultraviolet test, restricting use to indoors, and the coating will not cover major defects in the underlying deposit.

Recovery Technologies

Although waste recovery technologies may be considered a part of WM practices, a review of these technologies will be treated separately. With the high cost of raw materials and pending land disposal restrictions, reclamation of raw materials from waste streams is the preferred treatment option. Many studies have concluded that a large part of industrial operational costs can be subsidized by the scale of the recovered metals.

Electrolytic Recovery

Electrolytic recovery or electrowinning is one of the many technologies used to remove and concentrate metals from process water streams. This process uses electricity to pass a current through an aqueous metal-bearing solution containing a cathode plate and an insoluble anode. Positively charged metallic ions cling to the negatively charged cathodes leaving behind a metal deposit that is strippable and recoverable.

Walters and Vilagliano of the University of Maryland have demonstrated the electrolytic recovery of zinc from metal finishing rinse waters [18]. Results of the study indicated that zinc can be recovered from plating bath rinse waters. A batch electrochemical reactor with stainless steel electrodes was employed. The controlling factor in achieving high rates of zinc deposition appeared to be agitation. The study cited that mechanical mixing and nitrogen gas aeration were both effective. NaCl was added to maintain a minimal conductivity level. Higher current densities produced higher deposition rates, as opposed to lower current densities. A noticeable disadvantage was that corrosion could become a significant limiting factor, where electrodes would frequently have to be replaced.

Another study investigated the removal of copper from dilute, chelated rinse waters [19]. Copper removal efficiencies ranged from 80–85 percent over a broad pH range (3 to 11). Optimum removal efficiency was achieved when the influent copper concentration reached levels above 50 mg/L. A mathematical model was developed based on certain physical principles that influence the performance of most electrolytic cells. A mass balance analysis for a plug-flow reactor, approaches a first-order chemical reaction as:

$$R = 1 - e^{(-kal/v)} \qquad (2)$$

where:

R is the fractional removal of copper
k is the mass transfer coefficient (cm/sec)
a is the specific cathode area (0.526 cm^{-1})
l is the length of the recovery unit (cm)
v is the superficial fluid velocity (cm/sec)

This model will predict an exponential decrease in copper concentrations along the length of the device.

Several case studies were performed by Centec Corporation, at the Huntington Park area, for the purpose of evaluating electrolytic recovery or regeneration of spent drag-out and rinse tank solutions [20]. For one plant, six electrolytical recovery units were employed. Solutions from the drag-out tank were continuously recirculated through the electrolytic cells that removed the residual plating metals (copper, nickel, gold, and rhodium). These cells maintained very low metal concentrations in the drag-out tank. It was demonstrated that utilization of electrolytical recovery units can eliminate the need for precipitation units that generate hydroxide sludge. A credit was given to the lessor for the metals recovered from the cells.

Data collected from the metal finishing industry survey in Ontario, Canada, identified two types of commercially available, electrolytic cells, which are classified as low surface area (LSA) or high surface area (HSA) [21]. The LSA cell consists of anodic and cathodic plates, closely spaced (approximately 25 mm), to increase cell efficiency. Recovery of metal by using the LSA cell results in a sheet of metal that can be easily handled and reused in the originating plating bath. HSA cells use cathodes of a filamentous structure allowing for the metal bearing solution to be pumped through them. To recover the plated metal in sheet form, the metal contacted cathode, which functions as an anode, is placed in an electro-stripping tank. A stainless steel plate is used as a "starter cathode." The accumulated metal sheet is peeled off and reused. Recovery rates using electrolytic cell technology are shown in Table 5.

An innovative electrolytic technique used to increase the mass transfer rate for enhanced metal recovery from low-concentration rinse solutions employs a three-dimensional flow through carbon filter electrode. Use of this high mass transfer (HMT) electrochemical process has been demonstrated at Electronic Precision Circuitry (EPC), Incorporated [22]. Some of the advantages of the use of carbon fibers are as follows:

1. Chemical inertness to most acids, bases, and solvents
2. Good electrical and thermal conductivity
3. Excellent friction and wear characteristics
4. Low density and cost per unit volume

The HMT recovery process is designed to "electrically" reduce metal values, i.e., spent process solutions, to a level acceptable for discharge without the production of sludge. The recovery process employs both electrowinning and electrorefining. The electrically conductive carbon fibers are woven into fabric layers that are connected to two electrical distribution grids mounted within a plastic-coated frame. Due to its electrolytic effect on metal ions and the high hydrogen over-potential characteristics, these carbon filters can be operated at a high current density in solutions with low metal concentrations. As a result, overall power consumption is optimized, hence leading to lower operating costs. The two-step process cycle begins with the electrowinning of the metal onto the carbon filter cathodes (see Figure 2). Once the cathodes reach their design metal loading capacity, the carbon assemblies are transferred to another reaction cell where the metal is dissolved and replated onto stainless steel starter sheets that will be peeled off in scrap form. Projected annual operating costs for treating both rinses and process dumps, when employing the HMT recovery system, were lower than handling dump disposal alone.

Table 5. Recovery Rates Using Electrolytic Technology [21]

Solution	Estimated Recovery Rate	Recovery Technique	Reuse Application	Metal Value ($/kg)	Recovered Metal Value ($/yr)
Acid Zinc Concentrate from Ion-Exchange (2.5 g/L Zn)	5 kg/week	LSA	As Anodes	1.37	343
Nickel Plating, Drag-out	23 kg/week	LSA	As Anodes	10.12	11,638
Copper Plating, Off-line Drag-out	0.35 kg/week	LSA	Sold	1.81	32
Cadmium Plating, Drag-out	0.7 kg/week	LSA	Sold	3.52	123
Acid Copper Plating, Drag-out	23 kg/week	HSA	Sold	1.81	2,082
Acid Zinc Plating, Drag-out	16 kg/week	LSA	As Anodes	1.37	1,096
Bright Nickel Plating, Drag-out	30 kg/week	LSA	As Anodes	10.12	15,180

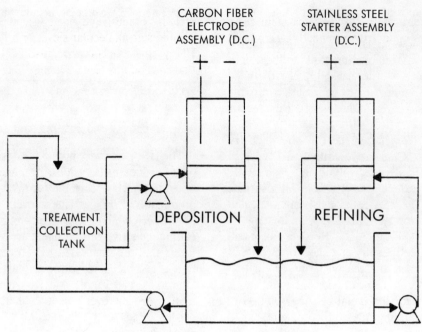

FIGURE 2. HMT treatment schematic [22].

Membrane Technologies

Membrane processes have experienced widespread application in the metals treatment industry for recovering valuable metals or purifying plating solutions. Both reverse osmosis (RO) and electrodialysis (ED) have been successfully employed to recover metals for reuse in plating baths and to remove toxic impurities from rinse solutions.

RO is a pressure driven membrane process where water molecules are forced through the microscopic pores of a semipermeable membrane. The production rate of an RO membrane is a function of dissolved solids concentration, temperature, pressure, pH, and chemistry of the solution to be treated. To date, no commercially available membrane polymer has demonstrated tolerance to all extreme chemical factors, such as pH, strong oxidizing agents, and aromatic hydrocarbons [11].

Many studies have been conducted evaluating the application of RO technology to the metal finishing industry. A study by Crampton successfully demonstrated the use of a cellulose acetate membrane to concentrate the drag-out from plating solutions [23]. Both the concentrate and purified water are recycled. However, this type of membrane is pH sensitive and is primarily used to treat Watt's nickel plating baths. Other drag-out recovery applications successfully treated copper, zinc, and chromic acid baths. Cartwright defines various membrane types and their associate applications [24].

Application—Membrane Type

Nickel—cellulose acetate
Copper sulfate—hollow-film polyamide, cellulose triacetate and spiral-wound thin film composite
Zinc sulfate—spiral-wound thin film composite
Brass cyanide—polyamide cellulose, triacetate hollow fiber
Copper cyanide—polyamide hollow fiber
Chromium (VI)—spiral-wound thin film composite
Other—thin film composite (TFC)

Table 6 shows RO test data for six common plating bath rinses.

Table 6. RO Test Data for Six Common Plating Bath Rinses [24]

Plating Bath	Bath Temperature °C/°F	Toxic Containment	Percent Rejection
Watt's Ni	60/140	Ni^{2+}	99+
Copper cyanide	60/140	Cu^+	85
		CN^-	46
Zinc cyanide	27/80	ZN^{2+}	96
		CN^-	80
Brass cyanide	27/80	ZN^{2+}	98
		Cu^+	97
		CN^-	
Decorative Cr	43/110	Cr^{6+}	90
Hard Cr	55/130	Cr^{6+}	92

Potential problems with RO systems stem from membrane fouling due to suspended precipitated solids coating the membrane surface; precipitation of salts in the concentrate; membrane deterioration due to chemical attack; and for higher strength wastes, high levels of dissolved solids in the permeate.

Electrodialysis (ED) is an electrochemical membrane process employed to remove ionic impurities from process solutions. A conventional system consists of an anode and cathode separated by permeable membranes [21]. In ED systems, an electrical potential is applied across the membranes to provide the driving force for ion passage through the membranes. Membranes used in this process are "thin sheets" of the same polymeric network used to make ion-exchange resins. Unlike RO systems, these membranes are tolerant of most chemical environments. Another distinction that ED systems exhibit over RO systems is the ability to separate water from the salt by selective removal of the salt from solution rather than concentrating the salt into a brine.

It has been reported that ED is most effective for treating concentrated metal finishing wastewaters, achieving recovery efficiencies ranging from 90–95 percent [25]. At 90 percent efficiency as many as 125 cell pairs are required with replacement and installation costs at $80/pair. The operating cost to recover one pound of metal is estimated to be one kilowatt hour of electricity. It is feasible to recover zinc, nickel, and cadmium with ED systems.

Another study demonstrated the application of ED for treating rinsewater containing fluoborate anions, resulting from solder, tin, copper, and nickel plating operations [26]. To achieve optimum results, a graphite anode was developed that is approximately one-fortieth the cost of the platinized titanium anodes typically used. Furthermore, use of electrolyte-containing anions that match those present in the feed solution hinders the contamination of the by-products associated with ED usage.

Ion Exchange

Ion-exchange (I/E) and resin adsorption systems offer versatile separation processes with wide applications to the metal finishing and printed circuit board (PCB) industries. Not only can I/E processes recover valuable metals for reuse, they can serve to provide deionized water makeup and purify waste streams for recirculation. The I/E process can be defined as a separation technology that removes various ionic species from solution via interchanging reversible ions, between the solution (aqueous waste stream) and an exchange (resin) [27]. The exchanger can be a natural material (e.g., soils, coal, lignin, metallic oxides, and bacteria), synthetic resin, insoluble salt, molecular sieve, or a liquid membrane. I/E systems can comprise adsorbent columns, membranes modules (e.g., Donnan Dialysis), or liquid-liquid extraction.

Success for treating aqueous metal-bearing waste streams depends upon the selection of a suitable resin. Some of the inorganic I/E resins used to treat these wastes streams are insoluble salts (e.g., phosphates, tungstates, silicates, zirconium phosphates, and oxides), heteropolyacids, complex salts (based on ferrocyanide), and new zeolites (prepared from thermal methods) [28]. Certain resins will remove cations (e.g., lead, nickel, cadmium, barium, and zinc), while others will remove anions (e.g., arsenate, selenate, chromate, uranium, and cyanide) as ferrous cyanide complexes.

Exchanger categories include both strong and weak cations and anions, as well as chelating ion exchangers. Some of the more common reactive groups are listed as follows [5]:

Reactive Groups	Composition	Exchangeable Ions
Strong Acid	Sulfonic	Cations
Weak Acid	Carboxylic	Cations
Weak Acid	Phenolic	Cesium and polyvalent cations
Strong Base	Quaternary amines	Anions
Weak Base	Tertiary, secondary amines	Anion of strong acids
Chelating	Iminodiacetate and oxime group	Cations (esp. heavy metals)

Ion-exchange systems have been used in many applications. Generally, system performance and mode of operation are greatly influenced by the characteristics of the waste stream that include pollutants to be removed, constituent concentration, pH, desired effluent quality, and recovery/regeneration considerations. Factors affecting system performance are process residuals (sidestreams), throughput capacity, and operational costs.

Ion-exchange systems can be operated in a batch (contact) or adsorber column (flow through) mode. During a batch operation, the resin is contacted with the waste stream in a mixing tank. Upon completion of the residence time, the resin is separated from the bulk solution and regenerated for further use. The continuous flow systems have some distinct advantages over batch operations in that they have greater applicability and higher efficiency, and consist of the following steps: service to breakthrough, backwash to expand and resettle the bed, regeneration to replace the depleted ion, and rinsing to remove the excess regenerant. Continuous flow systems can be designed to be operated in either cocurrent or countercurrent configurations. In cocurrent systems, the feed and regenerant both pass through the resin in a down flow mode. One disadvantage encountered when employing this configuration is the high cost of regeneration, especially for strong acid and base exchangers. Countercurrent systems offer a more efficient use of regeneration reagents. The flow is reversed or reciprocated which has the effect of concentrating more contaminant into the regenerant stream. Examples of this latter application are the acid purification unit (APU) and the reverse flow ion exchanger (RFIE).

Many studies have been conducted that demonstrate various applications of adsorbents (resins) and process modes for treating synthetic and industrial aqueous metal wastewaters. In one study, Tare et al. [29] examined the kinetics of metal uptake from solutions of multimetal-ligand systems in treating a surrogate printed circuit board semiconductor manufacturing wastes by a chelating I/E resin (Chelex 100™). Since the presence of strong complexing organics (e.g., citrate, EDTA, tartrate) and inorganic ligands (Cl, F, PO_4) in dilute aqueous metal bearing wastes adversely affect conventional precipitation processes, I/E resins were evaluated as a treatment alternative. Selectivity constants, for specific metal ions, were assessed at predetermined uptake rates. Results of this study indicated that metal uptake rates were strongly influenced by solution pH and waste aging. More specifically, Cr^{3+} and Pb were more slowly exchanged from complexed solutions aged 30 days than from fresh batches. There was no qualitative agreement between the metal uptake rates and their respective selectivity constants, suggesting that kinetics may have promoted the metal separations. Desorption was readily achieved by the use of hydrochloric acid.

In another study, an innovative treatment system was employed to treat rinsewaters and bath dumps containing copper and nickel resulting from a PCB manufacturing operation [30]. Both RFIEs and electrowinning were used to recover these metal values from the generated waste stream. Some of the advantages associated with these technologies are reduced equipment size due to smaller columns, lower startup costs, lower operating costs resulting from countercurrent regeneration of the exchanges, longer resin life when using a fine mesh resin that is less prone to fracture and oxidize with shorter contact times, the ability to treat concentrated solutions, and the elimination of a hydroxide sludge with recovery of metal sheets.

It has been recognized that I/E resins have difficulty in removing mercury (Hg) from brine solutions, such as chlor-alkali plant effluents. The problem was confronted with the idea of producing a selective adsorbent for the purpose of removing mercury from solution. The resin selected was a diatomite matrix adsorbent (AD-30) that has the ability to form strongly bound irreversible complexes with mercurials. The capacity for this resin was 12 mg Hg/g of adsorbent. This adsorbent contained sulfhydryl groups attached to an inorganic matrix via saline intermediates. It was also resistive to oxidation by ferricyanide, HCl and dissolved oxygen. Numerical simulations or models, which accounted for kinetic, pore diffusion and matrix effects, correlated favorably with adsorption column studies for removing organic mercurials [31].

A more recent process was developed by the Akzo Chemicals Company that removed mercury from a chlor-alkali plant effluent by using a Rohm & Haas Duolite 67–73 (a weak acid cation-exchange resin) with high specificity for mercury. This process was demonstrated to be successful in removing Hg from a process stream flow rate of 1.75 gpm/ft^3 to well below 5 ppb [32].

Another common application of an I/E system is the acid purification unit (APU). Typically, these systems have been employed to recover acids from aluminum anodizing solutions, acid pickling liquors and rack-stripping solutions. These systems have shown to be successful in recovering solutions with positively charged ions (e.g., aluminum and iron) since these ions pass rapidly through the strong base anion-exchanger resin. Lower concentrations of metal species are not recovered as efficiently using the APU. Table 7 presents minimum concentrations recommended for efficient metals removal using the Eco-Tech APU [5].

Liquid I/E is a relatively new process that combines both I/E principles and liquid-liquid extraction techniques for the recovery of metal ions from aqueous waste streams [33]. This application has shown promise for treating dilute metal-laden solutions. Like resin systems, the actual exchange takes place when the solution of metal ions, upon contact with an extractant, produces an unstable liquid-liquid dispersion. The process occurs via a three-step mechanism: (a) the complexed metal ion diffuses to the aqueous/organic interface; (b) the ion is stripped of its hydration layer and complexes with the extractant at the interface; and (c) the organic/metal complex diffuses into the bulk organic phase where, simultaneously, hydrogen ions are transferred to the aqueous phase. Thereafter, the stripping solution (extractant) is regenerated.

Table 7. Recommended Minimum Concentrations (g/L) for Efficient Metals Removal Using the Eco-Tech APU [5]

Solution	Iron	Zinc	Aluminum	Copper	Total Metals
Hydrochloric acid	30–50	130–150	—	—	—
Sulfuric acid	30–50	—	5	20	—
Nitric-hydrofluoric acid	—	—	—	—	30
Nitric acid rack stripping	—	—	—	—	75–100

Note: The APU can be used for solutions with lower concentrations of these metals, but the metal removal efficiencies will be lower unless a larger unit is used. Metal removal efficiencies average 55 percent for typical systems.

Chapman et al. [34] investigated physical and chemical rate phenomena which influence metal extraction efficiency and selectivity. Extraction efficiency is determined by the selection of a reagent that provides the optimum equilibrium phase distribution properties for the metal species of interest. Two major factors that influence the rate of metal transfer between the two immiscible liquid phases are contacting and solution pH. A generalized representation of the reaction can be expressed as

$$M^{2+} + 2HR \rightleftharpoons MR_2 + 2H^+ \quad (3)$$

where M^{2+} is an aqueous metal ion that is exchanged for hydrogen ions by an organic phase solute, HR, to form an organic soluble metal species MR_2. A more representative and thermodynamically correct mass action equilibrium constant K can be incorporated to form the following equation:

$$K = \frac{[MR_2][H^{2+}]}{[M^{2+}][HR]} \quad (4)$$

The reagent, HR, can be a common acid (carbonic or phosphoric acid) or a chelating agent. It was concluded that the solvent extraction of metals occurs by means of a heterogenous chemical reaction where the multicomponent nature of the process must be understood before extraction rates can be determined.

As previously mentioned, ion-exchange materials have been used in conjunction with electrolytic recovery applications for reclaiming metals from process wastewaters. However, electrolytic recovery from dilute solutions has experienced several limitations whereby: 1) electrical conductivity may be insufficient for adequate electrodeposition to occur with minimal energy requirements; 2) a low metal-ion concentration provides a low driving force for diffusion transport of metal ions to the cathode, limiting the reaction rate; 3) electrodeposition creates a diffusion layer adjacent to the cathode surface causing metal electrical potentials to become more negative which can allow for other competing cathodic reactions (e.g., hydrogen evolution) to become more dominating.

A study was conducted at General Motors (GM) Research Laboratories employing an innovative two-phase electrolytic septum consisting of an "interlectrode solution" and cation-exchange resins (beads) [37]. When a cation-exchanger is immersed into a solution containing metal cations (copper and nickel) the ion-exchanger tends to attract these cations in ratios determined by thermodyanmic (Donnan) equilibrium. In dilute solutions, multivalent ions such as Cu^{2+} and Ni^{2+} are preferentially selected over monovalent species. When exposed to an electrical field, these ion-exchangers exhibit two properties that promote electrodeposition. The first is that concentrated "mobile" counter ions within the resin provide ionic electrical conductivity substantially independent of the surrounding solution which has the effect of increasing the net conductivity of the solution's resin mix. As a result, uniform metal coverage on the cathode surface is enhanced. Secondly, as the current passes through the solution cation-exchange mix, a discontinuity occurs around the resin bead column, leading to concentration effects in the solution. Both the rate of reaction and current efficiency for electrodeposition are improved. Results obtained for five different proprietary resins indicate that superior electrochemical metal recovery was attainable by loading dilute solutions containing Cu^{2+} and Ni^{2+} with strongly acidic particulate cation-exchange resins.

Regeneration of spent I/E resins is a very important part of cost-benefit considerations when considering use of I/E systems for removing heavy metal contaminants from aqueous waste streams. Etzel and Tseng [36] conducted a series of studies examining the cation-exchange removal of heavy metals with a recoverable chelate regenerant. Results showed that the sodium salt of EDTA, NTA, or CIT could be used to regenerate a strong cation-exchanger exhausted with Cu, Zn, or Ni. Regeneration was pH dependent, providing the best results in the pH range between 6 and 9. The predominant reaction of displacement was through a metal chelate formation. At a pH greater than 10, a precipitate formed which fouled the resin and reduced exchange capacity. Since regeneration efficiency for the sodium salts varied as the number of Na ions were available for exchange per mole of chelating agent solution, EDTA and CIT performed better than NTA as reagents. Recoveries for both heavy metal and chelating agent solution from the spent regenerant were evaluated via an electromembrane process. Since data revealed a more efficient recovery of Cu over Zn or Ni, a final experiment was performed which cyclically exhausted and regenerated a resin with EDTA and recovered EDTA-Cu via electromembrane deposition, during five complete cycles. Both EDTA and Cu recovery averaged approximately 97 percent.

DeVoe-Holbein Technology

DeVoe-Holbein International, N.V. is promoting a patented technology which can be used in ion-exchange equipment [5]. This process is an adaptation of biological transformations coupled with adsorption technology, where coordinating compounds are covalently bonded to the surface of an inert carrier material that, purportedly, captures metal ions. Living cells actually remove, by selection, a variety of inorganics from solution (e.g., Cu, Zn, Co, Fe, Se, Cn, and Mn). Specialized cells, collectively known as Siderophores, can remove specific metals via molecular sites that bind certain metal species. Siderophores, which belong to two classes of microbes, *Enterobactin* and *Desferrioxamine*, are covalently linked to porous glass bead supports. In addition, a series of synthesized metal-capturing compositions with catechol are used as the active components (Vitrokele™). It has been claimed that these formulations have shown to be effective in achieving ≥ 99 percent removal rates for specific or grouped metal ions. Table 8 shows some results of capturing toxic heavy metals by DeVoe-Holbein compositions.

One of the claimed advantages of this technology is that it is capable of yielding a more concentrated regenerant than ion exchange. However, residual solids present in the feed can foul the adsorbents in a column. As a result, pretreatment may be required to remove these solids.

Case studies applying this technology have reported successes in: (a) the removal of chromium from boiler blowdown water, chrome plating waste precipitator effluent, and cooling tower wastewater; (b) zinc removal from a 5-gpm countercurrent rinse effluent from a zinc chloride electroplating line; and (c) treatment of spent baths, acids, and emulsifiers from a large job shop containing nickel, zinc, brass, chromium, cyanide, and precious metals.

Centralized Treatment

With increasing disposal costs and more stringent regulations, a limiting factor associated with the treatment, reduction, and recovery of metal-bearing hazardous waste streams is economical. Small-sized manufacturers, job shops, and industrial operations are not financially capable, from the standpoint of economy of scale, to implement measures that embrace waste minimization practices. Capital start-up costs and the extent of payback period prevent many small-scale operations from capitalizing on in-plant pretreatment recovery processes. An alternative to remedy these problems has been promoted as centralizing treatment to accommodate numerous smaller operations located within a defined geographical area (e.g., larger metropolitan or regional areas). Consideration for the consolidation of waste treatment operations recognizes the advantage of recovery as a means to offset disposal costs.

Table 8. The Efficient Capture of Some Toxic Heavy Metals of Importance to the Hydrometallurgical and Metal-Finishing Industries by DeVoe-Holbein Composition [5]

Toxic Metals	DeVoe-Holbein Composition	Influent Concentration[a]	Effluent Concentration	Capture Efficiency(%)[b]
Cadmium	DH-516	674 ppm	<1.0 ppm	≥99.99
Chromium	DH-524	694 ppm	<0.01 ppm	≥99.99
Copper	DH-520	38 ppm	<1.0 ppt	≥99.99
Lead	DH-501	42 ppm	<42.0 ppb	≥99.99
Mercury	DH-573	12 ppm	<1.0 ppb	≥99.99
Nickel	DH-507	0.10 ppm	<1.5 ppt	≥99.99
Zinc	DH-508	6.5 ppm	<0.8 ppb	≥99.99

[a]Effluent concentration at or below normal detection limits using either radioactive tracer or atomic absorption spectrophotometric determinations.
[b]Capture efficiency determined as percent reduction in influent concentration; values are greater or equal to those shown due to detection limits of effluent metal concentration.

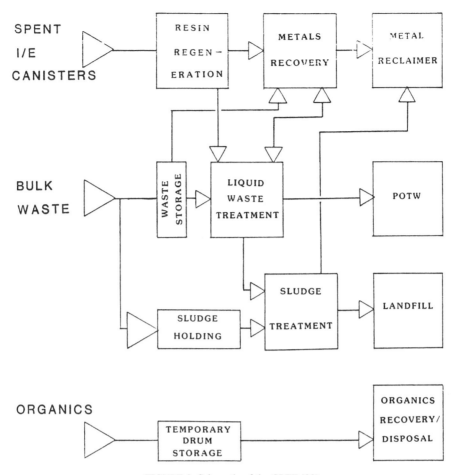

FIGURE 3. Schematic of the CRTF [38].

Interest in establishing centralized treatment facilities has spurred several demonstration projects to determine the feasibility of treating a large variety and quantity of metal/cyanide waste streams. A study was conducted in the Cleveland Metropolitan Area, having a pool of approximately 100 metal finishing plants, most of which were small job shops [37]. Determining factors for evaluating this approach are the volume and characteristics of the wastes to be considered for centralized waste treatment. This information was gathered from survey and site visits. Three categories of wastes were selected: batch dumps of exhaustible plating solutions, rinse waters, and sludge from in-plant pretreatment. Then, design and cost data were evaluated for a proposed centralized treatment facility at the Cleveland Resource Recovery Park (RRP). Two capacity designs (76 plant and 30 plant) were developed and compared. Each incorporated conventional treatment processes (e.g., cyanide oxidation, chromium reduction, precipitation, neutralization, clarification, thickening, dewatering) and recovery processes, emphasizing ion exchange. Capital operating costs were determined and, on that basis, projected waste treatment fees were calculated. On a per gallon basis, it would cost $0.085 for the 76-plant capacity versus $0.109 for the 30-plant. A significant cost limiting factor is the transportation cost associated with the servicing of the ion-exchange modules requiring removal and regeneration. However, as disposal costs rise, the attractiveness of considering centralized treatment is enhanced.

Another facility located in the Minneapolis/St. Paul area offers centralized recovery and treatment of wastes generated from 24 individual participating surface finishing and electronics manufacturing plants [38]. The centralized recovery and treatment facility (CRTF) collects and transports waste from individual firms to the centralized facility. Then, it recovers chemicals and metals and treats residual wastes for authorized disposal. The CRTF employs an array of physical/chemical processes including ion exchange and electrowinning to treat acids, alkalies, chromates, cyanide solutions, and metal hydroxide sludge.

Each of the participating firms has installed their own ion-exchange and carbon filtration waste collection system. The batch dump system with holding tanks must be large enough in capacity to hold a seven-day dump of the plant's waste. Trucks are then employed to transfer segregated I/E canisters and batch loads (tanker) to the CRTF. Figure 3 shows a generalized schematic of the centralized facility. Canisters containing spent resins are off-loaded, segregated by metals and cyanide, and then transferred to regeneration vessels where metals and salts are removed and waste cyanide converted into sodium cyanide. Metal values, such as copper, nickel, and zinc removed from the I/E resins are processed through one of three electrolytic metal recovery cells. As much as a 20:1 metal reduction is accomplished where metal deposited on cathode plates are stripped and sent to a reclaimer. The residual wastewater is treated, generating a hydroxide sludge which is further thickened and filter pressed. The filter cake is either processed further for metal reduction or shipped off-site to an approved hazardous landfill. Bulk shipments of acids, alkalies, chromates, and ammonia are processed similarly where metals are recovered, cyanide converted, chromates reduced, and corrosives (acids and bases) neutralized. Organic wastes are sent to recovery/disposal facilities. All treated wastewaters are sent to the local POTW via discharge permit arrangements.

Although the CRTF has only been in operation since 1985, participating plants have realized capital savings of 50 to 75 percent as compared with the installation of in-plant wastewater treatment systems. However, it is yet too early to provide meaningful operating data for each of the firms. Future audits and analyses must be performed before this demonstration is shown to be successful.

REFERENCES

1. Congressional Budget Office. *Hazardous Waste Management: Recent Changes and Policy Alternatives.* Prepared for Senate Committee in Environmental and Public Works (1985).
2. Freeman, H. M. and E. Eby. "A Review of Hazardous Waste Minimization in the U.S.," presented at the *International Congress on Recent Advances in the Management of Hazardous and Toxic Wastes in the Process Industries, March 8–13, 1987, Vienna, Austria.*
3. P.L. 98-616 (November 8, 1984).
4. Bussard, P. A. "New Directions in EPA's BDAT's Waste Minimization Programs," *Proceedings of the 9th AESF/EPA Conference on Environmental Control for the Metal Finishing Industry, Orlando, FL* (1988).
5. Palmer, S. A. et al. *Technical Resource Document: Treatment Technologies for Metal/Cyanide-Containing Wastes,* EPA/600/S2-81/106. Bedford, MA:Alliance Technologies Corp. (February 1988).
6. Resource Conservation and Recovery Act of 1976, as amended (42 U.S.C. 6905, 6912(a), 6921, 6922 and 6937), from 45 FR 33119 (May 19, 1980).
7. *Technical Assessment of Treatment Alternatives for Waste-Containing Metals and/or Cyanides.* Preliminary draft report under USEPA, OSW, Contract No. 68-03-3149. Springfield, VA:Versar, Inc. (1984).
8. Rothschild, B. F. and M. Schwartz. *Printed Circuit/Wiring Board Manufacture,* American Electroplating Society, Inc. (1983).
9. Umphres, M. B. and T. E. Pollick. "Treatment of Metal Finishing Wastes for Reuse/Recycle," in *Performance and Costs of Alternatives to Land Disposal of Hazardous Waste.* E. T. Oppelt et al. ed., from Air Pollution Control Association (APCA), International Specialty Conference, New Orleans, LA, p. 124 (1986).
10. *Control Technology for the Metal Finishing Industry: Evaporators, Summary Report.* EPA 625/8-79-002, IERL, ORD. Cincinnati, OH:USEPA (1979).
11. Grosse, D. W. "A Review of Alternative Treatment Processes for Metal-Bearing Hazardous Waste Streams: Treatment Technologies for Hazardous Wastes, Part IV," *Air Pollution Control Association,* 36(5):603 (1986).
12. Forrestal, B. *Factors Which Affect the Success of Metal/Chemical Recovery Installations in Metal Finishing Industry,* Project Report 71. Ontario, Canada: Ontario Research Foundation, p. 87-48-5259/CI (1987).
13. Mohler, J. B. "Art and Science of Rinsing," *American Electroplaters Society (AES) Lecture Series.* Orlando, FL:AES Inc. (1985).
14. *Control and Treatment Technology for the Metal Finishing Industry: In-Plant Changes, Summary Report.* EPA/625-18-82-008, IERL, ORD. Cincinnati, OH: USEPA (1982).
15. Apel, M. L., P. S. Fair and J. P. Adams. *Design and Application of a Spray Rinsing System for Recycle of Process Waters.* NTIS No. PB85106722XSP, Cincinnati, OH:USEPA (1983).
16. *Environmental Pollution Control Alternatives: Reducing Water Pollution Control Costs in the Electroplating Industry,* EPA/625/5-85/016. Cincinnati, OH:CERI, ORD (1985).
17. Kaplan, M. "Innovative Job Shop Adapts Non-Polluting Brass Substitute," *Plating and Surface Finishing,* 75(9):42 (1988).
18. Walters, R. and D. Vitagliano. "Electrolytic Recovery of Zinc from Metal Finishing Rinse Waters," *Proceedings of the 16th Mid-Atlantic Industrial Waste Conference, Lewisburg, PA* (1984).
19. Bishop, P. L. and R. A. Brenton. "Treatment of Electroless Copper Plating Wastes," *Toxic and Hazardous Wastes: Proceedings of the 15th Mid-Atlantic Industrial Waste Conference, Lewisburg, PA,* pp. 584–596 (1983).
20. Harrison, D. S. *Technical Assistance for Huntington Park Group Treatment Facility.* NTIS No. PB85222768XSP. USEPA Contract No. 68-03-2907-17. Cincinnati, OH:IERL, ORD (1984).
21. Forrestal, B. *Factors Which Affect the Success of Metal/Chemical Recovery Installations in Metal Finishing Industry,* Project Report 71. Ontario, Canada:Ontario Research Foundation, p. 87-48-5259/CI (1987).
22. Bailey, D. et al. "High Mass Transfer (HMT) Electrochemical Process: Integration of Metal Recovery and

Regulatory Compliance (A Case Study)," *Proceedings of the 9th AESF/EPA Conference on Environmental Control for the Metal Finishing Industry*, Orlando, FL (1988).

23. Crampton, P. *The Application of Separation Processes in the Metal Finishing Industry*, EPA/600/2-81-028m. Cincinnati, OH:USEPA, p. 1091 (1981).

24. Cartwright, P. S. "An Uptake on Reverse Osmosis for Metal Finishing," *Plating and Surface Finishing*, 71:62 (April, 1984).

25. McCoy and Associates, Lakewood, CO. "Land Disposal Alternatives Evaluated for Electroplating Wastes," *The Hazardous Waste Consultant*, 5:1–22 (July/August 1987).

26. Liskowitz, J. W. et al. "Removal and Recovery of Fluoborates and Metal Ions from Electroplating Wastewater," in *Project Summary*. EPA/600/S2-85/054. Cincinnati,OH:HWERL, ORD (1985).

27. Patterson, J. W. *Industrial Wastewater Treatment Technology, Second edition*. Stoneham, MS:Butterworth Publishers (1985).

28. Dofner, K. "Ion Exchanger Types," in *Ion Exchangers: Properties and Applications*. A. F. Coers, ed. Ann Arbor, MI:Ann Arbor Science Publishing Inc. (1972).

29. Tare, V. et al. "Kinetics of Metal Removal of Chelating Resins from a Complex Synthetic Wastewater," *Water, Air, Soil Pollution*, 22(4):431–439 (1984).

30. Pajunon, P. and E. Schneider. "Copper and Nickel Removal in Printed Circuit Board Processing by Ion Exchange and Electrowinning," *Proceedings of the 9th AESF/EPA Conference on Environmental Control for the Metal Finishing Industry*, Orlando, FL (1988).

31. Swan, G. A. et al. "Heavy Metal Removal from Aqueous Media," *American Institute of Chemical Engineers Symposium Series on Adsorption and Ion Exchange*, 71(152):96 (1975).

32. Rohm and Haas Co. *Duolite GT 73 Ion-Exchange Resin*, Product Bulletin (1986).

33. Gallacher, L. V. *Liquid Ion Exchange in Metal Recovery and Recycling*. EPA 666/2/81-028. Cincinnati, OH:USEPA (1981).

34. Chapman, T. W. et al. "Rates of Liquid-Liquid Ion Exchange in Metal Extraction Processes," *American Institute of Chemical Engineers Symposium Series on Adsorption and Ion Exchange*, 71(152) (1975).

35. Tison, R. P. "Electrochemical Metals Recovery from Dilute Solutions Using Ion-Exchange Materials," *Plating and Surface Finishing*, 75(5) (1988).

36. Etzel, J. E. and D-H. Tseng. "Cation-Exchange Removal of Heavy Metals with a Recoverable Chelant Regenerant," in *Metals Specification, Separation and Recovery*. J. W. Patterson and R. Passino, eds. Cholsea, MI:Lewis Publishers (1987).

37. Comfort, E. et al. "Centralized Treatment of Metal Finishing Wastes at Cleveland Resource Recovery Park—Part I: Design and Costs; Part II: Financing; Part III: Site Investigation," in *Project Summary*. EPA/600-S2-85/075. Cincinnati, OH:HWERL, ORD (1985).

38. Norgaard, G. E. "Centralized Waste Treatment Facility Opens in Twin Cities Area," *Plating and Surface Finishing*, 75(11) (1988).

2.24

Solidification/Stabilization Techniques: Promising Treatment Technologies for Remediating Superfund Sites

Joseph DeFranco[1]

After many months of debate, Congress passed the Superfund reauthorization legislation in November 1986 which extended the Superfund cleanup program for an additional five years. This legislation, the Superfund Amendment and Reauthorization Act (SARA), included many detailed programmatic requirements that were intended to spur the EPA toward increased site remediation activities. Passage of this legislation came only after thorough review of many important issues, such as site cleanup standards and the use of permanent treatment technologies, to name just a few.

A key requirement under SARA [1] is the need to select a remedial action "that is protective of human health and the environment, that is cost-effective, and that utilizes permanent solutions and alternative treatment technologies or resource recovery technologies to the maximum extent practicable (p. 00)." In effect, this new Superfund requirement recognizes the need to use permanent solutions wherever feasible and available at Superfund sites. It is rapidly becoming one of the key requirements that the EPA and state agencies are imposing in connection with the cleanup of any Superfund site, and it will certainly be a key criterion as sites are remediated in the future.

As required by law, the EPA has begun to pursue the development of various promising innovative and alternative treatment technologies for use in remediating Superfund hazardous waste sites. Permanent solutions are now being considered in the Agency's Records of Decision for Superfund sites in the cleanup stage. In addition, the EPA has initiated other programs aimed at commercializing the use of advanced treatment technologies that can provide permanent solutions and reduce both the short- and long-term potential for adverse health effects from human exposure to hazardous wastes.

One of the most promising innovative, alternative remedial technologies is a solidification/stabilization process used by Separation and Recovery Systems, Inc. (SRS) of Irvine, California, which has been previously developed by EIF Ecology of France (EIF). The SRS/EIF process is a chemical fixation technology developed in France in 1977 to detoxify organic hazardous wastes. This process is cost-effective, and most importantly, it renders the hazardous waste nontoxic, thus eliminating possible waste leaching and disposal problems in the future.

This chapter explores the Superfund policy considerations underlying the development of innovative, permanent remedial technologies. It also examines solidification/stabilization processes in general, and the SRS/EIF technology, in particular, in light of the new Superfund mandate for permanent waste treatment remedies.

1986 SUPERFUND AMENDMENTS

Much of the congressional focus during the Superfund reauthorization process dealt with the lack of cleanup goals in the original Superfund law. Critics of the EPA's initial implementation of Superfund objected to the "shell game" approach to cleanups, whereby toxic wastes were merely moved from one Superfund site to another site where they were contained in new or existing land impoundments.

A study conducted by the U.S. Office of Technology Assessment (OTA) found that the lack of cleanup standards was significantly inhibiting the effective

---

[1]Separation and Recovery Systems, Inc., 16901 Armstrong Ave., Irvine, CA 92714-4962

implementation of the Superfund program. This OTA study also indicated that there were specific statutory preferences in the law for "waste containment and groundwater treatment rather than waste removal and treatment." According to OTA, this institutional preference for land containment inhibited the growth of new treatment technologies, since the EPA's preferences for containment technologies created an uncertain market which eliminated important economic incentives for investing in research and development and for commercializing advanced technologies.

This study and other testimony before congressional panels on Superfund reauthorization proposals laid the groundwork for changes in how the EPA would be required to remediate Superfund sites in the future. Testimony suggested that less emphasis should be placed on cost factors and more on the need for effective cleanup solutions at Superfund sites.

In response to testimony during the reauthorization hearings, Congress directed the EPA in SARA to utilize permanent treatment technologies to the maximum extent practicable and authorized several new programs to encourage the growth of innovative, alternative treatment technologies. A key program established in SARA [1] is the Superfund Innovative Technology Evaluation (SITE) program, which attempts to remove impediments to the development and commercial use of alternative technologies. The SITE program includes both a development program for emerging technologies and a demonstration program of "the more promising innovative technologies to establish reliable performance and cost information for site characterization and cleanup decision making." The purpose of this demonstration program is to provide "performance, cost-effectiveness, and reliability data so that potential users have sufficient information to make sound judgments as to the applicability of [a treatment] technology for a specific site and to compare it to other alternatives." This program offers a significant opportunity to test promising permanent treatment options, such as innovative solidification/stabilization techniques, at specific Superfund sites and to carefully evaluate their role in meeting the permanent treatment mandates under SARA.

SOLIDIFICATION/STABILIZATION

Generally, solidification/stabilization technologies refer to treatment processes that transform hazardous wastes into a more manageable or less toxic form. In particular, solidification processes have referred to methods in which certain types of materials are added to hazardous wastes to produce a new solid material. On the other hand, stabilization techniques [2] have generally referred to processes "by which a waste is converted to a more chemically stable form (p. 14)." "Fixation" approaches can usually mean either solidification or stabilization techniques.

Solidification/stabilization processes are designed to achieve several objectives, such as improving the handling and physical characteristics of a hazardous waste or limiting the solubility of the hazardous constituents in any particular waste. Carlton C. Wiles [2] of the U.S. Environmental Protection Agency has noted that "solidification/stabilization technology has the potential for making a major contribution as one of the alternatives for managing hazardous wastes." In discussing solidification/stabilization technology, he has also stated that "lower permeability, lower contaminant leaching rates, and similar characteristics from [solidification/stabilization] may make banned wastes acceptable for land disposal after stabilization (p. 20)."

Most wastes chosen for solidification/stabilization treatment are liquids or sludges. By using a solidification technique, for example, these wastes would generally be mixed with a binding agent and cured or dried into a solid form. Several types of binder systems are available for use in solidification/stabilization processes. Most of these systems fall into two categories: inorganic or organic. However, combinations of both types of systems have also been used, for example, diatomaceous earth with cement and polystyrene, polyurethane and cement, and polymer gels with silicate and lime cement.

REVIEW OF SOLIDIFICATION/ STABILIZATION TECHNOLOGY

Overall, several solidification/stabilization technologies have been proposed for treating hazardous wastes. Solidification processes mechanically lock contaminants within a solid matrix structure, such as through "microencapsulation." Chemical stabilization processes reduce the solubility or chemical reactivity of the waste by changing its chemical state. Some commercial processes incorporate both solidification and chemical stabilization into one treatment technology which transforms the waste into environmentally safer forms. The waste is converted into a solid matrix to reduce the opportunity for volatilization, leaching or spillage. The SRS/EIF technology utilizes a combined physico-chemical process to permanently detoxify and encapsulate hazardous waste sludge.

SRS Technology

SRS has the exclusive license in the United States to market an innovative, alternative solidification technology developed by EIF Ecology of France. The SRS/EIF technology, which was developed over ten years ago for the cleanup of an organic sludge pit in France, utilizes a lime-based process to permanently fix waste into a matrix product. This process is carried out on-site and therefore does not require the costly transport of waste. It is unique because it can treat waste that contains high percentages of organics. Wastes previously treated by this process have included crude oil, refinery intermediates or final products, halogenated chemicals, PCBs, pesticides, sludges, tars, painting wastes, and acid sludges.

Soon after its development, the technology was selected by the French government to treat contaminated beach soil and debris from the Amoco Cadiz shipwreck off the coast of Normandy. Over the past ten years, the SRS/EIF technology has been used successfully to clean up ten waste sites in Europe containing a total of 200,000 cubic yards of waste material. The SRS/EIF technology was originally developed to address the problems associated with the cleanup of wastes containing a high percentage of organics.

In this regard, the solidification technology is unique because it permanently immobilizes hazardous wastes containing both high percentages of organics, and heavy metals. EIF Ecology set three objectives in developing this technology:

- Create a process chemistry which treats a wide range of organic constituents and allows for fixation of sludges containing a high percentage of organics.
- Detoxify the waste being treated, promote chemical reactions between the waste organic and inorganic constituents and the chemicals used, and not merely encapsulate the waste product.
- Use simple, readily available, and low-cost earth working equipment for waste removal and mixing with the fixation chemicals.

The SRS/EIF technology has met the stated objectives of its developers.

Process

The SRS/EIF technology is a lime-based process. The lime, which contains proprietary nontoxic chemicals, catalyzes and controls the reactions between the lime and the waste. A minimum organic waste content of 3–5 percent is necessary to facilitate the physico-chemical reactions of the technology. The sludge to be treated is placed in an on-site mixing pit and blended with the nontoxic lime and lime additives. The physico-chemical reaction, with lime as the base material, transforms the hazardous waste acid, organic sludges, and associated contaminated solids and liquids into a dry pulverized material.

The heavy metals and organics in the treated waste react to form a product matrix. The treated product is hydrophobic. Rain, surface water, and groundwater are repelled by the treated waste and do not move into the product. The permeability of the fixed product has been found to be 1×10^{-7} to less than 1×10^{-12} centimeters/second. In addition, the permeability of treated refinery impoundment sludges was recently reported to be less than 1×10^{-12} centimeters/second.

Test Data

Available test results on waste materials treated by the SRS/EIF technology have shown that the process is permanent and the leachability of the treated material continues to decrease with time. For example, test results on refinery acid sludge at a Baisieux, France waste site that was treated with the SRS/EIF technology indicate that the metals in the contaminated waste are, to date, immobilized. Furthermore, the results show that the levels of hexane and chloroform extractable organics in distilled water extracts have been reduced by approximately 96 percent or, in some instances, by more than 99 percent.

Tests conducted on acid sludge treated with the SRS/EIF technology at a D'Abscon, France waste site show equally impressive results (see Table 1). Chemical oxygen demand (COD) measurements of leachate from the samples taken shortly after treatment and every six months thereafter clearly show that the extract COD continues to decrease for a period of time after treatment. Similar test results have been found at another lube oil acid sludge SRS/EIF project near Bourron-Marlotle, France.

Additionally, the SRS/EIF technology has been used to stabilize 400 cubic yards of lube oil acid sludge during a test project at a Sand Springs, Oklahoma Superfund site. Table 2 shows the total constituent analysis for the untreated lube oil acid sludge. Table 3 shows the Toxicity Characteristic Leaching Procedure (TCLP) results for the treated product.

Overall, these test data indicate full compliance with applicable EPA regulatory threshold levels using the TCLP test procedures. The test results show that the SRS/EIF process effectively fixes the organic

Table 1. Commercial Project Results: D'Abscon, France Treated Product

Sample Date	COD of Extract, mg/L
05/14/84 (12 h after treatment)	2738
11/08/84	196
05/10/85	143
11/20/85	120
05/22/86	118
11/12/86	116
06/04/87	116

Table 2. Untreated Acid Sludge: Total Constituent Concentrations

Hazardous Constituent	Concentration, mg/kg
Volatiles	
Benzene	TR
Ethyl Benzene	TR–0.76
Toluene	0.79–1.5
Total Xylenes	3.2–7.0
2-Hexanone	1.9–2.8
Semivolatiles	
2-Methylnaphthalene	29–45
Phenanthrene	28–79
Pyrene	22–48
Benzo(A)Anthracene	ND
Chrysene	57–150
Major Fractions, %	
Oil	27.9–38.7
Water	9.8–18.5
Solids	53.6–57.2

ND = Not detected
TR = Trace

Table 3. Treated Acid Sludge: TCLP Extract

Hazardous Constituent	Concentration, mg/L
Volatiles	
Benzene	ND
Ethyl Benzene	ND
Toluene	ND
Total Xylenes	ND
2-Hexanone	ND
Semivolatiles	
2-Methylnaphthalene	ND–TR
Phenanthrene	ND
Pyrene	ND
Benzo(A)Anthracene	ND
Chrysene	ND
Phenol	TR–0.024
2-Methylphenol	ND–TR
4-Methylphenol	TR
2,4-Dimethylphenol	ND–0.024

ND = Not detected
TR = Trace

contaminants at levels below the detection limit of the TCLP procedure.

As part of this demonstration project, an environmental monitoring program was conducted at the Sand Springs treatment pits for the three components of concern: volatile organic compounds, sulfur dioxide, and hydrogen sulfide. For the purpose of this program, the primary areas of concern were the excavation of the sludge and the mixing process in the nearby blending pits. Air monitoring results indicated that the maximum level of sulfur dioxide and volatile organics during this program, with the water spray system properly positioned, was less than 10 parts per million (ppm) and typically around 1.0 to 2.0 ppm. The emissions of volatile organics from the blending pits during the exothermic reactions were generally found to be around 1.0 to 2.0 ppm and always less than 10.0 ppm.

Applicability of SRS/EIF Technology

Based on the test data obtained to date, it is clear that the SRS/EIF technology can be applied to a wide variety of hazardous waste contaminants. However, in order to be an effective option, the organic content of the waste must be at least 3–5 percent. Hazardous wastes containing as much as 80 percent organic material have been effectively treated with this process. This process is effective at immobilizing these organic contaminants, since they are reactants in the fixation reaction.

The ability of the SRS/EIF technology to treat organics makes it a good candidate for treating heavy tars with high organic concentrations. Unlike other solidification processes, the SRS/EIF technology treats both organic and inorganic wastes. This offers the significant advantage of possibly using one treatment technology at an abandoned hazardous waste site.

Advantages of the SRS/EIF Solidification

The SRS/EIF process has significant advantages over other commercially available solidification technologies. For example, the SRS treated waste is not merely encapsulated, but is permanently fixed in the waste. The process involves a chemical reaction between the lime and lime additives and the hazardous constituents, which permanently fixes the hazardous constituents in the matrix structure of the treated waste. This chemical reaction significantly reduces the toxicity of the hazardous waste.

Moreover, the hydrophobic character of the treated waste significantly reduces the extent to which water can penetrate the treated product. Thus, the fixation of the hazardous wastes in a nontoxic matrix and the hydrophobic characteristic of the endproduct provide double protection against contamination of surface water and groundwater.

The SRS/EIF process also offers a great deal of waste treatment flexibility, since it is capable of treating solids, semisolids, sludges, and liquids, including most organic and heavy metal contaminants, as well as contaminated debris. This process effectively cleans oil residues from the surface of debris without the problems associated with the use of many solidification mixing and piping systems. All the mixing is accomplished in a pit, thus allowing for the treatment of odd-sized contaminated debris. In addition, the SRS/EIF technology can be utilized in a variety of climatological conditions, including severe summer heat, rain, freezing temperatures, and even snow.

The SRS/EIF treatment process has several other significant physical advantages. In particular, unlike other approaches, the SRS/EIF process will only increase waste volumes by about 25 percent, a factor that is quite significant in comparison to other fixation technologies, which may increase waste volumes by as much as 50 to 100 percent. Furthermore, this process can treat up to 1000 cubic yards per day of sludge. One blending pit can process up to 350 cubic yards of waste per day. The endproduct is a clay- or dirtlike solid that can be easily moved. In contrast, other fixation processes form concrete blocks requiring additional time and resources for disposal.

With regard to incineration technologies, cost considerations favor the SRS/EIF technology over conventional incineration. The SRS/EIF total estimated turnkey system cost per unit of treated waste is approximately $70 to $95 per cubic yard. SRS/EIF quotes on a firm, fixed price basis per unit of waste treated. SRS/EIF will not request change orders for unforeseen problems that are encountered during a project.

A cost of $70–95 per cubic yard for the SRS/EIF technology compares favorably to incineration costs which can range anywhere from $250 to $600 per cubic yard. In addition, a key financial advantage to this technology is that it does not require any significant up-front capital investment or outlay to build or install. Normal earth moving equipment is generally the only major equipment which is required for the use of the SRS/EIF technology.

Regulatory Requirements

Waste stabilization technology has been specifically included in the EPA's National Contingency Plan—the federal regulation that guides remediation activities under the Superfund program. According to all data analyzed to date, the SRS/EIF technology is consistent with the federal Superfund regulatory guidance and constitutes full compliance with the cleanup mandates of the Superfund program. The SRS/EIF technology offers a viable remediation option which meets the spirit and intent of all cleanup requirements specified by the new Superfund law—SARA—for the following reasons:

- SRS/EIF technology is protective of human health and the environment.

- The estimated SRS/EIF process treatment cost of $70–95 per cubic yard of waste is quite cost-effective.

- The SRS/EIF solidification process is a permanent alternative waste treatment remedy.

The EPA has selected the SRS/EIF process for participation in the federal Superfund Innovative Technology Evaluation (SITE) program, and work is now underway to select an applicable Superfund site under this program.

Finally, the SRS/EIF technology can produce a treated sludge product that may not be required to be landfilled in accordance with the Resource Conservation and Recovery Act (RCRA) requirements because of its highly desirable coefficient of permeability. In contrast, incinerator ash may need to be disposed of in many instances in an RCRA specification landfill, or be solidified if it contains heavy metal contaminants.

* * *

The next decade will be a critical period to judge the effectiveness of our nation's approach to remediating hazardous waste problems. Estimates of the number of hazardous waste sites that will require cleanup in the future are quite staggering. For example, a study by the Office of Technology Assessment suggests that the number of National Priority List sites will rise to 10,000, at a cost of $100 billion to remediate.

As a result, the need for cost-effective, innovative hazardous waste treatment technologies is more imperative than ever before. A key to the nation's cleanup efforts will be the use of alternative technologies like the SRS/EIF solidification process. The use of SRS/EIF technology has been shown to be cost-effective and permanent. Its application to hazardous waste containing both organic and heavy metal contaminants makes it a viable remediation option at most Superfund and hazardous waste sites. The EPA's selection of the SRS/EIF technology for its SITE program demonstrates a further commitment by SRS to maximize the use of this process to its full potential.

REFERENCES

1. Superfund Amendments and Reauthorization Act (November 1986).
2. Wiles, Carlton C. "A Review of Solidification/Stabilization Technology," *Journal of Hazardous Materials*, pp. 5-21 (1987).